More praise for *Something New Under the Sun*

"McNeill has taken on the heroic assignment of synthesizing for the first time the environmental history of the entire planet during the twentieth century. . . . Essential reading."

—William Cronon, author of *Nature's Metropolis*

"McNeill writes deftly, usually pithily, often sagely—and he deals with matters on a global scale without losing coherence. . . . The book deserves a wide audience."

—Stephen J. Pyne, *The American Scientist*

"*Something New Under the Sun* is both an important, well-balanced work of history and a very readable book. It should appeal to a broader audience than historians and biologists on account of its style. McNeill quotes great poets and he himself sometimes produces imagery worthy of them."

—*Literary Review*

"[An] important and sometimes irreverent and witty book. . . . This book offers an important rethinking of the century just past. More than any other it explains how humans and the earth have lived in dynamic tension."

—*National Post* (Canada)

"John McNeill's *Something New Under the Sun* can fairly be described as learned, a rare thing for books about the environment. And, the author's strong opinions notwithstanding, it is admirably objective. . . . Instead of apocalyptic warnings, he offers dry wit and understatement."

—*The Economist*

"In a field that inspires great passion in its protagonists and frequent polarization of opinion, *Something New Under the Sun* is a breath of fresh air. It is also a monumental, important, and timely work of interdisciplinary scholarship written to be accessible to anyone interested in the relationship between our species and the planet that supports us." —*Guardian*

"An important, beautifully written book." —*Nature*

"Superb. . . . McNeill is persuasive because he is quiet, sceptical, detached, and self-critical, without a line of hyperbole, eco-freakery or environmental millenarianism. . . . With hardly a note of condescension to the reader, or compromise of scientific rigor, he makes everything easy to understand. . . . He writes a nearly perfect style: the pace is well-judged, the language clear, the images forceful but uncontrived."

—Felipe Fernandez-Armesto, *The Independent*

ALSO BY J. R. McNEILL

THE MOUNTAINS OF THE MEDITERRANEAN WORLD: AN ENVIRONMENTAL HISTORY
THE ATLANTIC EMPIRES OF FRANCE AND SPAIN
ATLANTIC AMERICAN SOCIETIES FROM COLUMBUS THROUGH ABOLITION
(coeditor)

SOMETHING NEW
UNDER THE SUN

W · W · NORTON & COMPANY · NEW YORK · LONDON

SOMETHING NEW UNDER THE SUN

AN ENVIRONMENTAL HISTORY OF THE TWENTIETH-CENTURY WORLD

J. R. McNEILL

First published as a Norton paperback 2001

For information about permission to reproduce selections from
this book, write to Permissions, W. W. Norton & Company, Inc.,
500 Fifth Avenue, New York, NY 10110

The text of this book is composed in New Caledonia, with the
display set in Torino.
Composition by Allentown Digital Services
Manufacturing by LSC Harrisonburg

Library of Congress Cataloging-in-Publication Data

McNeill, J. R.
Something new under the sun : an environmental history of
the twentieth-century world / J. R. McNeill.
p. cm.
Includes bibliographical references and index.
ISBN 0-393-04917-5
1. Human ecology—History—20th century. 2. Nature—Effect
of human beings on—History—20th century. I. Title.

GF13.M39 2000
304.2′8′0904—dc21

99-054900

ISBN 0-393-32183-5 pbk.

W. W. Norton & Company, Inc.
500 Fifth Avenue, New York, N.Y. 10110
www.wwnorton.com

W. W. Norton & Company Ltd.
15 Carlisle Street, London W1D 3BS

20 19 18 17

for Julie,

ONCE AGAIN

Contents

1. Prologue: Peculiarities of a Prodigal Century 3

PART ONE: THE MUSIC OF THE SPHERES 19

2. The Lithosphere and Pedosphere: The Crust of the Earth 21

List of Maps and Tables

MAPS

TABLES

Foreword by Paul Kennedy

At the beginning of the twentieth century, humankind in the West had already become aware that its collective economic activities were doing strange things to the environment. Salmon could no longer migrate upstream through chemical-tainted waters. The air surrounding industrial cities—and further afield, as the winds moved on—was full of particles of burned fossil fuels. Smog took the lives of thousands with respiratory problems each year. Huge gashes had been carved in the landscape to gain access to fresh coal supplies, and ugly heaps of slag blotted once-pleasant countrysides.

The twin sources of this environmental havoc were also known to observers in 1900. The first was that the world's human population, which had grown rather slowly for almost four million years, began to accelerate in the late eighteenth century and still showed no signs of slackening. The second was that human economic activity had also accelerated ever since the post-1760 Industrial Revolution had allowed the substitution of inanimate for animate energy. All this caused intellectuals like the Cuban José Martí and the Englishman H. G. Wells to wonder whether this immense surge in human activities could be sustained in the decades to come without degrading nature.

Had such writers survived another 100 years, they would have been amazed at the even-greater pace of change that the twentieth century witnessed. The world's population quadrupled in that time, the global economy expanded 14-fold, energy use increased 13 times, and industrial output expanded by a factor of 40. But carbon dioxide emissions also went up 13-fold, and water use rose 9 times. Not all of this meant bad news—indeed the productivity increases of the twentieth century raised the living standards of hundreds of millions of human beings out of their forefathers' ghastly poverty—but the sheer size and intensity of the transforming processes also meant that the history of the twentieth century really *was* different, in environmental terms, from that of any preceding periods.

The twin challenge to all intelligent world citizens is, first, to understand the sheer dimensions of environmental change (and the many results thereof) in this past century; and, second, to think sensibly about how these problems might be addressed before dangerous thresholds are breached by our unwitting, collective activities. This message of first comprehending and then responding to environmental change is the hallmark of Professor McNeill's brilliant and remarkably concise examination of the past century. As his title makes clear, at least in this case Ecclesiastes may have been wrong in claiming that there was "nothing new under the sun." For what McNeill demonstrates in the seven chapters of Part One is that in all of the "spheres" that surround us—the lithosphere and pedosphere, the atmosphere, the hydrosphere, and the biosphere—we humans have impacted our planet more deeply in the twentieth century than we did in all previous history combined. One statistic alone sums this up: according to McNeill's (rough) calculations, humans in the twentieth century used *ten times* more energy than their forebears over the entire thousand years preceding 1900.

But Professor McNeill is not a mere recorder of environmental change. What really interests him is the interaction between what he calls "the planet's history and the people's history"—which is why the chapters in Part Two of this book are equally, perhaps even more, important. Here he deftly analyzes the elements of population growth, migration, technological change, industrialization, international politics, ideas—and their many "feedback loops" into the realm of environmental policies.

McNeill is neither a Luddite, nor a dogmatic "no-change-at-all" environmentalist. But he does caution us to be prudent, and to *take action*, lest the ecological thresholds that global society is steadily approaching are in fact closer than we think.

Something New Under the Sun is a clearly written and immensely insightful book. It carries a message that is deeply sobering, and deserves the widest attention from publics and politicians alike.

Acknowledgments

Without the help and support I received on it, this book would not have been ready until well into the twenty-first century, and it would have been a longer and lesser book. Several of my colleagues in the History Department at Georgetown University critiqued the manuscript, in most cases in its entirety, which claimed a good chunk of their time in the summer of 1998: Tommaso Astarita, Carol Benedict, Emmett Curran, Catherine Evtuhov, Alison Games, David Goldfrank, Andrzej Kaminski, David Painter, Aviel Roshwald, Jack Ruedy, Jordan Sand, Jim Shedel, Judith Tucker, John Tutino. Their collective expertise saved me from countless embarrassments. Other Georgetown colleagues read sections of the manuscript with equally helpful effects: Janice Hicks of the Chemistry Department; Martha Weiss from Biology; Tim Beach, Murray Feshbach, and Richard Matthew of the School of Foreign Service; and Steve King of the Government Department. I feel fortunate to work at a university where colleagues cheerfully shoulder burdens such as those I placed on this group.

Several of my students at Georgetown in 1998–1999 read parts of the manuscript and—gleefully—found ways to improve it: Dan Brendtro, Eric Christenson, Julie Creevy, Brett Edwards, Katie Finley, Justin Oster, and Jill Wohrle. Students from another time, the early 1990s, hooked me on world environmental history by their enthusiastic work in the best class I have ever taught: Sean Captain, Brad Crabtree, Nancy Golubiewski, Elena Garmendia, and Terri Willard. They helped spur me on by asking about progress on the book as the years ticked by.

Friends, relatives, and colleagues—pleasantly overlapping categories—elsewhere who read all or parts of the manuscript and improved it include Peter Campbell, Bill Cronon, Rebekah Davis, Don Hughes, John Kelmelis, Greg Maggio, Bob Marks, Elizabeth McNeill, William McNeill, and Kent Redford. I am also grateful to Alison Van Koughnett and George Vrtis for their help in sending useful research material my way.

I was able to work full-time on this book for two consecutive years thanks to the generous support of several institutions. The Fulbright program made possible six months' work in New Zealand, spent in the delightful environment of the University of Otago History Department. I thank Rob Rabel and Erik Olssen for arranging and smoothing my stay on the riviera of the subantarctic, where serious thinking on the book began. The Woodrow Wilson International Center for Scholars granted me a fellowship in 1996–1997 and provided a bracing intellectual environment, especially at lunchtimes. Many of the Fellows from that year helped me along in my work, most notably Temma Kaplan and Wolf Fischer. The Wilson Center also provided research assistants who made my task far easier: Christian Kannwischer, Peter Kocsis, Angeliki Papantoniou, and Toshuko Shironitta. In 1997–1998 the Guggenheim Foundation provided fellowship support which allowed me to stay holed up in my attic and complete the first draft of the manuscript. And Georgetown University granted me leave of absence for these two years, as well as additional research support over the years.

On several occasions I have received useful questions or comments from listeners patient enough to hear me out on the subjects of this book. These include audiences at the University of Otago, Massey University, and the University of Canterbury in New Zealand; at the University of Wisconsin, Carnegie-Mellon University, and the University of Amsterdam; and the Futures Group of the United States Environmental Protection Agency.

I started this book because Paul Kennedy, the general editor of Norton's Global Century series, asked me write it. I thank him for the faith he showed early on, and his patience thereafter. I also thank Steve Forman at Norton, who kept the book from growing to twice its final length, shamed me repeatedly with (or is it "by"?) his improvements to my prose, and kept up a steady barrage of encouragement. Susan Middleton earned my gratitude with her relentless efforts to achieve consistency and precision in my prose.

I finished the book because Julie Billingsley worked long and hard to allow me to. She sustained me over the years in ways too numerous to count.

Preface

What has been is what will be,
  and what is done is what will be done;
  and there is nothing new under the sun.
Is there a thing of which it is said,
"See, this is new?"
It has been already, in the ages before us.
There is no remembrance of former things,
  nor will there be any remembrance of later things yet to happen.
                              —Ecclesiastes 1:9–11

Most verses of Ecclesiastes contain useful wisdom for the ages. But the above words are now out of date. There is something new under the sun. The ubiquity of wickedness and the vanity of toil may remain as much a part of life today as when Ecclesiastes was written, in the third or fourth century B.C., but the place of humankind within the natural world is not what it was. In this respect at least, modern times are different, and we would do well to remember that.

RATS, SHARKS, AND HISTORY. This book is a history of—and for—environmentally tumultuous times. It is the story of environmental change around the world in the twentieth century, its human causes and consequences. In the pages that follow I aim to persuade you of several related propositions. First, that the twentieth century was unusual for the intensity of change and the centrality of human effort in provoking it. Second, that this ecological peculiarity is the unintended consequence of social, political, economic, and intellectual preferences and patterns. Third, that our patterns of thought, behavior, production, and consumption are adapted to our current circumstances—that is, to the current climate (and global biogeochemistry), to the twentieth century's abundance of cheap energy and cheap fresh water, to rapid population growth, and to yet more rapid economic growth. Fourth, that these preferences and patterns are not easily adaptable should our circumstances change. This last proposition pertains to the future and so I will not, in a work of history, pursue it far. In addressing these themes I also aim to convince you that the modern ecological history of the planet and the socioeconomic history of humanity make full sense only if seen together.

These, I think, are interesting propositions because they imply that we, as a species, are unwittingly choosing a particular evolutionary gambit. In the very long view of biological evolution, the best survival strategy is to be adaptable, to pursue diverse sources of subsistence—and to maximize resilience. This is because in the long run there will be surprises, shocks, and catastrophes that kill off some species, no matter how well adapted they may be to one specific set of circumstances. If a species can survive the periodic shocks that kill off competitors, then evolutionary success is at hand: there is plenty of open niche space to inhabit. For individuals of such cagey species, survival of periodic shocks opens up *Lebensraum* in which to spread their genetic footprint in space and time. Over the long haul, *Homo sapiens* has enjoyed great biological success on the strength of adaptability. So have some species of rat.

Adaptability is not the only strategy for evolutionary success. Another is supreme adaptation to existing circumstances, which can work well for a while, a long while if circumstances are stable. Koalas eat only eucalyptus leaves, pandas only bamboo. Both animals pursued specializations that served them well as long as eucalyptus and bamboo were plentiful. Sharks are supremely adapted to hunting, killing, and eating good-sized sea creatures. Sharks have done well with this for over 200 million years because the oceans, despite considerable changes, always contained a goodly supply of shark food.

The rat and shark strategies have their (rough) analogues in human societies. In recent millennia, cultural evolution has shaped human affairs more than biological evolution has. Societies, like rats and sharks, unconsciously pursue survival strategies of adaptability or of supreme adaptation. Stable—not to be confused with peaceful—social orders, such as those of pharaonic Egypt, feudal Europe, or imperial China, rested on exquisite adaptation to existing ecological (and other) circumstances. While those circumstances persisted, such societies prospered, but in the long run they faced crises made more acute by their very success. Fine-tuned adaptation (shark strategy) is rewarded by continuous success only so long as governing conditions stay the same. This success easily translates into conservatism, orthodoxy, and rigidity. But it is not a bad strategy: it can work for centuries. Or it could in times past, when humans lacked the power to disturb global ecology.

In the twentieth century, societies often pursued the shark strategy amid a global ecology ever more unstable, and hence ever more suited for rats. We energetically pursued adaptations to evanescent circumstances. Perhaps a quarter of us live in ways fully predicated upon stable climate, cheap energy and water, and rapid population and economic growth. Most of the rest of us understandably aspire to live in such ways. Our institutions and ideologies too are by now built on the same premises.

Those premises are not flimsy, but they are temporary. Climate has changed little for 10,000 years, since the retreat of the last ice age; it is changing fast now. Cheap energy is a feature of the fossil fuel age, roughly since 1820. Cheap water, for those who enjoy it, dates to the nineteenth century except in a few favored settings. Rapid population growth dates from about the middle of the eighteenth century, and fast economic growth from about 1870. To regard these circumstances as enduring and normal, and to depend on their continuation, is an interesting gamble. Throughout the twentieth century, more and more people, more and more societies, (quite unconsciously) accepted this gamble.

It is not a foolish gamble. Indeed, where and when states and societies must seek security in a competitive international system, where and when firms seek profit and power in a competitive marketplace, and where and when individuals seek status and wealth in competitive societies, the gamble is tempting. Those who reject it will lose out so long as present circumstances persist. So it is not a bad gamble at all unless one is concerned with the long run, or unless one imagines our present circumstances are soon to change. To make an interesting gamble yet more interesting, our twentieth-century adaptation, our modern behavior, increases the probability that our present circumstances will soon change.

The same characteristics that underwrote our long-term biological success—adaptability, cleverness—have lately permitted us to erect a highly specialized fossil fuel–based civilization so ecologically disruptive that it guarantees surprises and shocks, and promotes just the sort of flux that favors the adaptable and clever. And we have prospered mightily as a species amid this flux, multiplying far faster than ever before. We have created a regime of perpetual ecological disturbance, as if we had organized a grand global plot to do in species less cagey than ourselves. But we are not so clever as that. The regime of perpetual disturbance is an accidental by-product of billions of human ambitions and efforts, of unconscious social evolution.

While shaking up global ecology in ways that allowed us to proliferate biologically, we increasingly organized our societies around the new circumstances we created. To take but one example, the modern United States was built on the premise of cheap energy. This was not a bad strategy: cheap energy was a reality for most of the twentieth century, and the United States adapted to that circumstance as thoroughly and successfully as any society. But that very success means that, should the circumstance change, any adjustment will be all the harder. The United States, while an unusually flexible society in many respects, is an ecological shark. Many influential and powerful Americans were aware of this, and crafted domestic and foreign policy to perpetuate cheap energy. Think of the Gulf War of 1991.

The biological success of the human species is probably not at serious risk. As a species we are much more ratlike than sharklike. The social order of many societies, on the other hand, probably is at risk. Acute social strains, typical of modern times, will remain our fate as long as acute ecological disruption, also typical of modern times, remains our practice. In these ways the planet's history and people's history remain tightly linked, perhaps more so now than at most times past. One without the other is scarcely comprehensible.

CONFESSIONS OF A HISTORIAN. This book is about people and the environment. It is not concerned with ecological changes that humans had no role in bringing about, nor with those that, whatever their causes, have little chance of affecting human affairs. I hope the book will interest those who want to understand history in its fullest context, and those who want to understand environmental change in its historical context.

All historians write with viewpoints and biases. Here are some of mine. This book is anthropocentric. The American humorist Robert Benchley allegedly wrote a history of the Atlantic cod fishery from the point of view of the fish. The British historian Arnold Toynbee published "The

Roman Revolution from the Flora's Point of View," in which he gave speaking roles to plants.[1] An environmental history of the world in the twentieth century written from the point of view of lemmings or lichen might be very interesting, but I lack the imagination for that. My book leaves out a lot of ecological changes simply because they have little to do with human history.

Historians are mainly interested in change. This book concentrates on times and places where large changes took place, neglecting countless examples of continuity. Broad expanses of Antarctic ice, of Mauretania's desert, of Canadian tundra, of the Pacific Ocean's deep waters are much the same as they were a century ago. I will say little about such cases, and much more about Indonesia's forests and London's air quality. This amounts to a strong bias in favor of examples of change. Historians of the Industrial Revolution or the Bantu migration do the same thing: change is usually more compelling than continuity.

Modern environmental writing typically evaluates changes as either good or bad, but it rarely reveals the criteria for judgment. I will refrain from such evaluation in many cases, because environmental changes usually are good for some people and bad for others, and indeed good for some species or subspecies and bad for others. Where possible I will try to explain for whom (or what) a given development was good or bad. And if it was bad for virtually all life forms, I will abandon all effort at Olympian detachment and label it degradation, despoliation, destruction, or the like. Salinization of the Murray-Darling basin in Australia is degradation. Readers are invited to consider for themselves whether or not Amazonian deforestation is good or bad—and urged to think: for whom?

The answer is complicated. It depends on whose interests one rates over others (Brazilian ranchers, peasants, rubber tappers, Amerindians, among others); on how one evaluates the role of deforestation in producing changes in the global atmosphere, and how one regards the prospect of global warming; on how much one values the forms of life going extinct in Amazonia, their genetic information, and their role in the web of life around them. It depends on how much of Amazonia is involved, how fast deforestation is proceeding, and on what land-use patterns and ecosystems follow in its wake. It depends on all this and more. In such complex cases I will try to outline the impacts of environmental change, but will allow the reader to judge if they count as good or bad,

[1]Toynbee 1965, 2:585–99

as improvement or degradation. For those who like their issues uncomplicated and their morality simple, this will disappoint. For those who like apples and oranges reduced to dollars for convenient accounting, this will disappoint. And for those who like to be told what to think, this will particularly disappoint. I hope that leaves a goodly number of readers.

SOMETHING NEW
UNDER THE SUN

1

Prologue: Peculiarities of a Prodigal Century

The disadvantage of men not knowing the past is that they do
not know the present.
                          —G. K. Chesterton (1933)

Environmental change on earth is as old as the planet itself, about 4 billion years. Our genus, *Homo,* has altered earthly environments throughout our career, about 4 million years. But there has never been anything like the twentieth century.

Asteroids and volcanoes, among other astronomical and geological forces, have probably produced more radical environmental changes than we have yet witnessed in our time. But humanity has not. This is the first time in human history that we have altered ecosystems with such intensity, on such scale and with such speed. It is one of the few times in the earth's history to see changes of this scope and pace. Albert Einstein famously refused to "believe that God plays dice with the world."[1] But in the twentieth century, humankind has begun to play dice with the planet, without knowing all the rules of the game.

[1]Frank 1947.

The human race, without intending anything of the sort, has undertaken a gigantic uncontrolled experiment on the earth. In time, I think, this will appear as the most important aspect of twentieth-century history, more so than World War II, the communist enterprise, the rise of mass literacy, the spread of democracy, or the growing emancipation of women. To see just how prodigal and peculiar this century was, it helps to adopt long perspectives of the deeper past.

In environmental history, the twentieth century qualifies as a peculiar century because of the screeching acceleration of so many processes that bring ecological change. Most of these processes are not new: we have cut timber, mined ores, generated wastes, grown crops, and hunted animals for a long time. In modern times we have generally done more of these things than ever before, and since 1945, in most cases, far more. Although there are a few kinds of environmental change that are genuinely new in the twentieth century, such as human-induced thinning of the ozone layer, for the most part the ecological peculiarity of the twentieth century is a matter of scale and intensity.

Sometimes differences in quantity can become differences in quality. So it was with twentieth-century environmental change. The scale and intensity of changes were so great that matters that for millennia were local concerns became global. One example is air pollution. Since people first harnessed fire half a million years ago, they have polluted air locally. Mediterranean lead smelting in Roman times even polluted air in the Arctic. But lately, air pollution has grown so comprehensive and large-scale that it affects the fundamentals of global atmospheric chemistry (see Chapter 3). So changes in scale can lead to changes in condition.

Beyond that, in natural systems as in human affairs, there are thresholds and so-called nonlinear effects. In the 1930s, Adolf Hitler's Germany acquired Austria, the Sudetenland, and the rest of Czechoslovakia without provoking much practical response. When in September 1939 Hitler tried to add Poland, he got a six-year war that ruined him, his movement, and (temporarily) Germany. Unknowingly—although he was aware of the risk—he crossed a threshold and provoked a nonlinear effect. Similarly, water temperature in the tropical Atlantic can grow warmer and warmer without generating any hurricanes. But once that water passes 26° Celsius, it begins to promote hurricanes: a threshold passed, a switch thrown, simply by an incremental increase. The environmental history of the twentieth century is different from that of time past not merely because ecological changes were greater and faster, but also because increased intensities threw some switches. For example, incremental increases in fishing effort brought total collapse in some oceanic fisheries. The cumulation of many increased intensities may throw some grand switches, producing very basic changes on the earth. No one knows, and no one will know until it starts to happen—if then.

This chapter examines the long-term histories of some of the human actions that change environments. The length of the long term varies from case to case, mainly because of differences in the availability of information. The actions and processes in question are sometimes easily measured, sometimes not. The accuracy of the data is also open to question. Despite these problems, it is possible to make some judgments about how peculiar the last century was, and in what respects it departed sharply from the patterns of the past.

Economic Growth since 1500

Most of the things people do that change environments count as economic activity. Economists habitually measure the size of economies by summing the total value of goods and services brought to market or otherwise officially noted. The addition yields a single figure, the gross domestic product, or GDP. This is a very imperfect procedure, especially for times and places where significant production (and delivery of services) take place outside of markets. Economic historians are keenly aware of the drawbacks of this measurement, and have tried to adjust their figures accordingly.

Five hundred years ago the world's annual GDP (converted into 1990 dollars) amounted to about $240 billion, slightly more than Poland's or Pakistan's today, slightly smaller than Taiwan's or Turkey's.[2] Up to 1500 the world economy had grown extremely slowly over the millennia, mainly because (as we shall see) population had grown only slowly and improvements in productive technologies came very slowly by recent standards. After 1500, leading technologies were applied to the Americas and other regions, shipping became truly oceanic, and international trade grew. By 1820, the world's GDP had reached $695 billion (more than Canada's or Spain's, less than Brazil's in 1990s terms). The Industrial Rev-

[2]In 1990 Geary-Khamis dollars. The estimate is from Maddison 1995:19.

olution, further improvements in transport, and further development of frontier lands increased the rate of growth after 1820 so that in 1900, world GDP reached $1.98 trillion (less than 1990s Japan's). Indeed the period 1870 to 1913 remains one of spectacular growth spurts in the history of the world economy, faster than any that went before, and faster than much of what followed. After three decades of repressed growth (1914–1945) the world economy surged again, so that in 1950, world GDP attained $5.37 trillion (as large as the United States's economy in 1991). A long boom followed, based on more-open international trade, fast development of technology, and rapid population growth. By 1992, world GDP was about $28 trillion. This miraculous period of economic history, with all its upheaval, invention, organization, and suppression, is reduced to index numbers and growth rates in Table 1.1.

TABLE 1.1 EVOLUTION OF WORLD GDP, 1500–1992

Date	World GDP[a]
1500	100
1820	290
1870	470
1900	823
1913	1,136
1929	1,540
1950	2,238
1973	6,693
1992	11,664

SOURCE: Maddison 1995:19, 227.

[a]GDP figures are given in index numbers relative to A.D. 1500.

The world's economy in the late twentieth century was about 120 times larger than that of 1500. Most of this growth took place after 1820. The fastest growth came in 1950 to 1973, but the whole period since the World War II saw economic growth at rates entirely unprecedented in human experience.[3]

---

[3]Notable earlier expansions presumably came with the retreat of the last ice sheets, with the initial inventions and spread of agriculture, and with the establishment of large trade networks under the aegis of ancient empires. But all these economic expansions were slow and modest by post-1820 standards.

Most of this economic expansion was driven by world population growth. The rest is owing to more productive technologies and organization (and perhaps harder work). Per capita figures (Table 1.2) show that while the world economy has grown 120-fold since 1500, average income for individuals has grown only 9-fold.[4] This of course is a global average, and disguises huge variations among regions, countries, and persons.

TABLE 1.2 PER CAPITA WORLD GDP SINCE 1500		
Year	*Per Capita World GDP* (*1990 dollars*)	*Index Numbers* (A.D. *1500* = *100*)
1500	565	100
1820	651	117
1900	1,263	224
1950	2,138	378
1992	5,145	942

Source: Elaborated from Maddison 1995:228.

On average, we have nine times more income per capita than our ancestors had in 1500, and four times as much as our forebears had in 1900. Despite gross inequities in the distribution of this income growth—the average Mozambican today has an income well under half the global average of 1500—it must count as a great achievement of the human race over the past 500 years, and especially over the past century. The achievement has come at a price, of course. The social price, in the form of people enslaved, exploited, or killed so that "creative destruction"[5] could make way for economic growth, is enormous. So is the environmental price. Historians in the past thirty years have appropriately paid great attention to the social price of economic growth and modernization; the environmental price deserves their attention too.

---

[4]If world GDP has grown 120 times since 1500 and per capita GDP has grown 9 times, then population has grown about 13-fold (see below) and thus accounts for the majority of the economic growth since 1500.

[5]This phrase is Joseph Schumpeter's famous description of capitalism, but it applies equally well to the building of socialist economies in the 20th century.

Population Growth since 10,000 B.C.

Population is much easier to measure than economic activity, so although estimates prior to 1900 must be treated with caution for most parts of the world, the following reconstruction is more reliable than the previous one.

When humans first invented agriculture (around, say, 8000 B.C.), global population was probably between 2 and 20 million.[6] We were outnumbered by some other primates, such as baboons. But with agriculture came the first great surge in human numbers. Population grew much faster, probably between 10 and 1,000 times as fast as before, but nonetheless very slowly, by tiny fractions of a percent per year. By A.D. 1, the globe supported around 200 or 300 million people (roughly equivalent to today's Indonesia or United States). By 1500, world population had reached 400 or 500 million. It had taken about a millennium and a half to double, and grew at a rate well under 0.1 percent per year. After 1500, world population continued to grow quite slowly, reaching 700 million around 1730. At this point it began to rise more quickly, beginning the long boom still in progress today.[7] By 1820, human population reached a billion or so. Our spectacular biological success since then is sketched by the figures in Table 1.3.

TABLE 1.3 WORLD POPULATION SINCE 1820

Year	Population (billion)	Annual Growth Rate (%)
1820	1	—
1850	1.2	0.5 (1820–1849)
1900	1.6	0.6 (1850–1899)
1950	2.5	0.8 (1900–1949)
1990	5.3	1.8 (1950–1989)
2000	6.0	1.5 (1990–1999)

Source: Cohen 1995:79 and app. 2.

[6]Cohen 1995:77.
[7]Why this increase in the growth rate happened in the 18th century is unclear, but it may have had to do with the retreat of some lethal diseases, as well as eventually with improvements in sanitation, public health, and diet. A recent and useful discussion appears in Livi-Bacci 1992.

Since the eighteenth century our numbers have grown extremely quickly by previous standards. And in the period since 1950, population has increased at roughly 10,000 times the pace that prevailed before the first invention of agriculture, and 50 to 100 times the pace that followed. If twentieth-century rates of population growth had prevailed since the invention of agriculture, the earth would now be encased in a squiggling mass of human flesh, thousands of light-years in diameter, expanding outward with a radial velocity many times greater than the speed of light.[8] Clearly we will not keep the twentieth-century pace up for long. We are in the final stages of the second great surge in human population history. Demographers expect at most one more doubling to come. The twentieth century's global population history will be peculiar not only in light of the past, but in light of the future as well.

Another way to conceive of the extraordinary demographic character of the modern era is to estimate how many people have ever lived, and (with estimates about life expectancy) how many human-years have ever been lived. Such estimates require extra caution, of course. Some European demographic historians have made the heroic assumptions and subsequent calculations.[9] They figure that about 80 billion hominids have been born in the past 4 million years. All together, those 80 billion have lived about 2.16 trillion years. Now for the astonishing part: 28 percent of those years were lived after 1750, 20 percent after 1900, and 13 percent after 1950. Although the twentieth century accounts for only 0.00025 of human history (100 out of 4 million years), it has hosted about a fifth of all human-years.

Like the long-term course of economic growth, our population history also represents a triumph of the human species. It too, of course, has come at a price. In any case, it is an amazing development, an extreme departure from the patterns of the past—even though we tend to take our present experience for granted and regard modern rates of growth as normal. Bizarre events that last for more than a human lifetime are easy to misunderstand.

The long-term trajectories of economic growth and population growth followed one another closely for millennia. Only around 1820 did they

[8]This cheerful vision is adapted from Cipolla 1978:89.

[9]Biraben 1979:16; Bourgeois-Pichat 1988. See also Livi-Bacci 1992:32–3. Westing 1981 estimates 50 billion as the total number ever born, Keyfitz 1966 estimates 69 billion, and Haub 1995 prefers 105 billion. Some of the discrepancy derives from different starting points, e.g., 300,000 years ago for Westing and 1 million for Keyfitz.

begin to diverge sharply, with economic growth outstripping population growth—hence the rising per capita incomes. What made this possible were new technologies and systems of economic organization that allowed people to make far greater use of energy.

Energy History since 10,000 B.C.

Before the Industrial Revolution began, we had at our disposal the muscle power of our bodies and of some domesticated animals; the power (very inefficiently harnessed) of wind and water; and (for heat but not for power) the chemical energy stored in wood and other biomass. The Industrial Revolution changed everything because it brought engines that could convert into mechanical power the biomass energy stocks accumulated in the earth's crust over hundreds of millions of years: fossil fuels.

Physicists agree that the total quantity of energy in the universe is constant. On earth, energy is held in rough balance: what arrives from the sun as radiant energy is equivalent to what dissipates into space as heat. Energy can neither be created nor destroyed. Yet we commonly speak of energy production or consumption. The word "energy" is imprecise; the stuff hard to measure. The following reconstruction aims to be precise about what is meant by energy, but its quantitative elements deserve as much or more caution as the section on economic growth.

All our energy, ultimately, is nuclear energy, in that it comes from a nuclear fusion reaction in the sun.[10] It exists on earth in several forms, the important ones for people being mechanical (or kinetic), chemical, heat (or thermal), and radiant. The problem for us is to get energy in a useful form in the right place and the right time for whatever we might wish to do. We do this by means of converters, which change energy from one form to another, making it easier to store, transport, or use for work. Many economic operations make use of several converters. Each conversion involves some practical loss, in that a proportion of the preconverted energy is dissipated (usually as heat) or otherwise rendered into a form that is useless, impossible to capture. Hence converters have effi-

---

[10]This neglects gravitational energy, which keeps the earth in orbit at a convenient distance from the sun, and the nuclear fission that takes place within the earth's core and powers volcanoes.

ciency ratings. Human beings, for instance, are about 18 percent efficient: for every 100 calories I eat as food (chemical energy), only about 18 are converted into mechanical energy; the rest are lost for practical purposes, mostly as heat. Horses' efficiency is only about 10 percent.

Before the Industrial Revolution, the only important converters were biological ones.[11] The first human societies used only their own muscle power, derived from chemical energy stored in plants and animal flesh. Eventually, with a few tools, the deployment of this muscle power grew more efficient. The use of fire helped a great deal in heating, of course, and, when cooking was invented, rendered some otherwise inedible energy sources edible. But until roughly 10,000 years ago, for mechanical energy our ancestors depended on their own bodies in what one might call the "somatic energy regime."

Agriculture allowed people greater control over the plant converters we call food crops. Shifting agriculture probably increased energy availability 10-fold over that available through hunting and gathering, and settled agriculture another 10-fold. This translated into greater population densities. Then, as big animals were domesticated, people acquired more muscle power, more mechanical energy, in more concentrated form. Oxen for haulage and horses or camels for transport marked great improvements. Oxen could plow heavy soils, opening up new food possibilities, which in turn allowed for more people and more oxen in a positive feedback loop that extended and strengthened the somatic energy regime. Societies that did not domesticate large animals labored at a disadvantage. New crops, wheels, and horse collars improved the energy efficiency of societies over subsequent millennia, but even at the outset of the Industrial Revolution in Europe (c.1800), more than 70 percent of the mechanical energy used was supplied by human muscle.[12] The fundamental energy constraints remained the amount of arable land and the amount of water to produce crops.

Agriculture and animal domestication did create an energy surplus. Controlling that surplus, applying it as one wished, and enjoying the returns from it constituted the stuff of politics—directing the somatic energy regime. If applied judiciously, in war or irrigation for instance, surplus might create a windfall of increasing returns that made someone rich or powerful indeed—pharaohs, for instance. Since people are more

[11]This potted history is drawn from Cipolla 1978:35–69, Debeir et al. 1986, and Smil 1994.
[12]Smil 1994:226. Cipolla 1978:53 estimates 80–85% for plant, animal, and human energy sources combined, the balance deriving from wind and water power.

efficient than horses and far better than oxen as converters of chemical into mechanical energy, big domesticated animals were something of a luxury in preindustrial times. Slavery was the most efficient means by which the ambitious and powerful could become richer and more powerful. It was the answer to energy shortage. Slavery was widespread within the somatic energy regime, notably in those societies short on draft animals. They had no practical options for concentrating energy other than amassing human bodies.

An interesting feature of the somatic energy regime was its success in storing energy. In the form of heat or light or even electricity, energy is hard to store. Wind and direct solar power remain hard to store even with late twentieth-century technologies. Chemical energy in the form of plants is also hard to store, although with favorable conditions and appropriate techniques some crops can be stored for a few years, albeit with considerable wastage.

The vagaries of weather and crop pests caused the supply of food to vary greatly from season to season and year to year in preindustrial societies. This created a problem for society as a whole, and for its rulers, in that the available energy supply fluctuated uncontrollably and unpredictably over time. For rulers, the stock of human and domestic animal populations served as an energy store, a flywheel in the society's energy system. They could be put to work whether the primary energy source— plant crops—was bountiful or scarce. The stock could be built up in fat times and drawn down in lean times, but at virtually all times rulers could lay their hands on people and animals for their enterprises.

For ordinary people, livestock served the same purpose. They were a store of energy, one that could be raided when necessary to even out energy flows despite the inevitably uneven supply of staple foods. This provided households a flywheel in their domestic energy systems, proportional in size to the quantity of animals they owned (or could buy when needed).

The limits of the somatic energy regime were stringent. In a burst of effort, the human body can muster 1000 watts of power.[13] The most any society could devote to a given task, say ditch digging, dam building, or fighting, was—with people and animals as the main sources of mechanical power—a few million watts. The Ming emperors and Egyptian pharaohs had no more power available to them than does a single modern bulldozer operator or tank captain. Expanding their territo-

---

[13]Watts indicate the rate of energy use over time. One joule (the basic unit of energy) per second equals one watt.

rial domain might increase rulers' total energy supply, a goal vigorously pursued, but it could not raise the total that they could apply to a single task since it was usually impossible to concentrate more than a few thousand bodies on a given construction project or battle.

The Industrial Revolution first augmented and then quickly outstripped human muscle power. Wherever it spread, it ended the somatic energy regime, replacing it with a much more complex set of arrangements that one might call the "exosomatic energy regime," but might better be called the fossil fuel age: to date the lion's share of energy deployed since 1800 has come from fossil fuels.

From ancient times forward, notably in Persia, China, and Europe, sails, windmills, and watermills added slightly to the somatic energy supply of agrarian societies.[14] Incremental improvements followed for many centuries. But in the eighteenth century, steam engines tapped hundreds of millions of years' worth of photosynthesis, burning coal to convert chemical into mechanical energy. Coal of course had found uses for centuries, mainly as a fuel for heating. But the steam engine's capacity to convert that heat into mechanical energy capable of doing work opened up new possibilities.

The first steam engines were notoriously inefficient, losing more than 99 percent of their energy. But gradual improvements by 1800 allowed efficiency of about 5 percent and a capacity of 20 kilowatts of power in a single engine, the equivalent of 200 men. By 1900, engineers had learned how to handle high-pressure steam, and engines became 30 times as powerful as those of 1800. On top of this, steam engines, unlike watermills and windmills, could be put anywhere, even on ships and railroad locomotives. This created another positive feedback loop, in that it allowed transport of coal on a massive scale, providing the fuel for yet more steam engines. Nineteenth-century industrialization rested on this fact. World coal production, about 10 million tons in 1800, shot up 80- or 100-fold by 1900.[15]

By 1900 another major departure was underway: internal combustion engines using refined oil. A Scot, James Young, figured out how to refine crude oil in the 1850s, and an American, Edwin Drake, proved in 1859 that oil could be drilled through deep rock. The oil age had begun, albeit in a small way. Internal combustion engines, developed mainly in Germany after 1880, furthered the transition. They weighed less than coal-fired steam engines, they were much more efficient, especially at small

[14]Sørensen 1995:392–404 gives a world history of wind power.
[15]Smil 1994:186. Cipolla 1978:56 offers 15 million tons for 1800 and 701 million tons in 1900.

scales. On larger scales they could deliver much more power than steam engines. The provision of electricity needed power on such a scale; and automobiles required lightweight and efficient engines.

So from 1900 forward, biomass, coal, and oil provided large quantities of energy. In terms of usable energy, fossil fuels overshadowed biomass from the 1890s forward, even though the great majority of the world's population used no fossil fuels directly. Production and use of all three fuels grew throughout the twentieth century, although oil use grew much faster so that in proportional terms the other two declined. Some estimates of world fuel production and the usable energy derived therefrom appear in Tables 1.4 and 1.5. Not only did fossil fuels largely replaced biomass in the global energy mix in the twentieth century,[16] but the total en-

TABLE 1.4 WORLD FUEL PRODUCTION, 1800–1990

	Production (millions of metric tons)		
Type of Fuel	1800	1900	1990
Biomass	1,000	1,400	1,800
Coal	10	1,000	5,000
Oil	0	20	3,000

Source: Elaborated from Smil 1994:185–7.

Note: These figures do not reflect the energy yield of these fuels: a ton of oil gives 5–10 times as much energy as a ton of firewood, and perhaps twice as much as a ton of coal.

ergy harvest skyrocketed. The electrification of the globe, begun around 1890 and still in train, boosted demand and use of energy. Electric motors are highly flexible and have countless uses. Electricity also is good at providing light and heat. Lenin famously defined communism as electrification plus Soviet power, and rural electrification was a major achievement of Franklin Roosevelt's presidency.

The worldwide energy harvest increased about threefold in the nineteenth century under the impact of steam and coal, but then by another thirteenfold in the twentieth century with oil, and (after 1950) natural gas,

[16]The global *commercial* energy mix in 1994 was 40% oil, 27% coal, 23% natural gas, 7% nuclear, and 3% hydropower. Geothermal, wind, solar, and other energy sources combine to provided less than 1% (WRI 1996:276–7). Adding biomass (noncommercial) would scarcely affect the figures: commercial energy sources delivered about 25 times as much usable energy as biomass in the 1990s.

TABLE 1.5 WORLD ENERGY USE, 1800–1990			
	1800	1900	2000
Total (millions of metric tons of oil equivalent)	250	800	10,000
Indexed (1900 = 100)	31	100	1,250

Source: Elaborated from Smil 1994:187.

and, less importantly, nuclear power.[17] No other century—no millennium—in human history can compare with the twentieth for its growth in energy use. We have probably deployed more energy since 1900 than in all of human history before 1900. My very rough calculation suggests that the world in the twentieth century used 10 times as much energy as in the thousand years before 1900 A.D. In the 100 centuries between the dawn of agriculture and 1900, people used only about two-thirds as much energy as in the twentieth century.

This astounding profligacy, too, counts as something of a triumph for the human species, a liberation from the drudgery of endless muscular toil and the opening up of new possibilities well beyond the range of muscles. Even on a per capita basis energy use grew spectacularly, four- or fivefold in the twentieth century. In the 1990s the average global citizen (an abstraction of limited utility) deployed about 20 "energy slaves," meaning 20 human equivalents working 24 hours a day, 365 days a year. The economic growth of the last two centuries, and the population growth too, would have been quite impossible within the confines of the somatic energy regime.[18]

This energy intensification came at a cost. Here I mention two aspects of that cost. First, fossil fuel combustion generates pollution. So does, and always has, biomass burning. But because fossil fuels have more applications, their development has meant far more combustion in total, and far more pollution. Chapters 3 and 4 will address this theme. Secondly, fossil fuel use has sharply increased the inequalities in wealth and power

[17]This means total energy use has increased 80-fold since 1800. A compatible estimate comes from Starr 1996:244, who offers a 50-fold increase since 1850. Livi-Bacci 1992:28 accords well with Smil's data.

[18]It is theoretically possible that the higher yeields of modern agriculture could be obtained with careful techniques without the high energy inputs that are now conventional. But in practical terms these have been essential, and so it is fair to say that our population growth would have been impossible without high fossil fuel use.

among different parts of the world. The requisite technologies and corresponding social and political structures developed first and most thoroughly in Europe and North America. Other parts of the world generally remained dependent on biomass for heat and muscles for mechanical energy until 1950 or so. Indeed, the poorest countries remain so still. The average American in the 1990s used 50 to 100 times as much energy as the average Bangladeshi and directed upwards of 75 energy slaves while the Bangladeshi had less than one. Harnessing fossil fuels played a central (though not exclusive) role in widening the international wealth and power differential so conspicuous in modern history. This is a good thing if one prefers to see some people comfortable instead of almost all locked in poverty, but it is a bad thing if one prefers equality. In any case, inequality in energy use peaked in the 1960s. Thereafter the transition to intensive energy use spread around the world.

The exhaustion of fossil fuels on the global scale is not imminent. Predictions of dearth have proved false since the 1860s. Indeed, quantities of proven reserves of coal, oil, and natural gas tended to grow faster than production in the twentieth century. Current predictions, which will be revised, imply several decades before oil or gas should run out, and several centuries before coal might. We can continue to live off the accumulated geological capital of the eons for some time to come—if we can manage or accept the pollution caused by fossil fuels.

Conclusion

The human species has shattered the constraints and rough stability of the old economic, demographic, and energy regimes. This is what makes our times so peculiar. In the nineteenth century the world began a long economic boom, which climaxed in the twentieth century, when the world economy grew 14-fold. It expanded less than about fourfold in per capita terms, because world population multiplied fourfold in this century. Energy use embarked on a boom which began with a fivefold growth in the nineteenth century. That boom climaxed (to date) in the twentieth century with a further 16-fold expansion.

Why has all this happened now? The main answer is human ingenuity. Part of the answer is luck. First the luck: in the eighteenth century a large part of the disease load that checked our numbers, and our productivity too, was lifted. Initially this had little to do with medicine or public health measures, but reflected a gradual adjustment between human hosts and some of our pathogens and parasites. We domesticated or marginalized some of our killer diseases, quite unintentionally. This was luck. So was the ending of the Little Ice Age (c.1550–1850), which may also have had a minor role in permitting the great modern expansions.

Most of the explosive growth of modern times derives from human ingenuity. From the 1760s forward we have continually devised clusters of new technologies, giving access to new forms of energy and enhancing labor productivity. At the same time we have designed new forms of social and business organization that have helped ratchet up the pace of economic activity. Both machines and organization—hardware and software—lie behind the breakthrough of modern times.

The great modern expansion, while liberating in a fundamental sense, brought disruption with it. The surges in population, production, and energy use affected different regions, nations, classes, and social groups quite unevenly, favoring some and hurting others. Many inequalities widened, and perhaps more wrenching, fortune and misfortune often were reshuffled. Intellectually, politically, and in every other way, adjusting to a world of rapid growth and shifting status was hard to do. Turmoil of every sort abounded. The preferred policy solution after 1950 was yet faster economic growth and rising living standards: if we can all consume more than we used to, and expect to consume still more in the years to come, it is far easier to accept the anxieties of constant change and the inequalities of the moment. Indeed, we erected new politics, new ideologies, and new institutions predicated on continuous growth. Should this age of exuberance end, or even taper off, we will face another set of wrenching adjustments.

The twentieth century would appear equally unusual if one charted the long-term history of freshwater use, timber use, minerals use, or industrial output. All of these boomed after 1900. So did the generation of solid waste and of air and water pollution. Countless indicators of, and causes behind, environmental change would show much the same extraordinary story. The following pages will explore these stories, not from the perspective of the long term, but within the prodigal century itself.

THE MUSIC OF THE SPHERES

Followers of the sixth-century B.C. Greek philosopher Pythago-ras thought that the earth stood at the center of the universe's ten perfect spheres and that the spheres' motions made a har-monious music too subtle for human ears. For many millions of years, the earth's land, air, water, and living things have functioned in a complex evolving harmony, punctuated by oc-casional collisions with asteroids. Human action has added a new voice to that harmony, originally a soft one easily compat-ible with the others. Eventually that voice came to clash with the music of the spheres.

In the twentieth century, humankind rearranged atoms and altered the chemistry of the stratosphere. We made our impact felt on the smallest and the largest scales. This section of the book takes up environmental changes we wrought after 1900. It treats the rocks and soils that make up the land surface (lithosphere and pedosphere); the waters—salt and fresh—that make up most of the earth's surface, and percolate far below it (hydrosphere); the air and the lower elevations of space (atmosphere); and the community of all living things (biosphere).

This scheme of organization disguises the intimate links between the spheres. Just as history is a seamless web, so in ecology everything is connected to everything else. Coal scraped from below the earth's surface, when burned, releases gases and ash into the atmosphere. Rain washes some of this pollution out of the air and into waterways. Eventually some waterborne pollutants settle on the bottom of rivers, lakes, or seas and become embedded in sediments—back to the lithosphere again. The globe has a complex biogeochemistry in which elements, notably carbon, sulphur, nitrogen, but many others as well, cycle around among the spheres. My scheme of organization may give these cycles short shrift. But tracing the history of changes to biogeochemical cycles would conceal, by disconnecting, the cumulative impact of pollutants in the Mediterranean Sea or the air of Osaka. To most cultures, the world presents itself as land, air, water, and life, not as carbon and sulfur, and so I will treat it as such. One has to chop up the seamless web of ecology to write its history.[1]

[1]Readers can find historical biogeochemistry ably presented in Turner et al. 1990:393–466.

2

The Lithosphere and Pedosphere: The Crust of the Earth

Not every soil can bear all things.
—Virgil, *Georgics*

Humankind moved mountains in the twentieth century and for the first time became a significant geological agent. Our most consequential impacts occurred in the soil: we simultaneously corroded and enriched the substrate of civilization, so that some could soon bear nothing at all, while others eventually seemed able to bear, if not all things, at least a lot more of many different things.

This chapter considers major changes to the lithosphere and pedosphere, their causes and their consequences. Human action has altered the earth's surface biologically, chemically, and physically. Here I will entirely neglect our impact on the fungi, bacteria, rodents, and worms that inhabit the soil,[1] and focus on chemical and physical changes. They touched

---

[1] On these see Pimentel et al. 1995:1118–9.

human affairs—politics, economics, health, nutrition—in many ways in the twentieth century.

The Basics of the Earth's Crust

The lithosphere is the earth's outer shell of rock. It floats on molten rock, like a layer of scum on simmering soup. The shell is about 120 kilometers thick. In geological time the lithosphere moves: its various plates shift around, and parts of it are forced down into the magma and melt, while elsewhere magma cools, solidifies, and becomes new lithosphere. Humans notice only the sudden jolts in this slow cycling: earthquakes and volcanic eruptions. Compared with these slow but grand natural movements, the human imprint on the lithosphere seems faint.

The pedosphere is the soil, the earth's skin, a membrane between the lithosphere and the atmosphere. It consists of mineral particles, organic matter, gases, and a swarm of tiny living things. It is a thin skin, rarely more than hip deep, and usually much less. Soil takes centuries or millennia to form. Eventually it all ends up in the sea through erosion. In the interval between formation and erosion, it is basic to human survival, the source of sustenance for plants, the foundry of life.[2]

Soil Alchemy

Alchemists used to work hard trying to make gold out of baser metals. In modern times, farmers and agronomists have sought to do analogous things with the soil: make bad soil good, good soil better, and profit from it.

Without much understanding it and usually without intending to, people have long altered soil chemistry. Since the dawn of agriculture, human farming has reduced the nutrient supply in many of the earth's soils. This

[2]Hillel 1991:23–30; Rozanov et al. 1990:203–5; Stanners and Bourdeau 1995:147–8. Soil fulfills many other useful functions besides generating biomass. It filters toxins and pathogens, keeping them out of groundwater, and through the action of microorganisms neutralizes many pollutants. It helps regulate interactions between the geosphere, biosphere and atmosphere, and houses more carbon than either the atmosphere or the total biomass living aboveground.

happened on a very modest scale before cities, because most of what plants extracted from the soil soon returned to it after shorter or longer stays in animal and human alimentary canals and tissues. But with cities, human societies systematically exported nutrients from farming and grazing land. Some were returned, especially where human excrement ("night soil") was collected and distributed to farmers as fertilizer, a practice mentioned by Homer in *The Odyssey* but done most consistently in China and Japan. But much was not, and instead flowed into sewers, rivers, and the sea. In the 20th century, with its pell-mell urbanization and its vast expansion of farming and grazing, the scale of nutrient export became many times greater than ever before.

Nutrient depletion, especially of nitrogen and phosphorus, limits plant growth and hence crop yields. The history of agriculture is a struggle against this fact. Particular crops, if grown repeatedly (such as sugar, cotton, and maize), drain soils of specific nutrients. Crop rotation, an ancient practice, limited nutrient loss. Systematic use of legumes (whose associated bacteria fix atmospheric nitrogen in the soil) helped a great deal. When oceanic transport became cheap enough, in the nineteenth century, richer societies could import fossil manure in the form of Peruvian and Chilean guano to replenish the nutrient supply on their farms. But the world's agriculture would nonetheless produce far less (and the world have far fewer people) if chemists had not figured out ways to distill "superphosphate" from rock and to extract nitrogen from the air.

In 1842 an English gentleman farmer, John Lawes (1814–1900), first applied sulphuric acid to phosphate rock, producing a concentrated superphosphate which could be spread upon soil. Lawes had invented the first artificial fertilizer, and soon founded the first chemical fertilizer company. Britain and Europe had only a modest supply of suitable rock, so phosphate in Florida (after 1888) and Morocco (after 1921) was soon mined, shipped, chemically treated, transformed into superphosphate, and distributed to richer farmers in North America and Europe. The USSR developed phosphate mining above the Arctic Circle in the Kola Peninsula (1930) and then in Kazakhstan (1937) to build its fertilizer industry. After World War II, China and Jordan also developed significant phosphate deposits, as more recently did Thailand.[3] These places, together with some oceanic guano deposits, supplied the twentieth-century world with its increasingly heavy doses of superphosphate.

[3]Smil 1990:431. The world's largest producers of phosphate rock in 1994 were Morocco, Chile, Thailand, and Russia (U.S. Department of the Interior 1995:10–11).

Supplying nitrogen to the world's soils was more difficult, something that only lightning and certain microbes inhabiting the roots of legumes could do before the twentieth century.[4] Although air is plentiful compared with phosphate rock, getting nitrogen from it proved vexing to legions of late nineteenth century scientists. Then in 1909 an academic chemist, Fritz Haber, figured out how to extract nitrogen from the air through ammonia synthesis. Karl Bosch, an industrial chemist, oversaw the creation of mass production of nitrates with Haber's methods, so the process became known as the Haber-Bosch ammonia synthesis.

Haber was born in 1868 to a German-Jewish family in Breslau in what is now Polish Silesia. His father ran a dyestuff firm, for which Haber worked until he convinced his father to buy large quantities of chloride of lime, a treatment for cholera, hoping for vast profits during an 1892 epidemic in Hamburg. But the epidemic did not spread, Haber the elder was stuck with the chloride of lime, and advised his mistaken son to leave business for academia. By age 30, Haber became a distinguished chemist in Karlsruhe, doing some of the basic work that allowed the "cracking" of hydrocarbons, essential to the efficient refining of crude oil.

He was also an ardent German patriot. In 1900, German farmers used a third of the nitrates exported from the Chilean guano deposits. Without that imported nitrogen, Germany could not feed itself. Nitrogen fixation, Haber hoped, would resolve a general agricultural bottleneck and solve a German geopolitical problem at the same time. The Haber-Bosch synthesis helped Germany forestall hunger until late in World War I, despite an Allied blockade. It additionally provided nitrates for use in explosives. Haber won the 1918 Nobel Prize in chemistry; at the ceremony it was said he obtained bread from air. The Haber-Bosch ammonia synthesis became commercially lucrative soon after the War.[5]

Fritz Haber spent World War I creating poison gas for the use of the German military (which so distressed his wife that she committed suicide in 1916). After Germany's defeat, Haber labored from 1919 to 1926 trying to extract gold from seawater so as to help Germany meet its postwar reparation payments. His patriotic services did him no good after the Nazi accession to power in 1933; Haber resigned his post (director of the world's most prestigious physical chemistry research laboratory) under

[4]Lightning creates ammonia from atmospheric nitrogen, some of which rain carries into the soil and biosphere.

[5]German nitrate producers formed an export syndicate in the 1920s which met with some success, at least in Egypt (Friedrich 1993).

pressure, emigrated to Britain, and died within a year. But he more than anyone else shaped the world's soil chemistry in the twentieth century and allowed agriculture to flourish despite myriad forms of soil degradation.[6] "Whoever could make two ears of corn . . . to grow upon a spot of ground where only one grew before, would deserve better of mankind . . . than the whole race of politicians put together." So wrote Jonathan Swift in *Gulliver's Travels* (1726). Nearly two centuries later Haber did it.[7]

The Haber-Bosch process took a lot of energy, but wherever energy was cheap the manufacture of nitrogenous fertilizer proved feasible. However, economic conditions (notably the Depression of 1929–1938), slowed the development of nitrogen fertilizer use until after World War II. By 1940 the world used about 4 million tons of artificial fertilizer, mostly nitrogen and superphosphate, but also potassium fertilizers, derived from potash. By 1965 the world used 40 million tons, and by 1990 nearly 150 million.[8] This development was and is a crucial chemical alteration of the world's soils with colossal economic, social, political, and environmental consequences.

For one thing, artificial fertilizers allow perhaps an extra 2 billion people to eat. Without pumped-up yields, the world's population would need about 30 percent more good cropland—a tall order.[9] For another, it systematically widened the gaps between rich and poor farmers around the world between 1950 and about 1985, both within societies and among them. Before the 1960s, poorer countries used little artificial fertilizer, and their farmers found it increasingly difficult to compete with the growing grain surpluses of North America and Australia. After 1970, much of the increase in fertilizer use came in the poorer countries, but mainly on larger farms that could more easily afford the costs. Smaller farmers went to the wall. In Japan, Korea, and more recently in China, high levels of artificial fertilizer use allowed greater labor efficiency in rice paddies; millions of ex-peasants became urban workers, staffing the economic miracles of these lands. But in India, Mexico, and the Philippines, and other countries where large farmers made far more use of artificial fer-

---

[6]Haber's importance is discussed in Smil 1994:182, 189–90 and Smil 1993:165. A biography is Goran 1967. Bosch became chief of the giant firm IG Farben in 1925 and won the Nobel Prize in chemistry in 1931.

[7]Hillel 1991:129, 132.

[8]Figures are from Brown et al. 1996:9 and Solbrig and Solbrig 1994:215.

[9]Smil 1993:165 calculates that nitrogenous fertilizer alone supports one in three or one in four of the earth's inhabitants. East Asian and northwestern European agriculture are the most dependent on chemical fertilizers (together with some small sugar islands like Mauritius).

tilizers than small ones, displaced farmers contributed more to urban social strains than to economic miracles. So Lawes and Haber helped to shape the social structures of contemporary societies and the international division of labor.

Fertilizers mostly miss their targets and become water pollutants. Estimates vary, but usually somewhat more than half of fertilizers applied end up in the waters of agricultural communities and their downstream neighbors. This contributed mightily to the eutrophication of rivers, lakes, and seas, especially in Europe and North America. (Eutrophication means excessive nutrient supply; see Chapter 5). Moreover, even when fertilizers stayed in the soil, long-term "chemotherapy of the land" often led to problems in micronutrient supply, handicapping instead of helping farming.[10]

The impact of chemical fertilizers did not stop with chemistry. They strongly influenced the choice of crops after 1950: those that respond well to fertilizers (maize, for example) spread far and wide, replacing those that do not. Hence the trend toward more and more people eating fewer and fewer varieties of food: two-thirds of our grains now come from three plants: rice, wheat, and maize. Chemical fertilizer use also made food production thoroughly dependent on the fossil fuels needed for fertilizer production: our food is now made from oil as well as from sunlight. On the large scale, it has recast the global nitrogen and phosphorus cycles and strongly favored all species that thrive on heavy diets of these nutrients. What such basic planetary changes may mean for humanity remains unclear.

Soil Pollution

While the chemical industries of the twentieth century replenished key soil nutrients, they also helped contaminate soils. As a rule, soil pollution developed wherever chemical and metallurgical industries emerged, chiefly in Europe, eastern North America, the Soviet Union, and Japan.

---

[10]The phrase is Wes Jackson's quoted in Opie 1993:257 and Olson 1987:220–21. Shortages of zinc can result from high phosphorus loads; heavy use of nitrogen and potassium often leads to manganese shortage. Macronutrients are nitrogen, phosphorus, and potassium, needed in quantity by all plants; micronutrients, iron, zinc, and others too numerous to list, are needed in trace amounts—but needed.

One major source of soil pollution was the mining, smelting, refining, and use of metals such as lead, cadmium, mercury, and zinc. These metals proved helpful in modern chemical and metallurgical industries. Twentieth-century industrialization required them in quantity. But even in small doses these metals are dangerous to human (and other) life—although in still smaller doses, one of them, zinc, is essential. In the twentieth century they infiltrated soils (as well as air and water) on a magnified scale; lead and cadmium emissions, for instance, each increased about 20-fold from 1875 to 1975.[11] Most of these metals enter circulation first as air pollutants, some via water effluent, while some are dumped directly in the soil. Whatever their pathways, when such metals get into soils, they enter the food chain.

Japan provided a dramatic illustration. Surges in mining and metallurgy after the Meiji restoration (1868) brought acute heavy metal pollution. Copper contamination reduced rice yields around several mines at the end of the nineteenth century, but only on a local scale. Mining and smelting delivered heavy metals to rice paddies in several Japanese river basins early in the twentieth century, often provoking farmer protests. In 1926, in the Bamboo Spear Affair, farmers whose land suffered from copper contamination besieged the smelter of the Kosaka mine. In the Jinzu River valley a few cases of a bone disease, now called *itai-itai* ("ouch-ouch"), appeared before World War II. It turned out to be a consequence of cadmium poisoning. Hundreds more cases followed the war. Heavy-metal production and pollution took off after 1950, as the Korean War jump-started Japanese heavy industry. By 1973 Japan led the world in zinc production and ran a close second in cadmium output. Through various pathways, heavy metals contaminated irrigation water. When that water drained away from rice paddies, it left behind a residue of concentrated contamination. Cadmium, in particular, is easily absorbed by rice plants. By 1980 about 10 percent of Japanese paddies had become unsuitable sources of rice for human consumption because of cadmium pollution of the soil. Cadmium and other heavy metals, ingested through food, killed hundreds of Japanese and made thousands sick in the course of the twentieth century. The juxtaposition of mining, smelting, and rice paddies made heavy-metal soil pollution more serious in Japan than anywhere else.[12]

Elsewhere soil pollution affected society and ecology too, if less painfully than in Japan. Soils, and consequently the vegetables and peo-

[11]German Advisory Council on Global Change 1995:86.
[12]Asami 1983, 1988.

ple, of Polish Silesia showed unhealthily high concentrations of cadmium, mercury, lead, and zinc by the 1970s. Soils downwind of the smelters of Sudbury, Ontario acquired concentrations of nickel and copper about 400 times background levels, killing almost all vegetation. After 1970, forests and grassland soils around the world showed signs of elevated heavy-metal concentrations, although health dangers existed only in a few industrial districts. In the 1980s northern Indian soils, for instance, showed contamination levels one or two orders of magnitude lower than those of Hesse (Germany).[13] Urban soils in the twentieth century accumulated concentrations of trace metals 10 to 100 times greater than background levels. After the mid-1970s, airborne emissions and thus trace-metal deposition on soils declined, especially of lead and cadmium, a consequence mainly of regulations in the industrial world. But lead persists for 3,000 years in the soil, so this particular legacy of the twentieth century—which has had demonstrable ill effects on health—will long endure.[14]

Industrialization generated all manner of toxic wastes besides metals. Man-made chemicals existed only after the midnineteenth century and acquired environmental significance only after the midtwentieth. Roughly 10 million chemical compounds have been synthesized since 1900; perhaps 150,000 have seen commercial use.[15] Synthetic chemical production (by weight) expanded 350-fold between 1940 and 1982. The industries that used these chemicals generated quantities of wastes, much of which was hazardous. No historical data exist on toxic waste production (and definitions differ widely), but it is likely that levels increased slowly before 1940, but quickly after 1950 when the chemical industries thrived. A large proportion of toxic wastes ended up in landfills (50–70% perhaps). Smaller amounts, generally in urban areas and in enclaves of chemical, petroleum, and metallurgical industry, went directly into soils, dumped legally or otherwise along roadsides, in parks, and on private land.

Before 1980 such wastes generally attracted only passing notice: they were part of the cost of doing industrial business. In the United States in 1936, chemical companies dumped 80 to 85 percent of their toxic wastes,

---

[13]German Advisory Council on Global Change 1995:84.

[14]Asami 1983; Kitagishi and Yamane 1981; Logan 1990; Nriagu 1990a, 1996. Anthropogenic dispersal of trace elements (from mining, smelting, fuel combustion, etc.) far outstripped the work of volcanoes, forest fires, sea spray, and other natural forces by 1980. The relevant ratios: arsenic, 3 to 1; cadmium, 7 to 1; lead 25 to 1; mercury, 11 to 1 (Brown et al. 1990:439). On lead in soils and health, see Mielke et al. 1983.

[15]Tolba and El-Kholy 1992:249 estimates 100,000 in use currently; Prager 1993 estimates 80,000.

untreated, into adjacent pits, ponds, and rivers.[16] But in 1976 to 1980, in Love Canal (near Buffalo, New York), evidence accumulated of cancers and birth defects apparently deriving from chemicals buried there between 1942 and 1953. The Hooker Chemical Company, after burying a toxic stew, sodded over the land and deeded it to the local community, which built a school and residences on the site. By 1980 the federal government evacuated thousands and Love Canal was fenced off, officially a national disaster area. Soon hundreds of other communities made the equation between health problems and chemical dumps. Toxic waste dumps—of which in 1980 there were 50,000 in the United States alone—became a focal point of environmental politics, legislation, lawsuits, and a modest cleanup effort associated with the so-called Superfund, a federal effort to finance environmental restoration of hazardous waste sites. Disposal of toxic chemicals became an increasing problem in the United States and Europe, where legislation discouraged former patterns of casual disposal.

Exporting hazardous wastes for disposal elsewhere became an international business in the 1970s. Mexico buried and dumped U.S. wastes, Southeast Asian countries accepted some of Japan's, Morocco and some West African countries took wastes from Europe and the United States. Perhaps the biggest importer before 1989 was the former East Germany. One of the ironies of the reunification of Germany (1989–1990) is that West Germans reacquired hazardous waste problems they had (temporarily) exported. By the late 1980s the international trade in toxic waste involved millions of tons annually, and the spectacle of rich countries paying poor ones to take their poisons aroused political resistance. In 1987 a shipload of toxic incinerator fly ash from the United States toured the Atlantic looking for a country willing to take it. A spate of protocols and conventions in the late 1980s and 1990s aimed to regulate this international toxic trade, much of which was illegal.[17]

Modern armed forces are great users of synthetic chemicals. In both the Soviet Union and the United States the biggest single contaminator of soils after 1941 was the armed forces, often beyond regulation during World War II and the Cold War. Soils and groundwater in many of the military reserves in the United States and USSR, and in some foreign

---

[16]Colten 1994.

[17]J. Clapp 1994 suggests the volume by the late 1980s was 30 million to 45 million tons, of which 20% went to Third World countries. Prager 1993 has slightly lower estimates. A high-ranking economist at the World Bank insisted the international trade in toxic wastes was economically rational and should be encouraged, not forbidden.

bases, became so contaminated that cleaning up looked impossibly expensive.[18]

Before 1950 the cumulative effect of chemical poisoning of soils came to little. After that, more and more locales became contaminated. Soil pollution problems metastasized after 1975, following the migration of heavy industry from Europe, North America, and Japan to Korea, Taiwan, Brazil, and elsewhere. Everywhere, soil pollution centered on urban and industrial zones. To a smaller extent it affected rural soils, through agricultural chemicals and the airborne deposition of nitrogen, sulfur, and trace metals. Soil contaminants working their way into water, the food chain, and human bodies changed millions of lives for the worse, and shortened thousands.

Earth Movers

Natural forces move a lot of rock around. Volcanic eruptions, tectonic movements, scouring glaciers, and natural erosion have long sculpted the face of the earth. In the twentieth century, humankind came

TABLE 2.1 AVERAGE ANNUAL TRANSPORT OF ROCK AND SOIL

	Billion Tons
Wind erosion	1.0
Glaciers	4.3
Mountain building	14
Oceanic volcanoes[a]	30
Humankind[b]	42
Water[c]	53

Source: Hooke 1994.
[a]This refers to the midoceanic ridge upthrusts of new rock.
[b]Hooke offers estimates of 40 billion and 45 billion tons, depending on assumptions.
[c]The figure for water transport includes (nonhuman) sediment delivery to lakes, seas, and oceans (c.14 billion tons) and silt moved around within watersheds (39 billion tons).

[18]In the USA, the military released or stored 33 billion cubic meters of hazardous wastes at more than 10,000 sites. The estimated cleanup costs are $170 billion to $370 billion, and the job will take 75 years (USDOE 1995). On the Soviet legacy in the former East Germany, see the German Advisory Council on Global Change 1995:175.

to rival these forces as a geological agent, mainly through mining and accelerated erosion. Table 2.1, composed of rough estimates, suggests how far humanity has come as a geological force as of the 1990s.

At the beginning of the twentieth century, human geological impact probably came to less than a tenth of its 1990s proportion, putting it on a par with glaciers.[19] In some places, such as England, people in 1900 already moved far more rock and soil than did nature. But England, with its giant coal industry and comparatively feeble natural forces, was unusual.[20]

MINING. People have long scratched and dug the earth in search of useful metals and fuels. The scale, even at the height of Roman (1st–3rd centuries A.D.) or Song Chinese (10th–12th centuries) operations, remained small until after 1820. Industrialization provoked a frenzied quest for metallic ores after 1870, and steam engines led to a surge in demand for coal. Coal extraction (Table 2.2), for example, at 10 million tons in 1800, grew eighty- or a hundredfold in the nineteenth century and another six- or sevenfold

TABLE 2.2 WORLD COAL OUTPUT, 1850–1995

Year	Output (million metric tons)	Index (1900 = 100)
1800	10	1.3
1850	76	10
1875	283	37
1900	762	100
1925	1,358	178
1950	1,800	236
1975	3,257	427
1995	5,000	656

Sources: Headrick 1990:60; Erickson 1995:78. Similar figures appear in Smil 1994:186.

[19]This very rough guess assumes that the population was one-fourth as great in 1900 as in 2000, that the global economy was about one-fourteenth as large, and that the world economy involves much more mining now than it did then.

[20]Sherlock 1931:238.

in the twentieth. Modern coal mining first centered in Britain, which at one point exported coal to India and Argentina. But soon coal miners burrowed into the earth all round the world; the United States led from the 1890s to the 1950s, then the USSR, and after about 1980, China. Iron ore followed much the same trajectory, mined on large scales first in Britain, then in Germany and the United States by 1890, and in the USSR by 1930. By the 1990s, China, Brazil, Australia, and Russia led the world in iron-ore extraction. Gravel, sand, and other construction materials made up a huge share of mining activity, focused wherever urban growth required building materials. By 1980 or so, quarrying moved more earth than did natural erosion.[21] By the 1990s, U.S. miners moved about 4 billion tons of rock per year, and the world figure was perhaps four to five times that.[22]

All this mining corroded the lithosphere with a warren of underground shafts and chambers, and after the appearance of the requisite earth-moving machinery, pockmarked the earth's surface with thousands of huge open pits, mainly in the United States, Russia, Germany, and Australia. It also generated mountains of waste rock and slag, and filled rivers with slurry and silt.[23] The coal districts of Britain erected small sierras of slag, one of which slid down onto a village in Wales in 1966. In the Sacramento River valley, hydraulic gold mining (1880–1909) increased the silt load 10-fold; the Perak River in Malaysia (tin mining) and the Clutha River in New Zealand (gold) ran dark with silt from hydraulic mining at roughly the same time.[24] Modern mining can alter landscapes and lives for miles around, as it did on New Caledonia.

A cigar-shaped island the size of New Jersey or Kuwait, New Caledonia lies almost equidistant from New Guinea, Australia, and New Zealand in the southwest Pacific. Captain James Cook named it after the Roman term for Scotland, with which it has nothing in common except mountains. French missionaries went to work among the Melanesian population in 1840, and France annexed New Caledonia in 1853, adding to its Pacific empire. Twenty years later, prospectors found deposits of nickel,

[21]Nir 1983:70.

[22]Hooke, cited in Monastersky 1994. Ryabchikov 1975:142 estimated that humankind moved 3,000 km² annually with plows, extracted 100 billion metric tons of ore, fuel, rock, and gravel.

[23]Between 1950 and 1990, mining in the USA "destroyed" an area of fertile soil the size of New Jersey (Arnold et al. 1990:77).

[24]Meade et al. 1990:266. Hydraulic mining, practiced in New Zealand and Australia too, involved blasting away loose rock and soil with high-pressure hoses. In California it was made illegal after 1909, by which time it had moved eight times more earth than the excavation of the Panama Canal.

a hard, corrosion-resistant metal that was eventually found to be useful in making aircraft, armaments, and in nuclear power generation. New Caledonia happened to contain between a quarter and a third of the known oxidized nickel in the world under its mountain summits.

Early mining used picks and shovels, and immigrant labor from Japan, Java, and Vietnam. By 1926 New Caledonia led the world in nickel production. For much of the postwar half-century it ranked behind only Canada and the USSR. Indonesia overtook it in 1994. Between 1890 and 1990, miners employed by the Société le Nickel (SLN) moved half a billion tons of rock to get 100 million tons of ore and 2.5 million tons of nickel. This took the form of opencast mining, which after World War II involved beheading mountains.

Mining in the age of fossil fuels rearranged landscapes quickly and thoroughly. This is a Chilean copper mine at Chuquicamata in 1944. Copper was one of the important raw materials at midcentury, in large part because copper made good wire for carrying electric current. As industry and homes converted to electricity, mines like this one grew very busy around the world, from Japan and New Caledonia to Zambia and Utah. Wireless communications and fiber optic cable reduced the demand for copper after 1980, closing down some of the world's copper mines.

The environmental and social effects proved profound. To get at the nickel, miners decapitated the ridges. Streams filled with silt and debris, making fishing and navigation impossible. Floods and landslides destroyed lowlands, dumping gravel on arable land and shearing away coconut groves. Silt smothered offshore corals in one of the world's largest lagoons. Many Kanaks (as the Melanesians of New Caledonia are called) lost their livelihoods, their homes, and their lands in the first decades of nickel mining. Smelters, built locally because of the high cost of shipping raw ores to European markets, filled the air with smoke and noxious gases. Kanaks and missionaries complained to no avail. In the 1930s, sulfuric fumes dissolved the roof on a mission church. All this combined with wage labor, cash cropping, cash taxation, and the influx of immigrant miners to disrupt Kanak life. But the environmental and social change had just begun.

After 1950, bulldozers, hydraulic shovels, and 40-ton trucks replaced picks and shovels, and the scale of production increased 10-fold by 1960 and 100-fold by 1976, driven by Japanese industrial expansion and boom times in the world armaments business associated with the Cold War. The environmental and social dislocation intensified, contributing to an independence struggle and political violence that wracked New Caledonia in the 1980s. In that decade, the French government began to impose environmental regulations on the SLN's active mines, but the pollution, erosion, and siltation from abandoned mines will continue for decades, if not centuries.[25]

The history of nickel mining in New Caledonia is an extreme example of modification of the lithosphere, and of consequent environmental degradation and social disruption. Had the mines been in France itself, matters might have been different. But with imperialism and industrialization after 1880, more and more of the world's mining took place in parts of the world where constraints were few and ineffective. Recent parallels occurred around other great mines of Melanesia: the Panguna mine in Bougainville, Papua New Guinea (after 1960); the Ok Tedi mine in central New Guinea (after 1980); and the Freeport mine in Irian Jaya (after 1980).[26] Variations on the New

---

[25]Dupon 1986; Winslow 1993. In 1994, nickel accounted for a quarter of New Caledonia's GNP (U.S. Department of the Interior 1995:587).

[26]Hyndman 1994. Mining at Ok Tedi loaded the Fly River, New Guinea's largest, with heavy metals. Panguna mine tailings completely killed the Jaba River on Bougainville, where pollution and erosion were among the political issues behind a rebellion in the late 1980s in which the locals sought independence from Papua New Guinea. The copper and gold mine at Freeport

Caledonia theme occurred in Chile, Australia, Zambia, Siberia, Utah, and elsewhere.[27]

THREE PULSES OF SOIL EROSION. Soil erosion is as old as the continents. Accelerated soil erosion is that which human action provokes, and it is as old as agriculture—and on a trivial scale much older still. Nowadays, people induce about 60 to 80 percent of all soil erosion. That proportion, it is safe to say, is at an all time high.[28]

Soil erosion is a multiple menace as well as an ancient one. Soil loss from farmers' fields lowers crop yields. Lost soil must go somewhere else, and often it finds inconvenient lodging from the human point of view. Eroded soil ends up in reservoirs and lakes, affecting aquatic life. It silts up shorelines, harbors, and river channels, requiring dredging. Before modern dredging machines, siltation frequently forced the abandonment of ports, as at ancient Miletus in Asia Minor. Eroded soil often takes centuries to make its way to the sea.[29]

In the long haul of human history, soil erosion has surged three times.[30] The first pulse came when agriculture in the Middle East, India, and China emerged from the river valleys and spread over former forest lands. This of course happened slowly, say between 2000 B.C. and 1000 A.D., as states, economies, and populations grew—and as iron tools made clearing forests easier. Wherever existing vegetation was cut or burned to make way for crops or animals, faster erosion resulted. Typically these higher rates declined if stable farming or grazing systems developed, but rates rarely dropped to previous levels under natural vegetation. And in any case, stable agrarian systems normally were not stable for long, but came and went under the impacts of pandemics, war, migration, and climate change. So while this ancient pulse of soil erosion peaked long ago, it is still not a spent force.

---

polluted local soils and waters with heavy metals. In all these cases locals suffering the effects of mining's ecological disturbances have turned to political violence, leading to the use of Australian, Papuan, and Indonesian troops to protect the mines. (Papua New Guinea and Bougainville were administered by Australia until 1975.)

[27]Utah has the world's biggest man-made hole, the Bingham Canyon copper mine (Goudie 1985). Ripley et al. 1996 detail the environmental impact of Canadian mining; Young 1996:105–30 does the same for Australia.

[28]For estimates see Alexander 1993:230, Judson 1968:373, and Lal and Pierce 1991a:2. My colleague Tim Beach, a soil scientist, regards the 60–80% figure as too low.

[29]Meade et al. 1990.

[30]This is adapted from Dregne 1982; see also Butzer 1975.

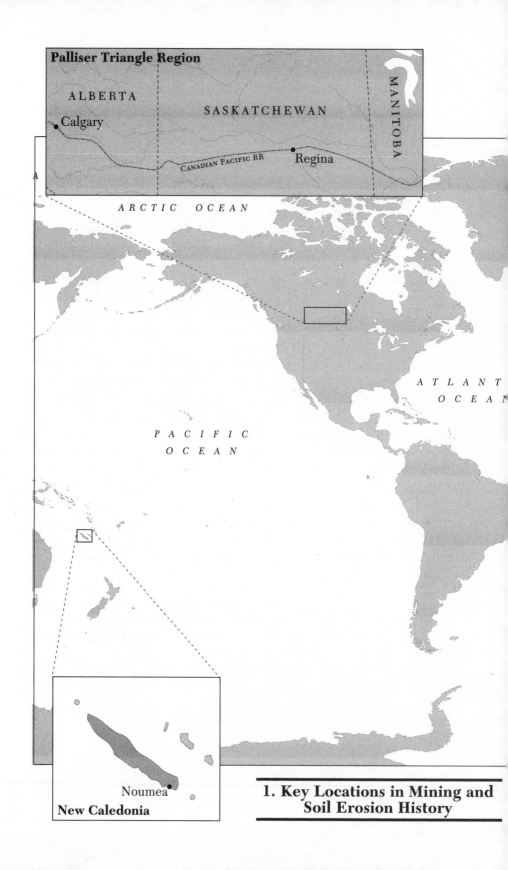

Palliser Triangle Region

ALBERTA

Calgary

SASKATCHEWAN

MANITOBA

Canadian Pacific RR   Regina

ARCTIC OCEAN

ATLANTIC
OCEAN

PACIFIC
OCEAN

Noumea

New Caledonia

**1. Key Locations in Mining and
Soil Erosion History**

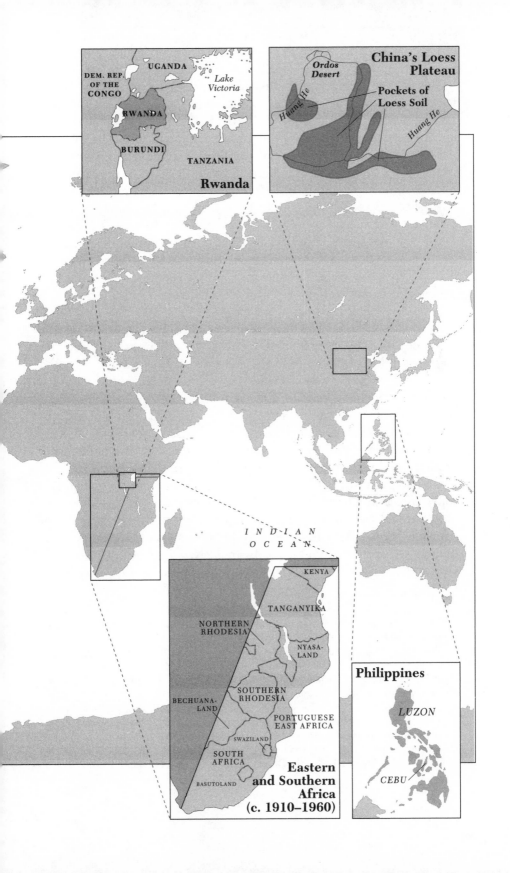

China's loess plateau typifies the first pulse of soil erosion history. Home to some 40 million people in an area about the size of France, the loess region is located in the middle reaches of the Huanghe (Yellow River) valley. It is one of the world's most easily eroded landscapes. The soil consists of very deep windborne deposits, blown in from Mongolia over the course of 3 million years. It is loose and easily dislodged. Rains come in intense summer cloudbursts. Before cultivation, when forests covered most of the loess plateau (prior to 3,000 years ago),[31] erosion carried off about 1.1 billion tons of soil per year. The Yellow River was called the Dahe, or Great River. Over the next 2,000 years, intermittent cultivation cleared most of the plateau, erosion increased, and the Yellow River acquired its present name. By the early twentieth century, soil loss reached 1.7 billion tons annually, and by 1990, 2.2 billion.[32]

The second global surge in soil erosion came with the frontier expansion of Europe and the integration of world agricultural markets. This pulse began with the European conquest and Euro-African settlement of the Americas after 1492. The conquest provoked a demographic catastrophe in the Americas, and in thickly settled mountainous regions of the Andes and Central America, agricultural terraces fell apart and soil erosion spurted. Grazing animals, introduced by Europeans, flourished in the Americas, and their hooves loosened yet more soil. This pulse of soil erosion weakened when population stabilized around 1650 to 1700.[33]

After 1840, however, it strengthened again when some 60 million Europeans migrated overseas or to Siberia. Many settled in cities, but millions worked the land throughout temperate North America, South America, South Africa, the Maghreb (Mediterranean Morocco, Algeria, Tunisia, and northwestern Libya), Australia, New Zealand, and the North Caucasus.[34] For most of these regions, the European invasion brought the

---

[31]This is the conventional view, supported by Fang and Xie 1994 and Ren and Walker 1998, questioned in Menzies 1996:556–8.

[32]Wen 1993:73–5; Ren and Zhu 1994. Since the river changed course in 1855, its delta expanded 50 km into the Yellow Sea. Deltaic growth has slowed since 1970 because of new dams that trap sediment (Milliman et al. 1987). The middle reaches of the Huanghe are 50% silt by weight according to Lal 1990:145.

[33]Smith and Baillie 1985. For the long-term erosion rhythms of Mexico, see Heine 1983.

[34]In the USA around 1910, for example, one-eighth of immigrants lived on the land, mainly Germans, Scandinavians, and Britons but also a sixth to a tenth of Irish, Italians, and Poles. Overall, in the period 1870–1920 some 10–15% of American farms were run by immigrants, and in Wisconsin, Minnesota, and the Dakotas as much as 60–65% (Conzen 1990).

first serious and protracted soil erosion.[35] Most of those who left their homes were peasants, and most of them came from northern Europe—a matter of vital significance.

Their experience of land and agriculture was a very unusual one. Northern Europe, from Ireland to Poland, enjoys a mild rainfall regime and, for the most part, has low slopes and heavy soils resistant to erosion. Cultivators could leave fields bare at any time of year and let hoofed animals roam without any serious risks of erosion. The soft, gentle rains lacked the energy to dislodge soils. But the same systems of cultivation and grazing, when transported to landscapes with lighter soils, steeper slopes, and more intense rainfall, led to devastating erosion in the Americas, South Africa, Australasia, and Inner Asia. Had the conquerors and colonizers of the modern world come from a different environment, one that did not invite negligence of soil conservation, then this second pulse would have been faint.

Moreover, European settlers usually had the power to shunt local populations onto marginal lands. Consequently, steep lands in particular, but those with unstable soils too, came under the plow or digging stick, often for the first time, opening them to a burst of erosion. This process was especially characteristic of the zones of European settlement in southern, North, and East Africa. In South Africa, Rhodesia (now Zimbabwe), and Kenya, for example, European power allowed the appropriation of the better lands for white farmers. African farmers had worked these lands with hoes and digging sticks (but not plows) for centuries. But after about 1890, European farmers introduced plows and commercial agriculture, planting wheat, tobacco, coffee, and other crops. This provoked a spate of erosion in the white highlands of Kenya and the "commercial" (white-owned) lands of Rhodesia. Meanwhile Africans farming on their own account were shunted onto less promising land—steeper or dryer or both. In South Africa, Basutoland (now Lesotho), Northern Rhodesia (now Zambia), and Kenya, people and livestock huddled into smaller areas than before, making it hard to refrain from farming unstable soils. Black farmers in South Africa created an organization concerned with soil erosion (among other issues) in 1918. Erosion in many parts of South Africa reached alarming proportions by the 1930s. Authorities took notice, spurred by the example of the American response to the Dust Bowl, and tried to impose conservation schemes. Jan Smuts, the dominant political

[35]Butzer 1975:70.

figure in South Africa, gave voice to a widespread view, at least among whites, when he said of South Africa "erosion is the biggest problem confronting the country, bigger than any politics."[36] Government estimates suggested erosion had reduced agricultural productivity by a quarter in 25 years. However, official antierosion measures proved unpopular—occasionally to the point of rebellion—because they usually involved forced labor or compulsory culling of cattle. In Lesotho, such schemes, poorly designed and grudgingly implemented, may have promoted rather than retarded erosion.

In southern African cases, accelerated soil erosion derived from a complex mix of social forces, not least the politics of settler societies. White settlement, culture, and inappropriate technique played a role: plows and furrows invited erosion far more than cultivation by hoe had done. In Basutoland at least, successful missionaries converted Africans from their animist beliefs, and inadvertently removed cultural constraints on tree cutting, promoting deforestation and erosion. Moreover, the incentives and pressures of cash cropping mattered. Population growth (after about 1920) played a modest role. But the political decisions to restrict Africans to native reserves played a much larger one.

Another factor, a controversial one, was the role of southern and East African culture, specifically communal land tenure and cultural attachment to cattle. Unless well regulated, communal land tenure provided incentives for individuals, families, and lineages to maximize short-term yield through overgrazing or hasty soil husbandry. They could keep the fruits of their efforts while spreading the costs among their neighbors. Since cattle were a preferred store of wealth and sign of status, Africans had a general incentive to maximize their herds and overgraze unless societal rules discouraged it. Social regulation, which may have worked in more tranquil times, proved difficult to maintain under the various pressures of colonialism, marketization, and long-distance labor migration, all of which disrupted southern African life between 1890 and 1960.[37]

---

[36]Jacks and Whyte 1939:21. The novelist Elspeth Huxley, of Kenya, also regarded soil erosion as a particular menace, and regarded antierosion measures as more important than politics. See her obituary in the *Economist,* 18 January 1997:86.

[37]Anderson 1984; Beinart 1984; Khan 1997, SADCC 1987; Showers 1989; Stocking 1985; Whitlow 1988; On parallels in Northern Rhodesia (Zambia), see Pletcher 1991. Jacks and Whyte 1939:247 offer the following: "From Cape Town to Cairo, European influence has been responsible for the rapid, and in places now uncontrollable, biological deterioration of the land." Antierosion schemes remain unpopular in many places because they interfere with grazing or cultivation, as in Ethiopia. See Campbell 1991.

In many other parts of the overseas European frontier, but especially in North America and Australia, land was cheap, sometimes simply there for the taking. This helped lure migrants. But it also discouraged them from investing money, time, or care in their lands, since they could expect to move on easily enough if they destroyed one parcel of land.

Furthermore, canals, railroads, steamships, and telegraphs knitted world markets together as never before, so after 1870 it made economic sense for new settlers to plow up the North American prairie and to run tens of millions of sheep on the lower slopes of New Zealand's Southern Alps. In such regions, overseas migrants had an ecological impact far out of proportion to their numbers, because they produced goods above and beyond their own requirements in order to sell to burgeoning urban populations far away.[38]

The Palliser Triangle of western Canada shows this process clearly. This is a semiarid wheat belt in the prairie provinces, including southern Alberta and Saskatchewan as well as the southwest corner of Manitoba. In 1857 the Royal Geographical Society sent John Palliser, an Irishman, to reconnoiter. He found the land inhospitable and unsuited to any settlement. It belonged to nomadic Indians and the buffalo before the Canadian Pacific Railway went through in 1885. A trickle of hopeful settlers followed, often enticed by the railroad's propaganda. When after 1897 the prairies enjoyed a run of rainy years, the trickle of settlers swelled to a flood. Population grew about 15-fold between 1901 and 1915. Rudyard Kipling, who passed through in 1907, thought that he was present at the creation of a new "Nineveh."[39] High wheat prices around the world during World War I, and bumper crops in 1915 and 1916, inspired new railroads, towns, and more settlers.

Settlers came mainly from humid lands in eastern North America and Europe, although not a few moved north from the U.S. prairies. They all sought to preserve soil moisture in the summer by leaving fields fallow, an idea widely promoted by professors and agronomists. But the prairies are a windy place, and by the 1920s this practice combined with dry years led to serious wind erosion. Thousands of farm families, mainly in Alberta,

---

[38]Of course, the opening of these agricultural frontiers connected to urban populations sometimes meant that existing fields could be retired, and could perhaps thus lower erosion rates in some parts of the world. No doubt this happened, but only on a small scale, because with growing populations around the world more land in total was tilled and grazed. The net effect of abandonment of fields came to little (except where terracing was involved), because newly cultivated lands typically eroded far faster than those already long in use.

[39]Quoted in Jones 1987:30.

gave up. Drought hit Saskatchewan in the 1930s. Dust storms darkened the skies. Three or 4 million hectares (the size of Belgium) of prairie lands were "completely destroyed." Dust blew east to Ontario, and in 1934 to the Atlantic. Social and economic distress matched that of the well-known Dust Bowl of the American plains (c.1931–1938) and spurred the success of unorthodox politics on the western prairies in the form of the socialist Canadian Commonwealth Federation and in Alberta the right-wing populist Social Credit movement. The Canadian dust bowl also provoked a major exodus, like that from Oklahoma and Kansas. The sad tale of farming in Palliser's Triangle was one of boom, erosion, and bust.[40]

Indeed, the widespread droughts of the 1920s and 1930s and the whipsaw effect of economic instability brought the second pulse of world soil erosion to a climax. The cycle of booms and busts in the world economy was pronounced in those decades, as economic globalization proved destabilizing and Keynesian fiscal policies to dampen fluctuations did not exist yet. These unrestrained cycles translated into agricultural expansions and abandonments, which, like the weathering produced by freeze and thaw, dislodged soils around the world. G. V. Jacks and R. O. Whyte, high apostles of soil conservation in the mid-twentieth century, thought "more soil was lost from the world between 1914 and 1934 than in the whole of previous history." An exaggeration, no doubt, but one with a grain, perhaps a clod, of truth to it.[41]

The European settlement frontier filled up attractive farmland by about 1930, except in Soviet Kazakhstan, where the process recurred in the 1950s and 1960s in the concentrated and accelerated form characteristic of Soviet projects. With experience, farmers and graziers often learned to limit erosion from the high levels of initial settlement. But almost nowhere in these frontier lands did farmers and graziers manage to reduce erosion either to levels that prevailed before their arrival or to northern European levels.

The second pulse, like the first, provoked its antithesis: sustained effort at soil conservation. Farmers had carefully preserved soils for millennia, and in some settings (such as Inka Peru) attained a high standard. But in

[40]Anderson 1975; Jones 1987; Stark 1987. On the American Dust Bowl, see Worster 1979. Wind erosion brought on by inappropriate agriculture also ravaged Kazakhstan in the 1950s and 1960s, and Australia with every drought since that of the late 1890s.

[41]Jacks and Whyte 1939:213. For other instances of accelerated erosion in this second pulse, see Bahre 1979:76–7 (on Chile), Barker and McGregor 1988 (on Jamaica), Barrett 1997:ch. 3 (on the North Caucasus), Beach 1994 (on Minnesota), Molina Buck 1993 (on Argentina), and Wilson and Ryan 1988 (on Ontario).

most societies, especially those where labor was more scarce than land, soil conservation had few followers and no official support. In the early twentieth century some concerns about soil loss emerged, not least in South Africa. But soil conservation was "thorny, it [was] packed with political dynamite, and it [would] always keep for another couple of years."[42] Then in the early 1930s, drought struck the U.S. southern plains, and land recently broken to the plow began to blow away. In 1934, red Nebraska dust blew into the corridors of power in Washington, D.C., into the lungs of legislators, and soil conservation soon became a major American policy initiative.

American soil scientists spread this gospel, especially within the British colonial empire but also in Mediterranean Europe and China. Simultaneously, Soviet authorities sponsored American-style shelterbelts to check wind erosion in Ukraine. Dozens of countries set up soil conservation agencies, often closely modeled on the American example. China organized against erosion in the 1950s. Sustained government research, outreach, and funding sometimes produced favorable results. In the loess hills of the upper Mississippi valley, for instance, 1990s erosion rates were half of those obtaining in the years 1925 to 1935.[43] In many settings, erosion control efforts reduced the toll. But nowhere could they check it. So this second surge, like the first, is not over.[44]

The third great pulse in world (accelerated) erosion history gathered in the 1950s and is still rolling along at its peak. In the global sense it is superimposed on its predecessors, but it mainly affects different regions of the world. After 1950 the populations of the world's tropics experienced an unprecedented outbreak of health and survival, as infectious disease came under greater control. Demographic growth, often together with state policies and land tenure patterns, spurred land hunger and land clearing, even on steep and marginal lands. Lowland peasants migrated to highland regions, mountain peasants invaded rainforests, and still others colonized semiarid land. Once again, ingrained agronomic knowledge and familiar animals and technologies often proved inappropriate to new settings.

[42]The words are by Elspeth Huxley, quoted in Dregne 1982:12.

[43]Argabright et al. 1996.

[44]On soil conservation history, see Dregne 1982, Grove 1990, Helms 1992, Reij et al. 1996, and Wen 1993. Lal 1990:132, Opie 1993:9, and Pimentel 1993:4 voice skepticism about the success and adequacy of soil conservation programs.

With the U.S.-led integration of the world economic system after 1945, tighter market links again played a key role in promoting erosion. Coffee, citrus fruits, bananas, and beef cattle claimed many fertile tropical lowlands, obliging subsistence food production to shift onto marginal lands. In some cases, such as the postwar expansion of coffee in southern Brazil, commodity crops colonized former forest zones directly.[45] In various ways cultivation spread onto mountain slopes and into rainforests where it had scarcely existed before. In tropical settings rainfall usually comes in downpours with exceptional erosive power, so that even with the best of intentions, appropriate know-how, and technology—a rare combination in any circumstances—preventing sharp surges in soil erosion proved very difficult.

The Philippine Islands are for the most part steep and subject to heavy monsoon downpours, and thus prone to erosion even without human influence. On Luzon, the northernmost of the islands, cash cropping encouraged forest clearance and cultivation from about 1880. After 1898, when American forces drove Spain from the Philippines, the quartermasters of the U.S. Army generated regular demand and good prices for food crops.

On remoter islands, such as Cebu in the central Philippines, subsistence needs and international politics, rather than the market, inspired land clearing. American conquest (1898) had proved the final chapter in a period (c.1860–1900) of demographic decline, but American occupation soon favored population growth and plantation development. Peasants took to the hills and cleared forests; rains stripped the soils. Cebu's uplands began to erode especially quickly after about 1920 with forest clearance driven by demographic pressure. The process climaxed during World War II, when combat and Japanese occupation drove many more Filipinos to uplands and forests. By 1950 the pace of erosion had begun to slow, in part because in many areas there was "no soil left to erode." Contour plowing on the slopes and agroforestry from the 1970s also helped stem—but by no means stop—the tide of erosion.

Then the timber companies arrived. Filipino politicians, in charge of their own country from 1946, added to their coffers by leasing timber concessions throughout the Philippines. Rapid and unusually thorough deforestation resulted, mainly after 1960, and sped erosion to the point

[45]Foweraker 1981; McNeill 1988.

where in 1989 the World Bank considered it the most acute environmental problem in the country.[46]

Another zone of highland terraces, Rwanda in east-central Africa, has an erosion history no less distinctive. This region, with its rich volcanic soils, abundant rain, and a comparatively mild disease regime, supported unusually dense rural populations in modern times. The western part of Rwanda is high rolling plateau leading up to mountains as high as 4,000 meters. These slopes receive torrential rains almost daily around the equinoxes. Before 1800 these slopes carried forest and minimal human population, but gradually pioneer cultivators made their way up into the lower hills. In the twentieth century, the migrants pushed farther west and farther up. Population pressure drove this movement (see Table 2.3).

TABLE 2.3 POPULATION DENSITY OF RWANDA, 1910-1996

Year	Density (people per km²)
1910	50–60
1932	51
1948	80
1978	180
1992	270
1996	260[A]

Sources: Bart 1993; Derenne 1988; Population Reference Bureau 1996.

[A]This figure is comparable to the density of Israel, El Salvador, or Haiti, and 10% lower the Belgium's.

So did politics. Belgian authorities, having acquired these German colonies in the Versailles settlement of 1919, attempted to reduce the frequency of famine, by obliging Hutu peasants to cultivate larger areas of land and to adopt crop rotations that left soil without plant cover during rainy seasons. Antierosion practices familiar to Hutu farmers were lost as the Belgians sought to maximize food production. As peasants shortened fallows and cleared new lands, erosion problems mounted. Belgian officials began to take note of rapid erosion in the 1920s and 1930s and, as their coun-

[46]De Bevoise 1995; Kummer 1991:41, 1994; Lewis 1992:174–6 and 182–4. Copper mining also added to Cebu erosion: it had, and has, the biggest copper mine in Southeast Asia, dumping 100,000 metric tons of tailings a day.

terparts did in British colonies, imposed forced-labor soil conservation schemes, still remembered with distaste. In 1951 the Belgian governor called erosion in Rwanda "a matter of life and death."[47] After the early 1950s the spread of bananas as a cash crop helped mitigate erosion. Unlike most of the world's cash crops, banana trees provide good soil cover with their broad leaves. Nonetheless, rapid soil loss continued in many parts of Rwanda.

After independence (1961), with population growth over 3 percent per year, peasant settlement of steeper and steeper slopes proceeded apace. Revived government concern about erosion brought back coerced labor for soil conservation schemes. In some parts of Rwanda, ever denser population allowed sufficient labor to attend conscientiously to soil conservation, and by the 1980s some slopes had stabilized. But others eroded away faster than ever. The civil war and aftermath (1994–1996) reduced rural population sharply, and perhaps provoked the sort of erosion characteristic of depopulated terraces. But perhaps not: Rwanda and Burundi still have the highest rural population densities in Africa.[48]

In at least one East African context, the Machakos Hills of Kenya, soil conservation schemes undertaken after independence worked well. Colonial land policy in Kenya had concentrated Africans on poorer lands, such as those of the semiarid and often steep Machakos District near Nairobi. From at least 1930 the Machakos Hills suffered from acute erosion, causing food problems and emigration to still drier lands. Between 1930 and 1990, population density tripled and cultivated area increased sixfold. But in the 1970s the Kenyan soil conservation service and the local Akamba farmers stemmed the tide of erosion. What made the difference, apparently, was greater security of land tenure, especially that of cultivators as against herders, and a more democratic approach to implementation in which Kenyan authorities worked closely with existing self-help groups in Akamba society. Swedish aid money helped too. Plentiful labor and secure tenure encouraged families to husband soil by leveling their plots, keeping animals away, channeling watercourses, and other means. Intensive farming stabilized the soils of the Machakos Hills, even as population density grew.[49]

[47]Bart 1993:23.

[48]Ibid. 339–45; Derenne 1988. In postwar Rwanda, population is growing again fast, spurred by "revenge fertility," the effort of Hutu and Tutsi to ensure their security and political position in the country's future. War between the Tutsi and Hutu has been a recurrent feature since at least 1959, making for rapid depopulations in many locales, which in terraced landscapes presumably have led to extra erosion through labor shortage.

[49]Moore 1979; Ondiege 1996; Tiffen et al. 1994. The success here may not be easily repro-

The timing, causes, and consequences of accelerated erosion differed considerably among the Philippines, Rwanda, and Kenya. They differed still more in cases such as Fiji, where the spread of sugar cane to sloping land after 1960 accelerated erosion; or Papua New Guinea's highlands, cultivated for 8,000 years but since 1930 with cash crops on more marginal lands, which doubled erosion rates; or Bolivia, Madagascar, Ethiopia, Nepal, Sri Lanka, Haiti, and Guatemala.[50] In most cases, however, two chief factors explain modern increases in erosion: migration or growth of population, and the intensification of market links. But the equations were always complex. Population growth had baleful effects on soil erosion in some settings and useful ones in others. Cash cropping, while usually a force for erosion, in Rwanda was not. Colonial policies (frequently continued after independence) and insecure land tenure often played important inadvertent roles as well.

Population, politics, and economic change drove the third pulse (as they did the second) of world soil erosion, but other forces more than scratched the surface of the world's soils. Technological changes in agriculture, specifically the adoption of heavy machinery, led to soil compaction after 1930, and especially after 1950, when tractor sizes grew rapidly. Some farm machines now weigh over 20 tons. In North America, soil compaction, which inhibits plant growth, cost several billion dollars a year in the 1990s.[51] Industrial air pollution and the heavy use of nitrogen fertilizers after 1960 led to the acidification of soils, especially in Europe. Irrigation (see Chapter 6) inadvertently brought on soil salinization, an ancient problem that by 1990 beset about 7 percent of the world's land. Saline seeps, unrelated to irrigation, emerged in North America's high plains, where they retired about 1 million hectares (equivalent to Lebanon) of farmland (1945–1990). Western Australia too suffered from "salt creep" after about 1950.[52] Most decisively, urbanization and road building buried

ducible elsewhere because the Akamba people had a history of private property in land, unusual in Africa. Kenya's soil conservation program, created in 1974, has probably been the best in Africa.

[50]See Blaikie 1985:177–9, Lal 1990:133–41, and Roberts 1989:168–69 (on many tropical locations); Grepperud 1996 (on Ethiopia); Ives and Messerli 1989 (on Nepal); McCreery 1989 (on Guatemala); Zimmerer 1993 (on Bolivia); and Randrianarijaona 1983 (on Madagascar).

[51]Cruse and Gupta 1991; Raghavan et al. 1990.

[52]Craswell 1993:268–9; Daniels 1987b; Hillel 1991:135–40; Young 1996:58–63. Soils in much of Europe have since 1960 become one or two orders of magnitude more acidic, lowering

soils throughout the 20th century. Between 1945 and 1975, farmland area equivalent to Nebraska or the United Kingdom was paved over.[53] All of these soil changes reduced vegetation cover, created more runoff and flooding and less infiltration, and hence promoted erosion.

Conclusion

The consequences of all these changes to the lithosphere and pedosphere are vast, but in important respects they canceled one another out. Soil degradation in one form or another now affects one-third of the world's land surface. The area now degraded by human action (about 2 billion hectares, or the area of the United States plus Canada) is a quarter again as large as the world's total cultivated area. An area of about 430 million hectares, seven times the size of Texas, has been "irreversibly destroyed" by accelerated erosion.[54] Erosion damaged some lands more than others. In China by 1978, erosion had forced the abandonment of 31 percent of all arable land.[55] African erosion rates, on average nine times as high as those in Europe, added to the food crisis there: Africa is the only continent where food production per capita declined after 1960. The United States in the twentieth century lost an amount of topsoil that took about 1,000 years to form, and currently loses 1.7 billion tons a year to erosion. In 1982, 40 percent of its arable land was eroding faster than the official tolerance level.[56] On one calculation, erosion in the United States in 1994 cost about $150 per person per year, twice as much per capita as in the world as a whole.[57] The United Nations's Food and Agriculture Organization (UNFAO) estimated in 1991 that erosion alone destroyed between 0.3 and 0.5 percent of the world's cropland every year, which helps explain why the pressure to clear forested land was so strong at the end of the century.[58]

---

their fertility (and increasing dependence on artificial fertilizers). Europeans have been inadvertently negating their great soil advantage over the rest of the world. In Europe, 36% of soils are free from major limitations; elsewhere in the world the percentage range is 10–25%.
  [53]Pimentel et al. 1993:280.
  [54]Lal 1990:130.
  [55]Ibid. 145–6.
  [56]NRC 1993a:221–2; Pimentel and Heichel 1991:115. The official soil-loss tolerance level is an estimate based on presumed soil creation rates.
  [57]Pimentel et al. 1995.
  [58]Cited in Solbrig and Solbrig 1994:231.

Humanity increased erosion two- to threefold over prevailing natural rates.[59]

But so what? In the same years when soil degradation reached its fastest pace, world food production climbed spectacularly. Around the globe, more food per capita was available by the end of the twentieth century than at any time in human history. The combination of intensive fertilizer use (mainly since 1950), genetic engineering of crops (mainly since 1970), and other magic tricks of scientific agriculture masked the effects of soil erosion and soil degradation. On the usual human time scales, soil degradation and erosion proved to be a local problem, not a general one. On geological time scales too, the impact of anthropogenic erosion seems minor, except in a few places. On average the earth's rocks have eroded, been deposited on ocean floors as sediment, consolidated into rock again, been thrust up above sea level—only to erode again—about 25 times in the long history of the earth.[60] Of course, from such Olympian heights no human problems matter.

Seen from an intermediate time scale, however, soil degradation and erosion may indeed carry important human consequences. Humankind has already played the fertilizer card in the earth's best agricultural lands, and further nitrogen and phosphate loadings no longer increase yields. The Netherlands, for instance, in the 1990s reduced fertilizer use because heavier doses did no good and did some harm. In other parts of the world, notably Africa, increased fertilizer use might work the same miracles as it did in Europe and East Asia, although much additional irrigation would be required. But there is a limit beyond which fertilizer cannot help make up for lost soil. Plant breeding, the other great source of improved yields in the twentieth century, has its limits too, although they are harder to discern. Clearly there is still some slack in the food system, and we can feed a few more billion people simply by using more land, more fertilizer, and breeding better crops. But equally clearly, doing so will be expensive, in environmental terms as well as in conventional terms, since food generated with more inputs will cost more money. So all that degraded, eroded, compacted, paved over, and contaminated soil may yet be missed—especially if freshwater or energy constraints begin to pinch, driving up the costs of irrigation and fertilizer. Today, as in 1900, we get 97 percent of our food from these vanishing soils.[61] As the story of the Machakos Hills in Kenya suggests, the loss is a matter of negligence, not necessity.

[59]Judson 1968:371. Lal and Pierce 1991a:2 say it increased by a factor of 2.6, which pretends to a precision that data scarcely permit.

[60]Judson 1968:373.

[61]See Shaler 1905:139 for the c.1900 figure.

3

The Atmosphere: Urban History

"This goodly frame, the Earth, seems to me a sterile promontory; this most excellent canopy, the air, look you, this brave o'er-hanging firmament, this majestic roof fretted with golden fire, why it appears no other thing to me than a foul and pestilent congregation of vapours."

So says Shakespeare's Hamlet (act 2, scene 2) in a characteristically bleak mood. From the point of view of life on earth, the air is a most excellent canopy. But lately one form of life on earth—humankind—has acquired extraordinary influence over the atmosphere and, in places, made it far fouler than either Shakespeare or a prince of preindustrial Denmark could imagine. More lately still, in the second half of the twentieth century, we simultaneously cleaned up the air in a few places and tinkered with some trace gases that govern fundamental conditions on earth. We have not made this goodly frame a sterile promontory, nor are we likely to, but we can certainly make it an awkward home for creatures attuned to conditions of the last few million years.

This chapter and the next consider the human history of the atmosphere in the twentieth century. The main story is how we made minute but consequential alterations in the atmosphere's minor constituents—things that are measured in parts per million or billion. For most of history, air pollution had existed only locally and had had only modest consequences. In the twentieth century, its scale grew exponentially, affecting the air of whole regions. With respect to some trace gases, pollution is global. Local air pollution history since 1900 mainly had concerned urban air quality, governed chiefly by fuel combustion. Regionally, the principal development has been the spread of acid rain.[1] Globally, the crucial trends have been the accumulation of greenhouse gases, which help regulate the earth's temperature, and the depletion of stratospheric ozone. These, then, are the subjects of these two chapters on the atmosphere. Their theme is the power of our energy-use habits to change the atmosphere on ever larger scales, and the ever greater consequences of the atmospheric changes we have wrought. For most of earth's history, microbes played the leading role of all life in shaping the atmosphere. In the twentieth century, humankind stumbled blindly into this role.

The Basics of the Atmosphere

The atmosphere is the thin gaseous envelope that surrounds the earth.[2] It is about 100 kilometers (60 miles) thick, although any outer boundary is arbitrary because it shades off gradually into space. Its air weighs about 5 quadrillion tons, about 0.0003 the weight of the oceans: hence it takes a lot less to pollute the air than it does the oceans. Air consists of thousands of gases, but two now predominate: nitrogen (78%) and oxygen (21%). In the very long haul, the chemistry of the atmosphere changes: Before there were plants on earth, there was not much oxygen. But in human times these changes have—until very recently—amounted to little.

The atmosphere is a dynamic place. It is a swirl of motion, driven by inequalities in air pressure, ultimately created by inequalities in temperature. Below 10 kilometers, about the elevation of Mt. Everest, there are

---

[1]Following the popular usage, I will use "acid rain" to mean all acid deposition, whether via rain, snow, mist, or as dry deposition.

[2]A useful primer is Turco 1997.

daily cycles, seasonal cycles, annual cycles, and a few longer and more ir-regular cycles. This is why weather is so complicated. In addition to churning about, air also exchanges heat, moisture, and gases with soil, water, and living things. So the very lowest altitudes are the liveliest. The outermost layer of atmosphere exchanges almost nothing with space, ex-cept for the all-important reception and reflection of the sun's rays, and is comparatively stable. Almost everything of interest to human history has happened in the lower elevations.[3]

To most of us the atmosphere seems infinite. But we do not need to do anything to the 5 quadrillion tons of nitrogen and oxygen to alter condi-tions on earth fundamentally. A few tweaks to the concentrations of key trace gases will do it. The three most important trace gases for modern history are carbon dioxide, ozone, and sulfur dioxide:

1. Carbon dioxide exists in concentrations now at about 360 parts per million. It is a "greenhouse gas," which means it keeps the planet warm by trapping solar rays reflected off the earth's sur-face. Without it and lesser greenhouse gases, the earth would be about 33° Celsius colder—frozen and lifeless. The human sources of carbon dioxide are fossil fuel burning and deforesta-tion.

2. Ozone is also a greenhouse gas and a component of urban smog; at low altitudes it is an unwelcome pollutant. But in the stratosphere ozone absorbs ultraviolet radiation from the sun, which protects the biota from potential harm. Ozone exists in concentrations of 1 to 15 parts per billion (ppb) as a low-altitude pollutant, but 500 to 1,000 ppb in the stratosphere.

3. Sulfur dioxide is the main ingredient in acid rain, which can damage forests and aquatic life as well as corrode metals and stone. It rarely exists in concentrations greater than 50 ppb. Its main human sources are fossil fuel burning and the smelting of metallic ores.

Other consequential trace gases include methane, chlorofluorocarbons (CFCs), and two nitrogen oxides. Some of their relevant characteristics appear in Table 3.1, provided for reference.

---

[3]The exception is stratospheric ozone depletion. The lower 10 km comprise the tropos-phere; between about 10 and 45 km up is the stratosphere. See Salstein 1995 for details.

TABLE 3.1 SOME IMPORTANT TRACE GASES IN TWENTIETH-CENTURY HISTORY

Gas	Human Sources	Human Significance	Anthropogenic Emissions (%)[a]	Concentration in 1900 (ppb)	Concentration in 1990 (ppb)
OF GLOBAL SIGNIFICANCE[b]					
Carbon dioxide	Fossil fuel burning; deforestation	Greenhouse gas	≈100	290,000	350,000
Methane	Rice fields, livestock, garbage, fossil fuels, mining	Greenhouse gas	≈60	900	1,700
Chlorofluoro-ocarbons	Refrigerants, foams, aerosol sprays	Ozone destroyer, greenhouse gas	100	0	≈3
Nitrous Oxide	Fertilizers, biomass burning, deforestation	Greenhouse gas, ozone destroyer	≈25	285	310
OF LOCAL AND REGIONAL SIGNIFICANCE[b]					
Sulfur dioxide	Fossil fuel burning, ore smelting	Acid rain	≈65	?	0.3–50
Nitrogen oxides	Fossil fuel and biomass burning	Acid rain, smog	≈65	?	0.001–50
Ozone (in the troposphere)	Vehicle exhaust's interaction with sunshine	Greenhouse gas, smog	50–70	10[c]	20–40[c]

Sources: Graedel and Crutzen 1989; Salstein 1995.

[a]Anthropogenic emissions are given as a percentage of all emissions (human plus natural) as of about 1990.
[b]The reason some trace gases have global significance while others have local or regional significance is their various "residence times." Those that on average remain in the atmosphere only briefly do not disperse around the world, while those that linger longer do. Residence times vary greatly, from a few days (sulfur dioxide) to a century or more (carbon dioxide, nitrous oxide, CFCs).
[c]Concentrations of tropospheric ozone refer only to western Europe.

In the twentieth century, human action put more than gases into the atmosphere. Fossil fuel burning, metal smelting, and waste incineration released thousands of tons of potentially toxic metals into the air as dust. Some of this inevitably worked its way into the food web, to the misfortune of fish, otters, alligators, minks, raccoons, and eagles among others. Human health suffered from lead emissions, which came mainly from automobile exhausts. Estimates of the historical evolution of metal emissions into the atmosphere in modern times appear in Table 3.2. The figures for lead emissions reflect the spread of the automobile after 1920, and those for nickel the rapid rise of the armaments industry from the 1930s. Another noteworthy feature, however, is the general decline in metal emissions after 1980, a consequence of environmental awareness and regulation, and of new technologies and efficiencies.[4]

TABLE 3.2 WORLDWIDE METAL EMISSIONS TO THE ATMOSPHERE, 1850–1990

		Yearly Average (tons)			
Period	Cadmium	Copper	Lead	Nickel	Zinc
1850–1900	380	1,800	22,000	240	17,000
1901–1910	900	5,300	47,000	800	39,000
1911–1920	1,100	8,000	49,000	2,100	49,000
1921–1930	1,400	9,600	110,000	2,100	62,000
1931–1940	1,700	12,000	170,000	4,900	75,000
1941–1950	2,200	17,000	170,000	8,000	96,000
1951–1960	3,400	23,000	270,000	14,000	150,000
1961–1970	5,400	44,000	370,000	26,000	240,000
1971–1980	7,400	59,000	430,000	42,000	330,000
1981–1990	5,900	47,000	340,000	33,000	260,000

Source: Nriagu 1994.

[4]Nriagu 1994.

Air Pollution before 1900

For most of human history, people could not pollute the air except by kicking up a little dust. Then half a million years ago we harnessed fire. We torched landscapes, releasing carbon dioxide and other gases into the air. But despite our devotion to fire, our atmospheric impact came to little. Natural processes, including the steady work of quadrillions of microbes and the occasional major volcanic eruption, governed the atmosphere.

When humankind occupied caves and began to burn fuelwood for heating and cooking, indoor pollution made its debut. Many caves inhabited millennia ago retain a patina of smoke on their walls, and cave dwellers presumably suffered lung and eye ailments derived from exposure to smoke. Blackened lungs are common among mummified corpses from Paleolithic times. When people built their own abodes, they often failed to ventilate them (perhaps because they wished to keep mosquitoes at bay) and lived in a pall of indoor smoke.[5] Some of the health effects of pollution, then, have been with us for many thousands of years.

Outdoor air pollution of any serious consequences came only with cities. Early cities, like some modern ones, often exuded pungent smells on account of decaying flesh, food, and feces. Cities under siege, with no opportunity to export the sources of offensive smells, could become insupportable. Egyptian literature records an instance in which Hermopolis surrendered to Nubian besiegers when its air drove the inhabitants to prefer Nubian mercy to their own stench.[6] Urban smoke darkened marble in ancient cities, annoyed classical writers such as the Roman poet Horace, and provoked a spate of laws among the ancient Jews.[7] Smoke and soot, not trace gases, dominated air pollution history in early times.

Ancient metallurgy added new pollutants, some of which wafted across

---

[5]Indeed, millions of people still do, wherever biomass is burned indoors (see Brimblecombe 1995). I spent a memorable afternoon in 1982 in a small hut in lowland Nepal, shifting between indoors and out, motivated by the conflicting desires to not offend my hosts and to avoid the indoor smoke that made my eyes smart. My hosts, who were used to the smoke, found it curious that I could not sit comfortably in it.

[6]The document is the Victory Stela of King Pi(ankhy) from c.734 B.C. (M. Lichtheim, *Ancient Egyptian Literature. III. The Late Period* [Berkeley: University of California Press, 1980], cited in Brimblecombe 1995).

[7]Mamane 1987.

seas and continents—the first evidence of pollution on regional scales. In the ancient Mediterranean world, mining and smelting played a substantial role in economic life. Noxious fumes from silver mines in Attica damaged human health, according to Xenophon and Lucretius.[8] The main metallic pollutants were copper and lead. Swedish and Swiss bogs and Greenland ice cores reveal significant lead deposition in Roman times, roughly 10 times background levels. Greenland ice shows that emissions of copper into the atmosphere surged twice before the Industrial Revolution, once after the introduction of coinage in the ancient Mediterranean, and once with the intense marketization of the Chinese economy—and burgeoning copper production—of the Song dynasty (960–1279 A.D.). Inefficient smelting technology sent as much as 15 percent of smelted copper into the air. Total copper emissions in the Roman and Song eras came to about a tenth of those of the 1990s, even though copper production was less than a hundredth of modern levels. Regional, indeed hemispheric, air pollution is about 2,500 years old, and—for copper emissions at least—was as great in Roman and Song times as at any time before 1750.[9]

Urban air pollution varied with the size and density of cities, with their industrial activity, and especially with their fuel use. As urbanization gathered pace in China, the Mediterranean basin, and West Africa after about 1000 A.D., larger numbers of people lived amid smoke and soot. The philosopher and physician Maimonides (1135–1204), who had wide experience of cities from Córdoba to Cairo, found urban air "stagnant, turbid, thick, misty and foggy," and thought it produced in urban dwellers "dullness of understanding, failure of intelligence and defect of memory."[10]

But transport difficulties limited *urban* air pollution levels: most industries that required combustion—for example tiles, glass, pottery, bricks, and iron—were located near forests, as moving masses of fuel usually proved too expensive. Most industrial pollution fouled air where few people breathed it.[11] Port cities constituted partial exceptions, for

[8]See Weeber 1990.

[9]Hong et al. 1996. This paper is the first attempt to reconstruct historic copper emissions, and conclusions must be considered tentative. Results from German peat bogs show medieval and Roman maxima for lead and copper pollution too (Gorres et al. 1995; see also Nriagu 1990a and 1996). Historical geochemical analysis offers exciting possibilities for understanding the history of metallurgy and mining, a central thread in human history. Chemists can also find traces of past metallurgy and pollution in bird feathers, fishbones, deer antlers, and human hair.

[10]Quoted in Turco 1997:137.

[11]In the Sumava Mountains, a rural area of the Czech Republic, copper and lead emissions rose sharply after 1640 (Veseley et al. 1993). The great north Chinese surge in coal use (11th century) mainly supplied rural iron forges (Hartwell 1967).

ships could transport wood or charcoal more cheaply. Hence Venice could maintain a glassmaking industry fueled by distant timber. Chinese cities too may have been exceptionally polluted, because the well-developed water transport system permitted high fuel use, at least in the Song capital of Kaifeng.[12] But for the most part, urban air pollution derived only from domestic fuel, often dung or wood, but sometimes smokeless charcoal.

Port cities short of fuelwood might turn to coal as a domestic fuel. London did so in a small way in the thirteenth century and a large way in the sixteenth, ushering in a new chapter in its air pollution history. Domestic coal use made seventeenth-century London air

> a cloud of sea-coal, as if there be a resemblance of hell upon earth, it is in this volcano in a foggy day: this pestilent smoak, which corrodes the very yron, and spoils all the moveables, leaving soot on all things that it lights: and so fatally seizing on the lungs of the inhabitants, that cough and consumption spare no man.[13]

London air got no better in the ensuing centuries, as population growth added to the number of hearths and chimneys. The poet Percy Bysshe Shelley noted its infernal qualities:

> Hell is a city much like London—
> a populous and smoky city.[14]

His fellow poet, Robert Southey, liked London air no better, describing it in 1808 as "a compound of fen fog, chimney smoke, smuts, and pulverised horse dung."[15]

With the rise of coal as the key fuel of the Industrial Revolution after 1780, clouds of pollution sprouted elsewhere in Britain and eventually beyond. By 1870, Britain had perhaps 100,000 coal-fed steam engines, all churning out smoke and sulphur dioxide.[16] The English Midlands be-

---

[12]Hartwell 1967. Kaifeng (about 500 km south of Beijing) was the first city in the world, so far as I can determine, to convert from wood to coal as its energy base. The transition took place late in the 11th century when the city had about a million residents, but the "coal regime" did not last long because Kaifeng was destroyed by Mongol conquest (1126) and plague in the early 13th century.

[13]John Evelyn, *Fumiugium* (1661), cited in Brimblecombe 1987:47–8.

[14]In the poem *Peter Bell the Third*, pt. 2, stanza 1.

[15]Quoted in Grimmett and Currie 1991:5.

[16]Clapp 1994:20. British coal contained about 1–2% sulphur; some coal is 4% sulphur.

came known as the Black Country. Copper smelting in the Swansea valley, Wales, generated acid rain that devastated vegetation. People suffered too: nearly a quarter of deaths in Victorian Britain came from lung diseases, mostly bronchitis and tuberculosis, often aggravated and sometimes caused by air pollution, mainly particulates. Air pollution killed Victorian Britons at (very) roughly four to seven times the rate it killed people worldwide in the 1990s.[17]

Air Pollution since 1900

The history of air pollution in the twentieth century is a dark cloud with a silver lining. One thing above all else accounts for air pollution in modern times: fossil fuel combustion. In 1900, coal combustion caused most of the air pollution, filling the skies with smoke, soot, sulfur dioxide, and various other unsavory substances. Coal-derived pollution came from smokestacks and chimneys: that is, industry and dwellings. Since the 1960s, automobile tailpipes have challenged smokestacks and chimneys, and by 1990, road traffic had become "the largest single source of air pollution around the world."[18] Pollution history followed the history of industrialization and "motorization."

The spread of coal-fired industrialization from Britain brought air pollution in its wake. In the late nineteenth century, a second phase of the Industrial Revolution took hold, centered on steel, iron, chemicals—and copious coal. Successful smokestack industries emerged in Europe, especially in Belgium and Germany; in the United States, especially in Pennsylvania and Ohio; in Czarist Russia, especially in Ukraine; and in Japan, especially around Osaka. Smaller pockets of coal-fired industry developed in India, South Africa, Australia, and elsewhere between 1880 and 1920. Cities in North America and Europe acquired electric grids

[17]Clapp 1994:64–68. Apparently about a fifth of deaths from bronchitis "took their origin in atmospheric pollution" (68). If the same proportion held for pulmonary tuberculosis, air pollution killed about 1.4 million Britons in the years 1840–1900 via these two diseases. If air pollution raised the death toll from pulmonary tuberculosis only 10%, then the total number of shortened lives was more like 800,000. (Roughly 28 million people died in Britain of all causes between 1840 and 1900, a figure derived from vital statistics in Mitchell 1978.) On 1990s death rates from urban air pollution, see Hall 1995 and Murray and Lopez 1996, summary:28.

[18]Walsh 1990:217; see also WRI 1996:86.

powered by coal combustion in the 1910s and 1920s. In all these places, smoke, soot, and sulfur dioxide blanketed industrial neighborhoods.[19]

Many people objected, particularly women whose duties included keeping homes and linens clean,[20] but the prosperity that came with the smoke seemed well worth the price to those whose opinions counted. The city fathers and industrial unions of Pittsburgh, the captains of German industry, and the Russian ministers of state generally regarded belching smokestacks as signs of progress, prosperity, and power. As Chicago businessman W. P. Rend put it in 1892: "Smoke is the incense burning on the altars of industry. It is beautiful to me. It shows that men are changing the merely potential forces of nature into articles of comfort for humanity. . . ."[21] Japan's first large steel town, Yawata, expressed this view in its civic anthem:

> Billows of smoke filling the sky
> Our steel plant, a grandeur unmatched:
> Yawata, O Yawata, our city![22]

Coal-based industrialization did not stop in 1920. It continued to develop in Europe and North America, but grew much faster elsewhere. In the Soviet Union it progressed with extraordinary speed after 1929 under the first two five-year plans. After Stalin consolidated his hold on Eastern Europe (1946–1948), Soviet-style industry, an especially energy- and pollution-intensive variety, spread to Poland, Czechoslovakia, East Germany, and Hungary from 1948 to about 1970, and simultaneously to new enclaves in Siberia. Japan's reindustrialization, beginning around 1950, was still based on coal, although increasingly on oil as well. In the 1960s and 1970s, steel, shipbuilding, chemicals, and other energy-devouring industries emerged elsewhere in East Asia, notably South Korea, Taiwan, Malaysia, and after 1978 in China. Restraints on air pollution mattered in none of these cases before 1980.[23]

Domestic hearths as well as blast furnaces burned coal. As city popu-

[19]In 1911 the U.S. Geological Survey estimated that smoke damage cost the American economy $500 million annually, equal to the total amount paid in property taxes (Rosen 1995:354). A review of coal's impact on the German environment is given by Cioc 1998.

[20]On the role of women in the smoke abatement movement in the USA before World War I, see Stradling 1996, Platt 1995, and Tarr and Zimring 1997. On York (England), see Brimblecombe and Bowler 1992. On Yawata (Japan), see Morris-Suzuki 1994.

[21]Quoted in Rosen 1995:385–6.

[22]Tsuru 1989:19.

[23]Fang and Chen 1996; Srinanda 1984.

lations grew, more and more people required transportable fuel for space heating and for cooking. Urban households took to coal because where they lived they could not easily lay their hands on fuelwood. So more and more cities followed down London's path, making coal the chief domestic fuel. West European and eastern North American cities turned to coal after 1850. After 1890, others in northern China and the American Midwest followed.[24] Domestic hearths usually burned very inefficiently, sending plenty of soot and smoke up their chimneys—and into the dwellings they heated. This combination of coal use for industrial and domestic purposes created intensely polluted coal cities, such as London, Pittsburgh, and Osaka.

The second major force behind local and regional air pollution in the twentieth century was the automobile. Tailpipes emitted various pollutants, including some that reacted with sunlight to create smog, others that added to acid rain, and after 1921, lead.[25] The spread of automobiles came in three main surges. The first took place in the United States in the 1920s, when new assembly-line techniques made cars affordable for millions of Americans. As late as 1950, Americans drove more than half of the cars in the world. Then cars became ordinary in western Europe (1950–1975). The motorization of East Asia, led by Japan, is still in progress (1960–present). China in 1997 had only 2 million cars. Worldwide the total number of motor vehicles, less than a million in 1910, reached about 50 million in 1930, surpassed 100 million by 1955, and 500 million by 1985. In 1995 the world tallied 777 million cars, trucks, and motorbikes.[26] On top of this, people drove their cars more and more. In the United States, total vehicle mileage quadrupled (to 1.9 trillion) between 1950 and 1990. Fortunately, in the 1970s cars became more fuel-efficient, and new technologies, where adopted, reduced their tailpipe emissions.[27] Carbon monoxide and lead emissions fell sharply. But other

[24]Locke and Bertine 1986 analyzed magnetite to find traces of smoke and sulfur dioxide. They conclude that in New England and the mid-Atlantic regions of the USA, coal use became widespread after 1850, and around Lake Michigan only after 1890. This probably correlates with wood shortages: by 1890 the vicinity of Lake Michigan had been logged over.

[25]Road traffic generates carbon monoxide, nitrogen oxides and hydrocarbons in particular. In 1980, about two-thirds of carbon monoxide emissions in rich countries (members of the OECD, Organization for Economic Cooperation and Development) came from vehicles; the proportion for nitrogen oxides was 47%, for hydrocarbons 39%. About 13% of U.S. production of CFCs went into cars in the 1980s as well (Walsh 1990:217–8).

[26]Precise data appear in Walsh 1990; the 1995 figure is from the *Economist*, 22 June 1996.

[27]In the OECD, total mileage doubled in 1970–1990 to nearly 7 trillion km (5 trillion miles), but gas usage increased only 10% (OECD 1995:40, 46).

pollutants continue to spew forth—including about a fifth of the carbon dioxide added to the atmosphere.[28]

While industrialization and motorization proceeded rapidly throughout the century, total air pollution probably did not quite keep pace. Many cities, especially between 1945 and 1980, improved their air quality drastically. The reasons were threefold: economic, political, and geographic.

First, the economic reasons: The world's fuel mix grew a bit cleaner after 1920. Oil replaced coal in many applications (the first large one being the Royal Navy of Great Britain), beginning around 1910 but especially after 1950. This happened mainly for price reasons: as the great oil and gas fields of the Americas and the Middle East developed, along with the infrastructure of pipelines and supertankers, prices fell between 1945 and 1973. So for domestic heating, power generation, and other uses, oil and gas replaced coal. Less importantly, other forms of commercial energy emerged, such as hydroelectric and nuclear, further reducing coal's role. Price-driven fuel substitution reduced pollutants from factory smokestacks and domestic chimneys—but not from automobile tailpipes.

Second, political reactions against pollution produced results. Episodic citizen protest at industrial pollution was part and parcel of industrialization around the world, but it yielded very little in the way of pollution reduction until the 1940s. St. Louis was the first major city to adopt successful smoke abatement policies and technologies in 1940; they spread quickly after World War II. Political pressure checked pollution, generally through regulation and new technologies, in North America mainly after 1966, in western Europe and Japan from 1970, in South Korea from 1980, and in eastern Europe from 1990. Political reductions in air pollution affected smokestacks, chimneys, and tailpipes.

Geographic change also reduced the human consequences of pollution, if not its total volume. Energy-intensive industries, long concentrated spatially in a few places close to coal and iron ore, such as western Pennsylvania or Germany's Ruhr region, gradually dispersed around the world after 1960. Moreover, they tended to leave cities for industrial parks and "greenfield" sites, so that they polluted air that fewer people breathed. Of course, this relocation did nothing to reduce total pollution levels and had deleterious effects on ecosystems previously little touched by pollution. Only from the human, and particularly the city dweller's, perspective did

---

[28]See the *Economist* survey entitled "Living with the Car," 22 June 1996.

this amount to pollution reduction.[29] The combination of these economic, political, and geographic changes made for the silver lining of the dark cloud.

In some cases the results were spectacular, as in the reduction in lead emissions. In 1921 a chemical engineer named Thomas Midgley (about whom more in Chapter 4) figured out that adding lead to gasoline would make it burn better and prevent engine knocking. Two years later, leaded gasoline—a "gift from God" according to the first company to sell it— went on the market in Dayton, Ohio. Over the next half century, cars burned about 25 trillion liters (6 trillion gallons) of leaded gasoline, three quarters of it in the United States. Despite public concern and govern-ment inquiry beginning in the 1920s, General Motors and DuPont (whose joint subsidiary produced tetraethyl lead) managed to prevent regulation of lead additives in the United States until the 1970s. By then medical research showed that most Americans had elevated lead levels in their blood, and that the lead came from gasoline. Similar findings in the USSR resulted in a ban on leaded gasoline in major cities in 1967, a rare instance of the Soviet Union leading the world to a healthier environ-ment.[30] In 1970 the United States mandated that by 1975 low-lead gaso-line be available at gas stations. Japanese companies adjusted to U.S. law first, converting car engines to low-lead gas beginning in 1972. Japan eliminated leaded gas in 1987.[31] The United States made the transition from leaded gas in the late 1970s, slowed by lawsuits filed by interested manufacturers. Western Europe followed in the late 1980s, as did many other countries in the 1990s.

In the United States, ambient lead concentrations in the air declined by about 95 percent between 1977 and 1994. Tests soon found much less lead in the blood of American children. The highest ambient lead levels ever recorded came not from the United States but from Bangkok, Jakarta, and Mexico City in the 1980s. In Africa and China, where leaded gas remained in use in the late 1990s, urban children carried lead in their veins sufficient to produce all manner of health problems.[32] The decline

[29]This represents a partial return to the pattern of a thousand years ago, when polluting in-dustries clustered in forested areas so as to be close to their fuel supply. But it also meant more people driving cars greater distances to get to and from the new industrial areas.

[30]Thomas 1995:305.

[31]Japan's lead pollution increased about 1,000-fold in 1949–1970 with the introduction of leaded gasoline (Satake et al. 1996).

[32]Among these were "metabolic disorders, neuropsychological deficits, hearing loss, retar-dation in growth and development" due to low-level lead poisoning. See Nriagu 1990a,

of atmospheric lead pollution around the world lagged behind the United States and Japan.

Richer countries after 1945 sharply reduced their urban coal smoke, soot, and dust and after 1975 cut sulfur dioxide, carbon monoxide, and some other pollution modestly. They did this by shifting from coal to oil (c. 1920–1970) and by energy conservation and pollution regulation (mainly after 1970). Early in the century in the Scottish industrial hub of Glasgow, experienced newspaper editors left extra space for obituaries during smog sieges. As late as 1950, Glaswegians ordinarily inhaled about 2 pounds of soot each year. But by 1990 Glasgow had reduced smoke, soot, and sulfur dioxide pollution by 70 to 95 percent, a feat matched in several former coal cities. Although a considerable achievement, this silver lining was essentially confined to about 20 rich countries in North America, western Europe, Australasia, and Japan, and thus to about one-eighth of humanity.[33]

Measuring pollution totals is a methodological minefield. Ignoring important caveats, I estimate that emissions of local and regional air pollutants around the world were five times (more cautiously two to ten times) as great in the 1990s as in 1900.[34] In the Western world and Japan, air pollution levels declined after the late 1960s (and smoke levels after the 1940s). While these lands accounted for the lion's share of world air pollution over the century, after 1950 Eastern Europe, and after 1970 East Asia and to a lesser extent Latin America, came on strong, so that global air pollution continued to grow late in the century. Table 3.3 contains some of the data on which my rough estimate is based.

1990b; Rosner and Markowitz 1985; and USEPA 1995. On China, see Shen et al. 1996; on Africa, Nriagu, et al. 1996a and 1996b. Even Greenland ice cores show the results of the transition to low-lead fuel in North America and Europe. By 1990, Greenland air bubbles trapped in ice had less lead in them than at any time after about 1820 (Tolba and El-Kholy 1992:17). The chief manufacturer of tetraethyl lead was a company owned by General Motors, which thus had an incentive to keep building engines with the higher compression ratios that needed ever more lead in their gas. In the USA the lead used per vehicle mile climbed by 80% between 1948 and 1968 (Ponting 1991:379).

[33]On Glasgow, see MacDonald et al. 1993.

[34]The estimate is rough for two reasons: (1) the data are incomplete and their reliability variable; in particular, data on Latin America, Africa, and India prior to 1970 are extremely scarce. (2) There is the apples and oranges problems, i.e., how does one assess smoke vs. lead as a pollutant?

TABLE 3.3 SOME AIR POLLUTANT EMISSION RATES IN THE TWENTIETH CENTURY (ANTHROPOGENIC SOURCES ONLY, 1900–1990)

Pollutant	Place	Increase Factor	Source
Copper	World	5	1
Lead	World	7	1
Zinc	World	5	1
Nitrogen oxides	USA	9	2
Nitrogen oxides	World	14	3
Volatile organic compounds	USA	2.7	2
Sulfur dioxide	USA	2.2	2
Sulfur dioxide	Europe	2	4
Sulfur dioxide	Chengde (northern China)	10	5
Sulfur dioxide	World	5	6
Ozone[a]	Europe	5	4
Methane	World	3.5	7

Sources: 1) Nriagu 1996; 2) USEPA 1995; 3) Smil 1990; 4) Graedel and Crutzen 1990; 5) Jiang 1996; 6) Husar and Husar 1990; RIVM 1997, and UNEP 1997:225; and 7) Stern and Kaufman 1996.

[a]Tropospheric ozone. Graedel and Crutzen 1989 give two- to fourfold for Europe's increase, 1890–1989.

Coal Cities

Two cities that relied heavily on coal but weaned themselves from it are London, once known as the Big Smoke, and Pittsburgh, formerly nicknamed Smoke City.[35]

[35]The material on London is from Brimblecombe 1987, Eggleston et al. 1992 and Stradling and Thorsheim 1999; on Pittsburgh from Davidson 1979 and Tarr 1996.

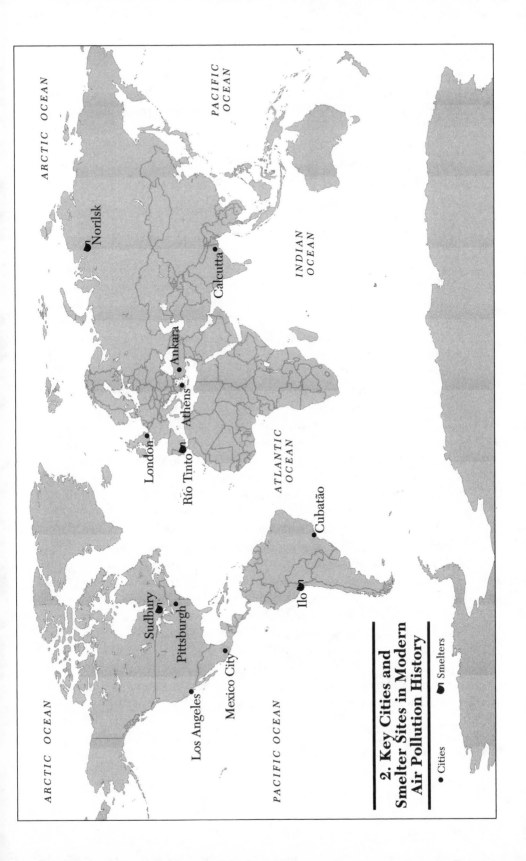

2. Key Cities and Smelter Sites in Modern Air Pollution History

• Cities ◆ Smelters

ARCTIC OCEAN

PACIFIC OCEAN

Norilsk

Calcutta

INDIAN OCEAN

Ankara

Athens

London

Río Tinto

ATLANTIC OCEAN

Cubatão

Sudbury

Pittsburgh

Mexico City

Ilo

Los Angeles

ARCTIC OCEAN

PACIFIC OCEAN

LONDON. Late Victorian London was the world's biggest city (6.6 million people in 1900), a sprawling metropolis with several hundred thousand chimneys and a few thousand steam engines, all burning coal. No one monitored air pollution carefully, but to judge by the frequency of London fogs (partly caused by pollution), London air was foulest around 1870 to 1900. One fog in 1873 caused people to walk into the Thames, unable to see where they were going. Several thousand people died prematurely on account of London fogs in this period—some 3,000 in the winter of 1879–1880 alone—generally from aggravated lung conditions. While no lethal fogs were recorded in London between 1892 and 1948, one in Glasgow in 1909 killed 1,063 Glaswegians, and 70 to 100 people died from severe pollution in Maastal bei Lüttich, Belgium, in a few days in 1930.[36] Smoke abatement made some progress in London and remained the focus of antipollution efforts to 1950. The spatial expansion of London, and more efficient industrial combustion, helped disperse and control pollution. But reformers did not touch the coal-burning domestic hearth, as sacrosanct to a pre-1950 Englishman as the automobile was to a modern American. In 1945, when some authorities sought to take advantage of the postwar reconstruction of London to create smoke-free zones, George Orwell defended coal fires in glowing terms as the birthright of free-born Englishmen.

Coal soon signed its own death warrant as London's fuel by killing 4,000 people in the fog of December 4–10, 1952. Chilly weather and stagnant air meant a million chimneys' smoke hovered over the streets of London for a week, reducing visibility to near zero. Healthy people found breathing uncomfortable; those with lung ailments often found it fatally difficult. In the twentieth century, only the 1918 influenza pandemic killed more Londoners. Public outcry and government inquiry followed, leading to the Clean Air Act of 1956, which sharply regulated domestic coal smoke. This helped London switch to gas and electric heat. Weaned from coal, London's smoke problem shrank to insignificance after 1956. Its sulfur emissions—even though not regulated until 1972—fell 90 percent (1962–1988) mainly because of the fuel shift. After the 1960s, London's air quality suffered more from tailpipe exhausts than from smokestack and chimney emissions. Ironically, the clearer air after the mid-1950s allowed more sunshine to penetrate to the city streets, where it reacted with tailpipe emissions to form photochemical smog. Londoners remain as attached to their right to drive cars as Orwell was to the right to a coal fire.[37]

[36]Zirnstein 1994:307.

[37]A 1991 fog, caused by tailpipe emissions and a temperature inversion, raised London death rates 10% for a week in December (*Economist,* 17 September 1994:91).

London had a long history of dirty air, dating back to the thirteenth century, when coal burning first became commonplace in London homes. London fogs, partly a consequence of coal smoke, earned a lasting reputation. During the week of December 4 through 10, 1952, London suffered the worst recorded air pollution disaster anywhere, bringing early death to 4,000 people. This is The Strand at midday, shrouded in a sulfurous haze. Within a few years, London converted to oil as its fuel for most purposes.

PITTSBURGH. In the United States, many cities built their energy systems around coal between 1850 and 1900. St. Louis and Chicago exploited the bituminous coal of southern Illinois, Pittsburgh and Cincinnati that of western Pennsylvania. They all had pollution problems, and from 1868 forward enacted smoke abatement laws. But no measures sufficed, and all remained smoky and sulfurous until 1940. Rebecca Harding Davis began her 1861 short story "Life in the Iron-Mills" this way:

> A cloudy day: do you know what that is in a town of iron-works? The sky sank down before dawn, muddy, flat, immovable. . . . The idiosyncrasy of this town is smoke. It rolls suddenly in slow folds from the great chimneys of the iron-foundries, and settles down in black, slimy pools on the muddy streets. Smoke on the wharves, smoke on the dingy boats, on the yellow river,—clinging in a coating of greasy soot to the house-front, the two faded poplars, the faces of the passers-by.[38]

This description reflected the Wheeling, West Virginia of Davis' youth. Journalist Waldo Frank wrote of Chicago in 1919: "Here is sooty sky hanging forever lower. . . . The sky is a stain: the air is streaked with the running of grease and smoke. Blanketing the prairie, this fall of filth, like black snow—a storm that does not stop. . . ."[39] Wheeling, Chicago, and scores of industrial towns in between were smothered in coal smoke for decades. Pittsburgh was the smokiest of all.

Pittsburgh first used coal in 1758, when it was a small fort perched at the edge of British settlement in the American colonies. Abundant local supply recommended coal even over wood. At the end of the Civil War (1865), half the glass and 40 percent of the iron made in the United States came from Allegheny County—the Pittsburgh area. The smoke inspired one 1866 visitor to describe the city of 100,000 as "Hell with the lid taken off."[40] Then the second industrial revolution came to Pittsburgh, and the steel industry took off. In 1884 this city of 300,000 people used 3 million tons of coal, 5 percent of the national total. For a brief interlude around 1887 to 1891, natural gas from the Appalachians drove down coal use in Pittsburgh and the skies cleared. But gas supplies ran out, coal returned, steel flourished, Pittsburgh grew, and smoke returned with a vengeance. Even Andrew Carnegie, whose coal-fired steel mills made him one of the world's richest men, complained of smoke in 1898. Smoke abatement laws had little effect. From the 1890s through the early 1940s, Pittsburgh became more like Hell with the lid clamped on.

[38]Jean Pfaelzer, ed., 1995, *A Rebecca Harding Davis Reader* (Pittsburgh: University of Pittsburgh Press):3.

[39]*Our America*:117, cited in Cronon 1991:12. Rudyard Kipling, who visited Chicago in the late 19th century, wrote "its air is dirt" (quoted in Cronon 1991:392).

[40]James Parton, cited in Davidson 1979:1037.

Andrew Carnegie, one of the great industrial entrepreneurs in world history, complained about the air pollution in Pittsburgh, but his steel empire did its part to fill the air with smoke and sulfur dioxide. Here are the Carnegie steel mills at Braddock, Pennsylvania, near Pittsburgh, on the Monongahela River. The photo is from about 1905.

For Pittsburgh's air, war was both hell and salvation. In 1940 and 1941, military orders boosted production, and the use of all available equipment, however old and inefficient, initially intensified Pittsburgh's smoke and pollution. But the example of St. Louis, where engineers, citizens, and politicians had come together to enact effective smoke prevention law in 1940, inspired Pittsburgh to imitation.[41] In late 1941 it too passed such laws. They were suspended for the duration of the war, but enforced on industry in 1946 and on homes in 1947, despite objections from coal interests, the United Mine Workers, and the railroads. Pittsburgh converted to cleaner anthracite coal, oil, and natural gas piped in from Texas. Locomotives and riverboats converted to electricity or diesel

---

[41]On St. Louis, see Tarr and Zimring 1997. Chicago, Cincinnati, Kansas City, and other coal cities followed St. Louis and Pittsburgh in the late 1940s.

In late October 1948 in Donora, Pennsylvania, not far from Pittsburgh, 20 people were killed by acute air pollution from local steel mills and zinc smelters. This is the zinc works of the American Steel and Wire Company, the main employer in Donora until it closed in 1956. The company, when sued for damages, maintained the killer smog was "an act of God." Donora was one of the few cities where cars sometimes stalled for lack of oxygen, but by the early 1950s, efforts to reduce air pollution in Donora, and the Pittsburgh area generally, began to show results.

fuel. By 1953 Pittsburgh's air was cleaner than at any time since the Civil War—excepting, perhaps, 1887 to 1891.[42] Tighter laws and more efficient fuel use continued to improve Pittsburgh's air in the 1950s and 1960s. Then, after the mid-1970s, the steel industry collapsed, mills closed, population declined, and Pittsburgh ceased to be a foremost manufacturing city—and the air got cleaner still. In 1985 a weekly magazine rated Pittsburgh as America's most livable city.

Both London and Pittsburgh underwent transformations that denizens of either city in 1900 or 1930 would have regarded as impossible. They did it at about the same time, which is partly coincidence but partly not. Both

---

[42]Ironically the worst recorded air pollution disaster in U.S. history happened just as Pittsburgh was cleaning up its air: in Donora, 30 km from Pittsburgh, where cars sometimes stalled for lack of oxygen. In October 1948, weather conditions trapped foul air from steel mills and zinc smelters, killing 20 people and making sick 6,000 (out of 13,000 in the town). See Snyder 1994.

transformations required the emergence of alternate fuels that permitted a cleaner energy system without any special economic sacrifice. Certainly some people suffered in the transition—coal deliverymen, chimney sweeps, laundry operators—but others benefited: pipe fitters, electricians, appliance salesmen. The coal industry survived in England and Pennsylvania until decades after the changes in the urban energy systems. Both transformations were eased by decentralization, the drift of population to suburbs abetted by the spread of cars. Both cities, although not their conurbations, slowly lost population: London's peak came in 1940, Pittsburgh's in 1950. Dozens of coal cities in the United States and western Europe followed roughly the same pattern, making for a great improvement in urban life for tens of millions.[43]

Like London, Pittsburgh, and most cities in the industrial world, New York in 1949 remained a coal-fired city, with thousands of chimneys and smokestacks. This is the view to the southeast from the top of the Empire State Building. By 1970, New York had followed St. Louis and Pittsburgh, and shifted away from coal as a fuel, and New Yorkers breathed easier.

[43]Eggleston et al. 1992; Powell and Wharton 1982; Schulze 1993: Stern 1982: On France, see Herz 1989. An exception was Belfast, which adhered to coal and suffered unreduced sulfur dioxide levels into the 1990s. No natural gas pipelines extended to Ireland.

Smog Cities

"Smog" is a term first coined by a London medical doctor in 1905 to mean smoke plus fog. It has evolved to mean any pollution-caused haze, in particular that created by the reaction between sunlight and nitrogen oxides or hydrocarbons. (I shall use it in this restricted sense). These pollutants, the precursors to smog, come mainly from vehicle exhausts. But they can also come from biomass burning or industrial combustion. In sunlight they tend to produce ozone, the most troublesome component of smog. Smog causes eye irritation and lung problems in people, and damages plants as well.

Geography matters in smog formation. It develops only where there is plenty of sunlight, and grows serious where topography and winds conspire to keep pollutants from dispersing. So sunny cities in bowls or basins, with mountains all around, are the most vulnerable. Mexico City fits that description all too well. Slightly less vulnerable, on geographic criteria alone, are Santiago, Los Angeles, Athens, Teheran, Chengdu (China), and Seoul.

LOS ANGELES. Photochemical smog made its debut in the human consciousness in Los Angeles, where the problem attracted notice in the early 1940s. During World War II, acute smog was mistaken for Japanese gas attacks; soon after it, smog became a regular character in Raymond Chandler novels and other Los Angeles literature.[44] Ever since, it has ranked high among daily complaints and political issues in southern California.

Here topography and history conspired. Los Angeles occupies a small coastal plain hemmed in by mountains on three sides. It has daily sea breezes that often blow yesterday's pollution back over the city, and frequent thermal inversions, which block the upward escape of pollution.[45] The American Southwest would not have big cities or smog without the cheap energy characteristic of the fossil fuel age. This, and cheap water, made possible greater Los Angeles' growth from 100,000 in 1900 to 1.4

[44]Brimblecombe 1995.
[45]Thermal inversions, or temperature inversions, exist when a layer of warm air stands above a colder one. This makes for an unusually stable situation, inhibiting the vertical mixing of air (and dispersal of pollutants).

By 1966, when this picture was taken, the Los Angeles smog had become justly famous. Exhausts from some 4 million cars fouled the L.A. air in 1966, affecting some 10 million people. Los Angeles was the first city in the world to combine large numbers of automobiles with abundant sunshine, the recipe for recurrent smog. Concerted effort at smog control since the 1960s has achieved only limited results. Los Angeles was built on the assumption that people drove cars, so ordinary Angelenos find it nearly impossible to live without them.

million in 1930 to 6 million by 1960.[46] American cities built after 1920 were shaped by mass ownership of automobiles, none more than Los Angeles, with its sprawl and freeways. In the 1940s, like several other American cities, it dismantled its system of public trains to make way for cars. Los Angeles' auto population quadrupled between 1950 and 1990 (to 11 million). Los Angeles is a city built for cars in a setting made for smog.

Smog became a political issue in the 1940s. The Los Angeles *Times* hired Raymond Tucker, the central figure in St. Louis's smoke reduction, in a press campaign for cleaner air. In 1947 the city created air quality boards that began to regulate refineries, factories, and finally even cars. By the 1960s, smog annoyed some 10 million people hundreds of days of the year and stunted tree growth as much as 80 kilometers (50 miles) away. After more stringent regulations in the early 1970s, ozone and smog

[46]Figures are for the Los Angeles–Long Beach conurbation (from Mitchell 1993).

decreased by about half in the Los Angeles basin, despite more cars and more driving. Yet in 1976, Los Angeles' air still reached officially unhealthy levels three days out of four. In the 1990s Los Angeles smog remained a regular health hazard, the most serious urban air pollution problem in the United States.[47]

ATHENS. Athens was built 2,500 years before cars. Yet it too acquired a serious pollution problem, which Athenians for decades have called *to nephos* ("the cloud"). Mountains surround the city on three sides; the sea borders the fourth. Thermal inversions are common in spring and fall. Like Los Angeles, Athens has a sunny climate, ideal for smog formation.

Modern Athens took shape after it became the capital of Greece in 1834. Its population climbed from 15,000 to 500,000 between 1830 and 1920, then quickly doubled as refugees from Anatolia poured in after a Greek military adventure there went awry. It reached 2 million by the early 1960s, and 3 million by 1980, by which time it occupied almost all the land between the sea and the mountains and accommodated one-third of the country's population.

Greater Athens has long hosted a large share of Greek industry, especially in the vicinity of Piraeus. Paint, paper, chemical, tanning, steel, shipbuilding industries, and more—half of Greek industry by 1960—squeezed into the capital area. Most factories were small-scale, unregulated, and energy-inefficient. Electrification, which grew exponentially after 1950, was based on fossil-fuel burning power stations in the western part of the city.[48]

Home heating added to the mix. Athens burned mainly wood and charcoal in 1920, but then took increasingly to imported coal until the Depression hit in 1931. After the further disruptions of World War II and the Greek Civil War (1940–1949), Athens slowly converted to an energy base of imported oil and domestic lignite, a dirty coal.[49] Before 1965, smokestacks and chimneys accounted for most of Athens' pollution, which took the form of smoke and sulfur dioxide. Much worse was yet to come.

[47]Particulates alone in the early 1990s killed about half as many people in southern California as did auto accidents (Lents and Kelly 1993; Levinson 1992:19–26; Turco 1997:148–63; WRI 1996:68–9). A new car in California in 1992 emitted only 10% as much pollutants as a 1970 new car.

[48]On Piraeus's evolution, see Sorocos 1985 and Leontidou 1990.

[49]Pelekasi and Skourtos 1992:24. In 1997, Athens began to use natural gas, piped in from Russia. Greece was the last of the countries in the European Union (EU) to make use of gas in its energy mix.

The automobile colonized Athens after 1955. The Athens subway, one of the world's first, consisted until 1997 of only a single line: for most Athenians there was no alternative to surface transport. In 1965 the city had 100,000 cars, and by 1983 a million. Since Greeks kept cars on the road for a long time, this vehicle fleet contained a hefty share of old clunkers, spewing out extra smog ingredients into the bright Athenian sunshine.[50] The unplanned growth of the city created a labyrinthine street pattern that contributed to traffic jams. The siesta tradition meant Athens featured four rush hours daily. At any given time after 1975 a large proportion of the engines running in Athenian cars were idling out of gear, waiting for traffic to clear. Until the late 1990s, Athenian buses, many of them imported from eastern Europe, were notorious polluters. All this combined to maximize tailpipe emissions per vehicle.

Occasional haze obscured visibility in Athens from at least the 1930s,[51] but *to nephos* appeared only in the 1970s. Smoke and sulfur dioxide levels actually dropped after 1977: their regulation amounted to a modest political challenge. But smog persisted, and indeed intensified after 1975 as Athens grew so prosperous that most families aspired to car ownership. The worst episode came during a heat wave in 1987, with excess mortality of about 2,000. Ozone levels in the late 1980s were twice those of 1900 to 1940.[52] Controlling cars proved a sterner political challenge than checking smoke and sulfur emissions.

To nephos entered Greek politics in the 1981 elections, when the socialist party (PASOK) implausibly promised to eradicate air pollution within three years, hoping perhaps to secure the gullible vote (a very thin slice of the Athenian electorate). PASOK won, restricted industrial fuel burning (1982), introduced low-lead gasoline (1983), and enacted rules prohibiting driving in the city center on alternate days, based on odd and even license plate numbers (1983).[53] Prosperous Athenians responded by

[50]See Pattas et al. 1994 on Athenian vehicle populations and fleet longevity; slightly different data appear in Katsoulis 1996). Cheaper maintenance and high purchase taxes made Greeks hang on to cars twice as long as the EU average. Three-quarters of the cars were still chugging away at age 20. In 1990 and 1997 the government approved incentives to try to modernize the car fleet, which seem to have worked.

[51]Papaioanniou 1967. Occasional dust storms affected the city before that. Klidonas 1993:85–6 notes the reduction in visibility in Athens during 1931–1984.

[52]Cartalis and Varotsos 1994.

[53]The history of Greek and Athenian air pollution regulation is found in Pelekasi and Skourtos 1992:90–122. Emergency levels of air pollution extend the restricted driving zone out to the suburbs. Taxi drivers objected strenuously. Dede 1993 considers Greek environmental regulation within a political economy framework.

buying second cars. Emission checks came in the early 1990s. These and further measures did not suffice, partly because of the unfortunate geography of Athens, and partly because enforcement of antipollution laws remained lax.[54] By the early 1990s Athens's smog was two to six times as thick as that of Los Angeles.[55]

Urban smog, in Los Angeles, Athens, or most anywhere else, persisted because most citizens preferred driving cars and breathing smog to limits on driving and less smog. It also persisted because of two things about which citizens had less choice: the weakness of public transit and the technology of automobile engines.

Megacities

Rural exodus and natural population increase in cities after 1950 combined to create a new chapter in the human experience. In 1950 only three conurbations in the world approached (or exceeded) 10 million inhabitants: London, New York, and Tokyo–Yokohama. By 1997, 20 cities surpassed 10 million, most in societies where law and regulation were uncertain, politics often unstable, and public finance always scarce. Urban growth outpaced basic infrastructure. Pollution in such circumstances was hard to control.

MEXICO CITY. The Aztecs in 1325 chose a magnificent setting for their capital Tenochtitlán, but the choice has come to haunt their descendants.[56] From the air pollution standpoint, Mexico City, in a steep-sided bowl, is a topographical error. Temperature inversions occur on 50 to 80 percent of days between November and May. At over 2,200 meters (7,300 ft), cars run inefficiently, generating more pollutants, and oxygen is scarce, intensifying the adverse health effects of ozone and carbon monoxide. There's plenty of sunshine to cook up smog.

[54]Klidonas 1993:92–4.

[55]Mantis and Repapis 1992; Sifakis 1991. See also the numerous papers in Moussiopulos et al. 1995, and the 10 technical papers in *Atmospheric Environment*, 1995, 29:3573–719. Carbon monoxide levels did, however, decline in Athens after 1986 (Viras et al. 1995). On health consequences, see Katsouyanni et al. 1990 and Pantazopoulou et al. 1995. Death rates among Athenians were notably higher on high pollution days (1975–1982).

[56]The following is based on Ezcurra 1990a, 1990b; Ezcurra and Mazari-Hiriart 1996; Levinson 1992:27–36; and WHO/UNEP 1992:155–64.

For a long time none of this mattered. The clarity of the air impressed the German scientist Alexander von Humboldt in 1803, and the air remained unusually transparent for another century and a half.[57] In 1900, Mexico City had 350,000 residents, many fewer than Pittsburgh. It then grew 60-fold in the twentieth century, as shown in Table 3.4.

TABLE 3.4 MEXICO CITY POPULATION, 1900–1997	
Year	*Population*
1900	350,000
1920	470,000
1940	1,800,000
1960	5,200,000
1980	14,000,000
1997	≈20,000,000

Sources: Ezcurra 1990b; Mitchell 1993a.

Mexico City also industrialized. In 1930 it had 7 percent of the nation's industry; by 1980 over 30 percent. This demographic and industrial growth reflected the centralizing ambitions of the leadership installed by the Mexican Revolution (1910–1920). By 1990 the bowl had 30,000 industries, of which 4,000 burned Mexico's high-sulfur fossil fuels.[58]

Mexico City also converted to motorized transport. The car fleet swelled from perhaps 100,000 in 1950 to 2 million in 1980 and over 4 million in 1994. Motor vehicles emitted about 85 percent of air pollutants in Mexico City in the 1980s, cars accounting for two-thirds, trucks and buses for the rest. The government, which owned the fossil fuel industries, for years discouraged energy conservation. The gasoline tax was one-tenth that of Japan and one-twentieth that of Holland or Italy.

[57]. . . except when early spring winds brought dust over the city from lakebeds drained at the end of the 18th century (Garfias and González 1992). In 1923 the writer Alfonso Reyes labeled Mexico as "la región más transparente del aire." In 1940 he complained that was no longer true. As late as the 1950s, Carlos Fuentes borrowed the phrase for the title of one of his novels, but it was quickly growing less apt as a description of Mexico City. Between 1937 and 1966, visibility decreased from 15 to 5 km in Mexico City, and worse was yet to come (Vizcaíno Murray 1975:109, 119).

[58]Garza Villareal 1985:133–94.

This combination produced what most observers considered the world's worst urban air pollution problem. In the 1970s sulfur dioxide levels in Mexico City usually ranged from one to four times the guideline of the World Health Organization (WHO), occasionally reaching 10 or 15 times "safe" concentrations.[59] They climbed only slightly in the 1980s: greater use of natural gas partly offset the city's growth. From 1975 to 1990, dust and soot grew thicker, from roughly twice to about three or six times WHO guidelines.[60] From 1965 to 1985, lead in the air doubled, reaching levels five times the Mexican legal limit, which prompted the introduction in 1986 of low-lead gasoline. The new gasoline unfortunately involved the use of additives that compounded the ozone problem.

Residents of Mexico City gasped and wheezed—and died[61]—from air pollution as much as any urban population anywhere. Schools occasionally closed for ozone alerts. Cautious parents equipped their children with surgical masks and tried to keep them indoors during peak pollution hours. Vegetation on the surrounding mountains suffered, affecting the water balance of the city. Tree rings from a fir grove southwest of the city showed sharply reduced growth from the 1960s; by 1993 pollution had killed one-third of the firs. In one acute pollution episode in 1985, birds fell out of the sky in mid-flight onto the Zócalo, Mexico City's great plaza.[62]

The policy response to air pollution in Mexico City included several measures: incentives to relocate industry (after 1978), the introduction of low-lead gasoline (1986), voluntary vehicle inspections (1988) quickly followed by mandatory vehicle inspections (1989), driving restrictions (1991), reforestation, and propaganda campaigns featuring cartoon birds that exhorted children to conserve energy and water. Even moss received legal protection. The United Nations twice awarded Mexico City prizes for its antipollution efforts.

By 1990 or so, the air stopped getting worse and by some measures improved. Sulfur dioxide, carbon monoxide, and lead concentrations no

[59]The WHO guideline is 30–70 μg (micrograms) per cubic meter; Mexico City's levels occasionally reached 900. No one knows what is "safe," a term that means quite different things to different people.

[60]The WHO guideline for mean total suspended particulates is 60–90 $\mu g/m^3$.

[61]Suspended particulate matter helped kill 6,400 annually in the early 1990s according to the World Resources Institute (WRI 1996:22). Health costs (treatment, lost wages, premature death) attributed to air pollution in Mexico City around 1990 came to U.S. $1.1 billion, or about 0.2% of Mexico's GDP. WRI 1996:24–5 says particulates killed 12,500 people per year in Mexico City and caused 11 million lost work days.

[62]*Wall Street Journal,* 4 March 1993:A1 (on trees); Pick and Butler 1997:202 (on falling birds).

longer regularly exceeded guidelines after 1990. But other pollutants, mainly ozone, caused the city's air to violate international standards on 90 percent of days during 1991 to 1995.[63] Extreme population growth and ordinary motorization, superimposed upon the Aztec choice of capital and the Mexican Revolution's commitment to centralization, made Mexico City's air pollution resistant to even the most vigorous antipollution policy. Centuries of history are not easily undone by public policy.

CALCUTTA. In some ways Calcutta suffered the worst of two worlds.[64] It became a Victorian coal city in the late nineteenth century and in the late twentieth evolved into a classic megacity, with a population (about 15 million in 1997) that grew faster than infrastructure and urban services. It achieved extraordinary success in regulating smoke after 1903, but eventually its growth outpaced its pollution control capacity.

Calcutta grew from about 1 million people in 1900 to almost 5 million by 1950, and then about twice that by 1980. These changes were spurred by natural population increase, rural exodus, and refugee settlement after the political violence that attended the partition of India in 1947–1948 and the war with Pakistan in 1970–1971. The city lies close to the coalfields of West Bengal, which made it a premier industrial city of India after 1880. It developed some dirty industries, including iron and steel, glass, jute, chemicals, and paper—powered until the 1940s by the high-ash (but low-sulfur) coal of Bengal.

Calcutta's setting, with frequent calms and thermal inversions in the winter months, often made for stagnant air. Visitors complained of the air in the eighteenth century, but it was the railway link to Bengal's coal fields (1855) that transformed Calcutta into a coal-smoke city. Jute and cotton mills led the way in fouling the air, but locomotives, steamships, home hearths, and coke making contributed to the haze. By the 1920s, Calcutta had 2,500 coal-fired steam boilers. In calm weather smoke lingered and death rates tripled.

Laws against coal smoke, first enacted in 1863, proved difficult to enforce, until the reforming zeal of the Viceroy of India, Lord Curzon (1859–1925), resulted in a smoke inspectorate (1903). Curzon told Ben-

---

[63]In 1995 domestic gas usage in Mexico City was (controversially) blamed for a third of the City's ozone. Nemecek 1995; Guzman et al. 1996. On 1990s Mexican air: Lacy 1993; Restrepo 1992; Simon 1997:77–82.

[64]Calcutta data from: Anderson 1995; WHO/UNEP 1992:91–98; Centre for Science and Environment 1982:74–90.

gal's chamber of commerce that smoke "besmirches the midday sky with its vulgar tar brush and turns our sunsets into murky gloom."[65] Between 1906 and 1912, numerous tall stacks sprouted in Calcutta and observable ground-level smoke declined by about 90 percent. In the 1920s the smoke inspectorate installed a rooftop observatory and by telephone scolded factory managers instantly when their smokestacks belched out thick smoke; smoke levels declined sharply again. "Probably no other industrial city in the world, before or since, has operated a pollution notification system so ambitiously conceived, so comprehensive in range, or so penetrating in its disciplinary effects."[66] The authoritarian power of the colonial government in India (seated in Calcutta until 1911) made smoke control easier there than in London or Pittsburgh, where coal interests retained great power. Curzon regarded smoke as a blemish on the escutcheon of European civilization and a poor advertisement for the benefits of imperialism. It had to go.

After the 1920s, with industrial smoke emissions under control, Calcutta enjoyed a breathing spell. But as demographic growth accelerated after 1950, smoke problems returned. In the 1970s Calcutta's sulphur dioxide concentrations more than doubled, reaching levels that averaged 25 percent above WHO standards, despite the low-sulphur coal. But Calcutta's most serious pollution problems came from domestic coal dust and soot, not sulfur dioxide. Its dust and soot levels billowed up in the 1970s, before levelling off at three to ten times WHO standards. For human lungs, breathing Calcutta's air after 1975 was equivalent to smoking a pack of Indian cigarettes a day. Nearly two-thirds of the population in the 1980s suffered lung ailments attributed to air pollution, chiefly particulates. Domestic use of coal probably damaged health at least as much as did industrial emissions, although industry used far more coal (2 million tons in 1990). Calcutta's dwellings almost all cooked with coal, and women and small children routinely inhaled soot and dust while preparing their meals. Men, who rarely cooked, breathed cleaner air.

Motor vehicles added little to the air pollution problems of Calcutta until the 1980s. After 1980 the fleet doubled every six years, reaching half a million in 1992. With little in the way of tailpipe emissions standards, and copious sunshine, ozone levels presumably grew rapidly. (They went unmeasured.) After 1991, when the Indian state lowered protection barriers and liberalized the economy, a newly prosperous class emerged in

[65]Quoted in Anderson 1995:313.
[66]Anderson 1995:323.

Calcutta as elsewhere, eager to drive private cars, so tailpipes contributed more and more to the air pollution cocktail of Calcutta. Population growth, the difficulty of regulating domestic fuel use, and eventual motorization brought the murky gloom of a century before back to Calcutta.

By and large the world's megacities all created air pollution problems as they grew. Those in rich, stable, and technically advanced societies lowered pollution levels sharply after 1970. Most of the rest got worse, as tailpipe pollution joined smokestack and chimney sources. The most acute problems plagued those cities cursed by geography: Beijing, Cairo, and Karachi suffered from desert dust as well as high levels of man-made pollutants; Mexico City and Seoul suffered from poor ventilation. In the early 1990s, Mexico City had the worst air of all megacities. Beijing, Shanghai, Seoul, and Cairo competed for a distant second place. Monitoring was so inadequate in Dhaka (Bangladesh), Lagos (Nigeria), and a few other megacities that no one could tell where they fit in the circles of air pollution hell.[67]

Recovering Cities

While Mexico City and Calcutta had limited or fleeting success in improving their air quality, some smaller cities did better.

ANKARA. Ankara was a sleepy provincial town of 30,000 when it became the capital of Turkey in the 1920s. The Turkish Revolution led by Mustafa Kemal Atatürk shared the centralizing tendencies of the Mexican Revolution, and Ankara doubled in population every decade from 1920 until 1980. Although it lies in a shallow bowl and has frequent wintertime temperature inversions, Ankara had no notable air pollution problems until the 1960s. But by 1970, when its population reached about a million residents, Ankara's emissions surpassed a threshold and developed growing sulfur dioxide, smoke, and soot problems, derived mainly from power stations and household use of lignite. High oil prices after 1973 drove Turkey to develop its lignite, which is high in sulfur and ash. Ankara's air became the worst in Turkey. By 1990 it had 4 million people, half a mil-

---

[67]The dust problems of Beijing and Karachi are in part man-made because of anthropogenic desertification. See WHO/UNEP 1992 for comparative efforts among megacities.

lion cars, and, in winter at least, air quality among the worst in the world. By the early 1990s, spurred in part by the longstanding Turkish quest to conform to European Union standards in hopes of admission to that club, Ankara began to tighten pollution controls and convert to natural gas, piped in from Siberia. Its air improved dramatically, despite continued urban growth.[68]

CUBATÃO. Nestled between the sea and a steep escarpment (Serra do Mar), Cubatão, Brazil, is close to São Paulo and closer to the port of Santos. In 1950 it consisted chiefly of banana plantations and mangrove swamp, but because of its hydroelectric potential and port access soon became a target for state-sponsored industrialization. This succeeded too well during Brazil's boom years of the 1960s and 1970s. Cubatão by 1980 became a city of 100,000, producing 40 percent of Brazil's steel and fertilizers and 7 percent of Brazil's tax revenues, but it earned the nickname Valley of Death. Its infant mortality was 10 times the average for São Paulo state; 35 percent of infants died before their first birthday. Soot and dust (total suspended particulates) averaged double the state's "state of alert" levels in 1980. Cubatão had no birds, allegedly no insects,[69] and its trees became blackened skeletons. Laboratory rats installed in a rented room in Villa Parisi, the poorest and most polluted part of town, managed to survive 1986 pollution levels, but with severe ill effects on their respiratory systems.[70] Acid rain killed vegetation on the slopes of the Serra do Mar, bringing on landslides. Communities had to be evacuated. Many observers thought Cubatão the most polluted place on earth—a title with many claimants.

After much dithering and denial, authorities responded to media harassment, citizen protest, and several deadly industrial accidents—especially after military rule (1964–1985) ended and questioners of the state development model grew less fearful. Under the impact of regulations, fines, and new technologies, pollution levels began to drop in 1984, and by 1987 came to only 20 to 30 percent of their previous levels. By the late 1990s, trees had returned to the slopes above the city and carp swam in the effluent pools of some of the chemical factories. The air of Cubatão was no worse than São Paulo's, which is to say it was bad but much better than it had been. In Cubatão the modernizing state cre-

[68]Türkiye Çevre Sorunları Vakfı 1991: 50–58, Levinson 1992: 42–48, Tuncel and Ungör 1996.
[69]According to newspaper reports cited in Findlay 1988 and Levinson 1992.
[70]Bohm et al. 1989.

ated the industrial pollution; when democratized and duly pressured, it also tamed it.[71]

Ankara and Cubatão addressed air pollution problems within 20 to 30 years of their first major appearance. London and Pittsburgh appear laggards by comparison. But relative to Mexico City or Calcutta, Ankara and Cubatão posed easy problems. Neither city had to regulate cars, infringe customary rights, or impose sacrifice on its citizenry.

Conclusion

After 1950, more and more people lived in more and more cities like Mexico City and Calcutta. In 1988 the WHO estimated that of 1.8 billion city dwellers in the world, more than a billion breathed air with unhealthy levels of sulfur dioxide and soot or dust.[72] The kind of civic, scientific, and political synergies that helped London, Pittsburgh, and dozens of other cities in western Europe, North America, and Japan to clean their air proved elusive in the megacities. They were growing too fast. Civic responsibility suffered, as many citizens identified more with their villages of origin than their city of residence. Law and regulation, even if on the books, proved hard to enforce. Economic development took precedence over other concerns. Perhaps most important, growth at such rates spawned any number of social pathologies that menaced the health of the poor and the security of the powerful: in this climate, reducing air pollution did not merit attention, effort, and resources. And so from Buenos Aires to Beijing and from Caracas to Karachi, urban air pollution clouds continued to gather.

[71]Findley 1988:52–68; Klumpp et al. 1996; Levinson 1992:37–42. Dean 1995:324–9 offers a less cheerful account of Cubatão's pollution history.

[72]Turco 1997:4. In 1995 Dietrich Schwela estimated 1.1 billion people breathed unhealthy air (WRI 1996:1).

4

The Atmosphere: Regional and Global History

This is the common air that bathes the globe.
                    —Walt Whitman, "Song of Myself"

The histories of urban air in the twentieth century were variations on earlier themes. In the case of Mexico City or Athens, pollution intensified. In the case of London or Pittsburgh, it abated. Regional and global air pollution, on the other hand, had scarcely existed in prior centuries.

Regional Air Pollution since 1870

Domestic hearths never, and cars rarely, caused noteworthy regional air pollution. That required the large-scale combustion of heavy industry generating pollutants that would hang around in the air for days or weeks (their residence times). The main pollutants responsible for regional-scale air pollution—sulfur dioxide, particulates, and nitrogen oxides—have residence times that allow them to stay aloft and spread with the wind. Regional air pollution grew acute where heavy industry enjoyed high prestige and political clout, where the objections of the landed interests counted for little, where dirty coal was the cheapest fuel, and wherever large-scale smelting took place.

In cases, a single smelter sufficed. The second Industrial Revolution (c.1870–1914) required plenty of copper, among other raw materials, and propelled a surge in copper mining and smelting in Spain, Chile, Japan, North America, and eventually southern Africa. In Spain, the Río Tinto mines, worked intermittently since Phoenician times but upgraded by new, British owners in the 1870s, soon yielded vast quantities of low-grade copper ore and all the sulfuric acid needed by Europe's emerging chemical industries. Ore was smelted on the spot, roasted over charcoal in open-air piles, drenching the vicinity in sulfuric acid rain. A British commercial agent acknowledged: "Throats and eyes are painfully affected and all iron corroded out. . . . Under these conditions vegetable life is impossible and animal life is rendered difficult." Pollution from Río Tinto smelters welded miners and peasants and their families together in one of the pivotal moments in Spanish labor history, a strike against the British mine owners (1888) in which 45 people were killed. Here pollution united whole communities, a characteristic of the industrial labor movement of the pre-1914 era.[1] But the scale, of pollution and protest, remained modest.

In the twentieth century, nickel joined copper as a key component in industrial and military production, and the scale of smelting dwarfed the efforts at Río Tinto. The nickel-copper smelters of Sudbury (Ontario) and the copper smelter near Ilo (southern Peru) produced poisonous airborne plumes that affected vegetation and lungs for miles around. Sudbury's open-roasting smelters created a blackened desert between 1888 and the 1920s, when the first smokestack was installed. Nearby farmers complained to little avail until 1972, when a smokestack taller than the Eiffel Tower began to disperse sulfur emissions over a broader landscape.[2] The Ilo smelter opened in 1960, and within four years farm-

[1]Kaplan 1981:38–47.
[2]Ripley et al. 1996:170–80; Dudka et al. 1995: Local vegetation remains thin because of decades of toxic metal deposition in the soil. The tall stack (381 m) does not disperse metal emissions much, only sulfur. The Inco smelter at Sudbury before 1972 emitted about 1% of the total sulfur released to the atmosphere, and contributed heavily to Canada's and the USA's acid rain. Farmers' agitation eventually (1974) got Inco fined $1,500, about one minute's worth of profits, Cox 1982. Quinn 1988 is a fine account of the damage caused by copper smelters in southeastern Tennessee. There much of the man-made desert has returned to life since 1930. Quinn 1989 covers some other smelters in the USA.

ers up to 200 kilometers away had organized a lawsuit seeking compensation for crop damage.[3] More damaging still were the Norilsk nickel smelters in northwest Siberia, part of a giant metallurgical complex built by gulag labor after 1935 and run by Stalin's secret police. Norilsk grew to be the largest city in the world above the Arctic Circle and a bulwark in the Soviet military-industrial complex. Its pollution killed or damaged a swatch of taiga forest half the size of Connecticut, and filled residents' lungs, contributing to severe health problems even by the standards of the late Soviet period. The men of Norilsk suffered the highest lung cancer rates in the world. In the 1980s the smokestacks of Norilsk spewed out more sulfur dioxide than did all of Italy.[4] Other Russian nickel smelters in the Kola Peninsula spread acid rain across northern Sweden and Norway, as well as east to Murmansk.[5]

What a single smelting complex could do badly, a cluster of heavy industries could do worse. From the late nineteenth century, several such clusters emerged, sometimes from state policy (Donetsk or Magnitogorsk in the USSR, Katowice in Poland); sometimes from the juxtaposition of coal, ores, and markets (the Ruhr, the English Midlands, the Great Lakes region of North America); and often from a mixture of the two (Osaka–Kobe in Japan, greater Los Angeles). These great industrial regions, the engines of twentieth-century economic growth at least until 1975, all produced enormous amounts of air pollution. The Ruhr and the "Sulfuric Triangle" make an interesting contrast, showing the impact of politics on pollution.

THE RUHR.[6] The Ruhr region in Germany is small as industrial heartlands go—about 1,500 km^2—but beneath it run some of the world's biggest coal seams. It lies to the east of the Rhine athwart the Ruhr and Emscher Rivers. In 1850 it was an agricultural area. By 1910 it produced 110 million tons of high-sulfur coal, employed 400,000 miners, and sustained the giant steel- and ironworks of Krupp and Thyssen, both crucial firms in the German military-industrial complex. Industry so important

[3]Vizcarra Andreu 1989. Ilo's emissions remained untreated as of 1994, but plans existed for desulfurization (Murley 1995:284).

[4]Peterson 1993:13: See Lincoln 1994:403 on Norilsk lung cancer rates.

[5]According to WRI 1996:206, the Norilsk smelter killed 350,000 ha (hectares) of forest and damaged 150,000 ha; Kotov and Nikitina 1996 offer estimates about 15% higher. See also Gytarsky et al. 1995, which has a map showing "total damage of vegetation" over an area of about 3,200 km^2, larger than Rhode Island or Luxembourg.

[6]This account comes from Brüggemeir 1994 and 1990, and Brüggemeir and Rommelspacher 1992.

to the German state escaped almost all regulation, so air pollution—smoke, soot, sulfur dioxide—attained gigantic proportions.[7] In 1900 the Ruhr was the biggest industrial region in Europe and probably the most polluted. Without it, Germany could scarcely have fought World War I.

International events made clear the extent of Ruhr air pollution in 1923. Furnaces went full blast during World War I, but in response to Germany's failure to pay war reparations, French and Belgian troops occupied the Ruhr in early 1923 and strikes shut down industry until autumn. Suddenly the skies cleared. Local harvests improved by half. Trees grew faster than before or after. Women enjoyed a respite from dusting homes twice a day. But the French weren't getting the coal they wanted, and Germany's currency was losing all value as the government printed money to support the strikers. The situation proved intolerable for both Berlin and Paris, so negotiations ended the strikes. Work—and pollution—resumed. Official inquiries into pollution, perhaps inspired by the bright summer of 1923, followed. But they concluded that pollution was inevitable, and that the Ruhr must adapt to it rather than try to limit it.

Pollution reached new heights in the 1920s, with new combustion technologies that permitted the use of lower-quality coal. One new plant, in Solingen, covered its surroundings in white ash within hours of firing up its boilers in 1929. Protests resulted in the installation of metal filters, but acidic gases disintegrated them within days. Even with better factory filters, Solingen's air was unfit for young lungs; its school had to shut down for 18 months to accommodate industry. Attitude, policy, and law followed this general pattern in the Ruhr because industrial firms and workers dominated the area from 1880 on. Profits and jobs mattered more than pollution; farmers, landowners, housewives, and others whose interests suffered could not compete politically with Krupp, Thyssen, or the industrial unions.

Profits, employment, and pollution burgeoned with Nazi rearmament after 1933. The Nazi affection for German blood and soil did not extend to German air, and certainly did not justify constraints on heavy industry, least of all during World War II. In 1944–1945 the Allies made Ruhr factories a special target, and although they found Ruhr smoke and haze made their bombers far less accurate than elsewhere over Germany, they eventually leveled a large share of Ruhr industry. German defeat in

[7]Wilhelmine Germany (i.e., 1871–1918) systematically changed laws to favor polluters at the expense of farmers, fishermen, and foresters, who once enjoyed a paternalistic protection in which, for instance, smelters had to cease operations during the spring growing season (Gilhaus 1995). See also Spelsberg 1984 on Wilhelmine air pollution.

At Duisburg, where the Ruhr River flows into the Rhine, tugboats, loading cranes, and switch engines go about their work against a background of industrial haze. Duisburg hosted iron, steel, and chemical industries, and was the busiest inland port in Europe. This picture was taken about 1952, by which time the Ruhr's industrial economy was on the road to recovery from damage sustained during World War II, and the air once more was highly polluted. By 1961, Ruhr air quality had become an important political issue in Germany, and by 1980, pollution of this intensity was a thing of the past.

war once again led to a momentary relaxation of pollution (1945–1948). But with the Cold War, recovery of the Ruhr's industry became essential. Europe needed German coal, iron, and steel during its reconstruction. By the late 1950s the German government subsidized Ruhr coal in order to combat imports, which boosted Ruhr production and pollution to new levels.

The Ruhr's air quality became a national political issue in 1961 when Willy Brandt, campaigning for the Chancellorship, said the skies over the Ruhr must become blue again. Brandt lost, but effective smoke and soot regulation followed in the 1960s, especially after he became Vice-Chancellor in a coalition government (1966–1969). Tall smokestacks went up, broadcasting the sulfur emissions downwind. This tactic, common in all industrial regions, distributed pollution over wider areas in less concentrated doses. Politically this often soothed matters and delayed effective emissions reduction. In the Ruhr it took until the early 1980s—and the emergence of a German Green Party—before the accumulated evidence of sulfuric acid rain generated the political will to clamp down on emissions.

The Ruhr's record of unbridled air pollution prior to the 1960s and considerable improvement in the subsequent 30 years is paralleled in the industrial regions of the United States, Japan, Sweden, and Britain. By and large, the industrial centers of the USSR and eastern Europe remained impervious to pressures to reduce pollution until the 1990s. Billowing smokestacks carried a totemic aura among Marxists: heavy industry promised both power for the state and proletarianization of society. Hence iron, steel and coal, and later chemicals, cement, and oil became priorities in the economic plans of the USSR, and after 1948 of its East European satellites. Among the many consequences were vast regions of intense air pollution, and hence an additional grievance against communist rule.

THE SULFURIC TRIANGLE. The triangle bounded by Dresden, Prague, and Krakow sits on rich seams of brown coal, high in sulfur and ash. Early industrialization took advantage of these deposits, and of water power from the Tatra and Sudeten Mountains. The first blast furnace in Europe opened in Gleiwitz (Gliwice in Polish) in 1796, inaugurating a long era of coal, iron, and steel production in what is now Polish Silesia. By 1900 this region supported a considerable industrial establishment, second in Europe only to the Ruhr.

Notable air pollution in this region dates from the late nineteenth century. Czech forests showed the ill effects of a high-sulfur diet from the 1920s. The Depression (1931–1938) checked industrial production and pollution, but then war stimulated it. Entirely under German control by late 1939, the region contributed heavily to the Nazi war effort. American bombing and Soviet artillery flattened most of its industry in 1944 to 1945. But its coal remained, and industrial development suited the interests and ideology of the communist parties in charge after 1948 in Czechoslovakia, Poland, and East Germany—and their overlords in Moscow. So heavy industry (steel, iron, coal, cement, chemicals, glass, ceramics) returned, expanded, and polluted skies as never before. Plant managers and party planners had even less to worry about in the way of resistance to pollution than had the lords of the Ruhr. The communists

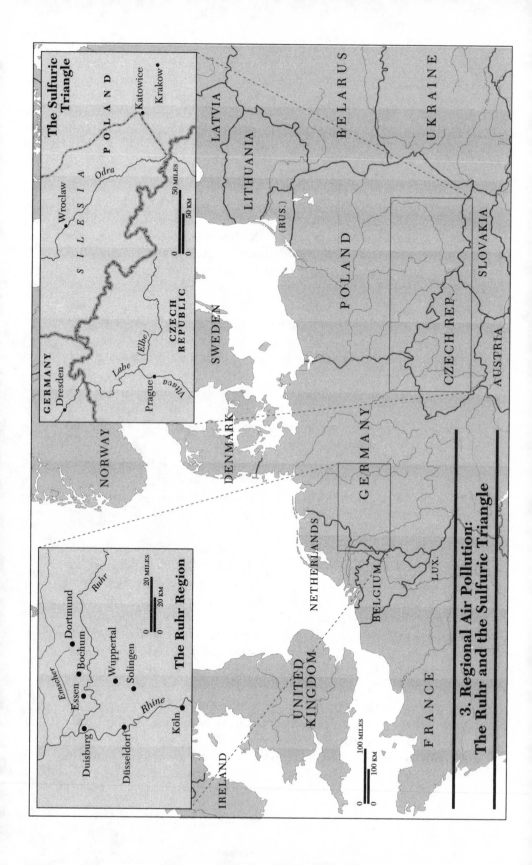

The Sulfuric Triangle

POLAND

Katowice

Krakow

S I L E S I A

Odra

Wroclaw

50 MILES

50 KM

GERMANY

Dresden

CZECH REPUBLIC

Labe (Elbe)

Prague

Vltava

NORWAY

DENMARK

SWEDEN

NETHERLANDS

GERMANY

BELGIUM

LUX.

FRANCE

AUSTRIA

CZECH REP.

SLOVAKIA

POLAND

(RUS.)

LITHUANIA

LATVIA

BELARUS

UKRAINE

IRELAND

UNITED KINGDOM

100 MILES

100 KM

**3. Regional Air Pollution:
The Ruhr and the Sulfuric Triangle**

The Ruhr Region

Ruhr

Dortmund

Bochum

Essen

Emscher

Duisburg

Wuppertal

Solingen

Düsseldorf

Rhine

Köln

20 MILES

20 KM

destroyed the landed interest quickly in Eastern Europe, and monopolized information and power so successfully that discordant opinions were rarely voiced in public until the 1980s. Moreover, the economic growth rooted in rapid industrialization (c.1950–1970) improved average living standards, so polluting smokestacks in the Sulfuric Triangle, as in Yawata (Japan), Pittsburgh, and the Ruhr, seemed a tolerable nuisance—for a while.

By the 1970s, pollution attained gargantuan proportions. Northern Bohemia had the most intensive pollution in Europe because of its extraordinary concentration of coal-fired power plants.[8] Coal provided three-quarters of Poland's energy and two-thirds in Czechoslovakia and East Germany. In the 1970s, two-thirds of Poland's sulfur emissions came from Upper Silesia, only 3 percent of the nation's territory.[9] Polish Silesia also acquired millions of tons of pollutants with the west wind from Czech and German provinces. By one calculation, the intense pollution in Upper Silesia and Krakow in the mid-1970s killed six to seven people per thousand each year. Life expectancy in Upper Silesia was shorter than elsewhere in Poland, genetic mutations and childhood developmental disabilities more common.[10]

Whereas the Ruhr after 1970 achieved some measure of control over industrial emissions, air pollution levels in the Sulfuric Triangle continued to climb, albeit more slowly, in the 1980s. Coal subsidies and fuel inefficiency remained standard. East Germany generated more sulfur dioxide per capita or per unit of GNP than anywhere else in the world. By 1990 Poland's Upper Silesian sulfur emissions were five times as great (per square kilometer) as the Ruhr's.

Authorities in Poland initially (c.1950–1970) denied that pollution existed in socialist economies, and eventually found it prudent to make environmental information state secrets. Consequently the citizen pressure, politicization, and media crusades that forced emission reductions in the Ruhr while leaving basic structures of society intact had no parallel east of the Iron Curtain. Since the state shielded polluters so thoroughly in Eastern Europe, citizen outrage at pollution, pent up for decades, was di-

[8]Two-thirds of Czechoslovakia's power plants were located here, and they generated 86 tons/km² of fly ash and 181 tons/km² of sulfur dioxide per year (Carter 1993a:77).

[9]By 1990 the proportion dropped to one-third (Carter 1993b).

[10]Trafas 1991. This includes only excess deaths from respiratory ailments, not cancers. Details on health costs of air pollution in Polish Silesia are given by Majkowski 1994. On Czech lands, see Bobak and Feachem 1995, who suggest that air pollution contributed about 3% to Czech mortality in the late 1980s.

rected at the state itself, not at individual firms. This contributed to the demise of the communist states in the years 1989 to 1991.

Even these world-shaking events could not change geology: the brown coal remained and continued to power industry and pollute air in the 1990s. In the former East Germany the events of 1989 to 1991 led to re-unification, investment, technical improvements, and quick changes in energy efficiency and pollution control, much of it achieved through greater use of natural gas. In Polish and Czech lands changes came more painfully. Czech air pollution levels declined by almost 50 percent be-tween 1985 and 1994. But improved air quality in both countries often came from plant closings (as for example at the giant Nova Huta steel-works near Krakow) more than from technical improvements.[11]

The air pollution experience of the Ruhr and the Sulfuric Triangle in the twentieth century shows the firm devotion of European states to in-dustrial growth; the devotion of Western firms to profit; the fervid com-mitment of the communist states to out-producing the West; and the low priority that air quality and human health commanded. But it also shows how, after 1970, more politically open societies could change under so-cial pressures, while more restricted and inflexible polities remain choked in the grip of pollution.[12]

The Air in Japan

In Japan, air pollution reached very high levels by 1970 and then subsided. Both the pollution and its abatement were linked tightly to Japan's politics—international, national, and local.

Japan shared many of the same commitments and characteristics that caused so much pollution in the Sulfuric Triangle and the Ruhr. From the 1870s it rapidly built dense concentrations of industry. Crowded cities and polluting industries lived cheek by jowl. Until the mid-1960s the state scarcely regulated the big firms on which its power, especially its geopolitical position, largely rested. In Japan, as in Germany, powerful na-tionalism translated into determined efforts to industrialize quickly, re-

[11]Carter 1993a and 1993b; Klarer and Moldan 1997; Stanners and Bourdeau 1995:ch. 4 et passim; Trafas 1991. The Czech data are from Bobak and Feachem 1995 and Moldan 1997.
[12]A careful treatment of this theme is given by Dominick 1998.

gardless of social and environmental costs. Japan and Germany can each lay plausible claim to the most polluted air (nationwide) in the twentieth century.[13] From the Meiji restoration (1868) until about 1965 Japan was "a polluter's paradise."[14] But by 1975 Japan's air pollution problem was under control, and by 1985 Japanese enjoyed cleaner air than almost any other industrial populace in the world. This was a dramatic turnaround, which like the Ruhr, Pittsburgh, or Cubatão, shows how reversible even intense air pollution can be.

Copper Mines and Smelters, 1885 to 1925. Mining and smelting gave Japan early experience with acute air pollution. The air and water pollution at the Ashio mine (see Chapter 5) provoked tense political struggles in the period 1890 to 1905. The Besshi copper mine, on Shikoku Island, caused pollution problems from the seventeenth century on. After 1885 the expansion of mining and smelting at Besshi intensified pollution and provoked 40 years of intense complaint and political tussle. To avoid the political problems of Ashio, the government encouraged mine owners to install antipollution devices. In 1910 the Minister of Agriculture, responding to agitation by farmers, coerced the proprietors, Sumimoto Interests, into restricting Besshi's operations during a 40-day span crucial to the maturation of rice. This practice, following a precedent set in Wilhelmine Germany, reduced but did not resolve antipollution struggles. After 1925, desulfurization equipment and a 48-meter smokestack diffused local pollution and defused political pressure. The Hitachi copper mine after 1905 followed the same course of increased local pollution and public protest, resolved by a 155-meter stack—tallest in the world—built in 1914–1915.[15]

Only a few communities succeeded in restraining air pollution, and usually their success consisted of the dilution and export of offending substances. Meiji Japan's intense commitment to industrialization, militarization, and imperial success legitimated the sacrifice of local and regional communities for the national interest; its centralized power structure made such sacrifice feasible. A fisherman whose livelihood was

[13]Britain is the only other contender. Poland, the Czech lands, and other communist countries did not industrialize early enough to develop world-class air pollution until late in the century.

[14]This phrase, and much of what follows, is drawn from Hashimoto 1989.

[15]On the pollution and politics of another important mine, the Kosaka copper mine, near Asita, see Okada 1990. After World War II the Hitachi mine (Ibaraki Prefecture) caused protests because its effluent poisoned local villagers (Miura 1975:286–305).

4. Key Locations in Japan's Air Pollution History

HOKKAIDO

Sea of Japan

HONSHU

Tokyo

Yokohama

Kyoto

Yokkaichi

Hanshin Region

Kobe

Osaka

Osaka Bay

Ube

SHIKOKU

KYUSHU

PACIFIC OCEAN

0	50 MILES
0	50 KM

destroyed by the Yawata steelworks reflected: "With the development of the Japanese nation, and the development of this region, it is us fishermen who have become the victims."[16] But fishermen did not run Japan.

THE HANSHIN REGION, 1890 TO 1970. The acute air pollution of mines and smelters affected small regions but presaged larger problems. The Hanshin region hosted more heavy industry than anywhere else in Japan except possibly greater Tokyo. (Hanshin includes Osaka, Kobe, and numerous smaller cities around and behind Osaka Bay.) Long a textile center, after 1880 the region bristled with new iron, steel, cement, and chemical plants. Population in Osaka and Kobe doubled in the 1880s, to about 620,000, and doubled again by 1900. Victory in the Russo-Japanese War (1904–1905) vindicated the Japanese economic strategy, and redoubled government commitment. Hanshin's industries spilled out amid residential and farming districts, causing acute social stress. Smoke and sulfur dioxide poured forth from thousands of smokestacks onto millions of people. Hanshin was the equivalent of the Ruhr: an essential industrial zone where national interest justified intense pollution.

Osaka began to monitor air pollution in 1912. Its smoke and ash concentrations rivaled those of St. Louis, Cincinnati, London, or Berlin.[17] Osaka was a coal city like London or Pittsburgh; the Japanese called it "Smoke City Osaka." As in other coal cities, murmurs of dissent arose against the orthodoxy of industrial development.

Nascent political pressure to curb pollution foundered when the demands and opportunities of World War I arose. Japan played only a small role in that conflict, but when it ended managed to acquire German colonies in the Pacific and some German concessions in China. The chaos in the Russian Far East after the Bolshevik Revolution gave Japan another opportunity to flex its imperial muscles, for which it needed more steel, ships, and weapons. Hanshin boomed. Its air pollution got worse in the 1920s, despite a 1925 law that required smoke prevention equipment on urban buildings. Administration and regulation could not keep pace with urban growth: Osaka doubled in population (to 2.4 million) in the 1920s, while expanding territorially. The whole Hanshin region was humming with the rapid growth of Japan's economy. Dozens of air pollution disputes led Osaka Prefecture to organize smoke prevention efforts, culminating in a 1932 law designed to increase combustion efficiency and re-

[16]Quoted in Morris-Suzuki 1994:203.
[17]Data from Miura 1975:244–56.

duce smoke. But the city of Osaka alone had 35,000 smokestacks in 1932, and only three smoke inspectors. With the inauguration of Japan's imperial adventure in China (1931), more metallurgical and chemical plants were built in Hanshin, and smoke levels doubled (1932–1934). Airplanes crashed on account of reduced visibility. But no pollution restrictions carried weight, as Japan's overriding priority became war production. Smoke, ash, dust, and sulfur bathed the region until American air power pounded Hanshin's industry to rubble in 1944–1945.

After 1945, Hanshin stagnated until the Korean War, when, as in the Ruhr, American geopolitical anxieties required the resurrection of industry recently destroyed by American bombers. Dustfall in Osaka, in 1945 only a quarter of 1935 levels, by 1955 surpassed prewar heights.[18] Automobiles, a small factor in Hanshin air pollution in the 1930s, added to the dark clouds from the 1950s, especially after 1970, when Osaka–Kobe–Kyoto had nearly fused into a single sprawling motorized conurbation. Smokestack and tailpipe pollution flourished until 1970, causing widespread health problems for a population approaching 10 million. Hanshin, like the Ruhr or Norilsk, was a place where polluting enterprises were protected in the national interest: a pollution reserve.

The Hanshin region was only one of several places severely polluted during Japan's economic miracle. Ube—a cement, chemicals, and coal center in the far southwest of Japan—suffered notoriously foul air. The Yawata steelworks, opened in 1901, by 1961 was blowing 27 tons of soot and dust out of its smokestacks every day and presumably contributed to the unusually high rates of respiratory disease observed in surrounding precincts. And Tokyo, where the air was polluted from the late nineteenth century, if not so badly as Osaka's, acquired an acute photochemical smog problem by 1970 due to growing auto use.[19] The Japanese economic miracle—like the German *Wirtschaftswunder*—came at a heavy air pollution price. But by the late 1960s the air all over Japan was beginning to clear.

THE JAPANESE ENVIRONMENTAL MIRACLE, 1965 TO 1985. Several convergent forces allowed Japan to change environmental course without derailing the engine of economic growth. The most important were a system of local government that responded to local concerns, an energy shift away from coal, widespread prosperity that encouraged cit-

[18]Data in Ibid. 37.
[19]See Hashimoto 1989 and Morris-Suzuki 1994:202–7 for several more examples.

izens to question the necessity of pollution, and the extraordinary rate of capital accumulation that allowed industry to spend on pollution control when obliged to.

Ube took effective action first. In the early 1950s an unusual coalition of academics, industrialists, and local bureaucrats formed. Epidemiologists from the local university's medical school demonstrated the health consequences of Ube's foul air. Kanichi Nakayasu, a leading industrialist who had visited Pittsburgh before World War II, went there again in 1954 and saw and smelled the difference. He convinced his peers that Ube could clear its air. Bureaucrats composed new regulations on emissions, chiefly of dust and smoke, and by 1961 Ube's air became a transparent shadow of its former self. Elite-driven reform worked. Its regulations became the basis for national law in 1962 that controlled—with indifferent success—smoke and soot.

Compliance with the 1962 smoke-and-soot law did not come naturally to most Japanese industrialists, but citizens and the state gave them a nudge. The 1962 law did nothing to curtail emissions of sulfur dioxide or heavy metals. Air pollution levels in Japan by the late 1960s were worse than ever before. Citizen action drove the next phase of reform. In 1967, asthma sufferers in Yokkaichi sued the giant petrochemical complex operating there since 1959. In 1972 they won large damages amid great publicity; the Mie Prefecture soon imposed tight sulfur emissions standards. At the Yawata steelworks, women pressured authorities into pollution abatement. Several prefectures canceled planned industrial complexes in the late 1960s, prompted by pollution and health concerns—an extreme departure from the previous mentality and practice in Japan.

By the late 1960s, new plants had to have pollution control technology to get local approval. Japan's limited supply of suitable terrain put local authorities in good bargaining positions vis-à-vis industrialists, allowing them to impose conditions that would not fly in countries with large expanses of flat ground. Indeed, the industrialist who valued good community relations (not a *rara avis* in Japan) found it prudent to police factory emissions more hawkishly than the law required. But the law was usually not far behind. Pollution control entered the main stage of national politics by 1965, well-publicized lawsuits continued, and people and prefectural authorities learned more and more about pollution and pollution control. A particularly bad smog episode in Tokyo in the summer of 1970 helped focus the attention of the national government, which later in that year passed a spate of antipollution laws and created a new agency to monitor environmental affairs. Standards grew tighter, industrialists invested in new antipollution technology, and Japan's skies got much, much cleaner. By 1975, Japanese businessmen were trying (unsuccessfully) to sell pollution control technology in Cubatão.

Japan even achieved quick control over car emissions. Tokyo's car fleet quadrupled to 2 million from 1960 to 1970. When the head of the American EPA appeared before Congress to sell the U.S. Clean Air Act of 1970, he used Tokyo, where traffic policemen wore masks to filter the air, as an example of how bad things could get. Tokyo brewed world-class smog and lived in a "permanent dusk."[20] MITI, the all-powerful Ministry of International Trade and Industry, reacted to public outrage by creating emission standards for the auto industry, following those the United States implemented in 1970. In both countries the car companies claimed the standards could not be met. The American companies sued the EPA. Japanese companies complained loudly but worked vigorously to meet MITI and EPA standards. Some companies met both, years before the law required it, which helped them surge into the American car market in the 1970s. By 1978 new Japanese cars emitted only about 10 percent as much pollution as had new 1968 models.[21] The Japanese aversion to old and used cars helped modernize the fleet rapidly and drastically reduced tailpipe emissions.

By the early 1990s, Japanese cities such as Osaka and Tokyo, and Japanese industrial districts such as Hanshin, suffered far less air pollution than they had in the 1960s or the 1930s.[22] Behind this remarkable turnaround stood some favorable conditions. Japan's struggle with smoke and soot ended well in large part because coal (half the energy mix in 1955, a sixth in 1975) gave way to oil and other energy sources. Antipollution technology made rapid progress, assisted by Japan's extraordinary capacity to import, modify, and assimilate useful technologies and laws. Japan, which had assiduously studied western legal codes in the early Meiji years, a century later adapted pollution control law from Germany, Britain, and the United States. Japanese society in the boom years also saved at uniquely high rates, allowing prodigious capital investment; a proportion of this could go to pollution control, occasionally above and

[20]See Nishimura 1989:v.

[21]Hashimoto 1989:42.

[22]See data on specific pollutants in *Environment*, March 1994, 36:36. Tokyo still had occasional ozone problems, but had reduced sulfur dioxide, carbon monoxide, suspended particulates, and ambient lead to well below WHO guidelines, which in cases involved 75–90% reductions in concentrations. Osaka's air pollution levels (c.1995) were about the same as Paris's or London's.

beyond legal requirements. It also went to energy efficiency, gospel in Japan after 1973. MITI funded appropriate research, and by 1983 mining and manufacturing used one-third less energy (per unit of production) than in 1973. Most importantly perhaps, after the 1947 constitution, Japanese municipal and prefectural officials became very responsive to citizen concerns, disciplined more by local elections than by Tokyo. This tempered the tendency to sacrifice local environments (and health) for the greater glory and progress of the nation—a strong tendency after 1868. The Japanese environmental miracle, like the economic one, derived from a mix of factors, several of which were distinctly Japanese.[23]

Acid Rain

Sometimes regional pollution extended across international borders. The major issue in transboundary air pollution was acid rain, derived from sulfur and nitrogen oxides. Unlike most pollutants, they last long enough in the atmosphere to travel thousands of kilometers. Fossil fuel combustion in each of the great industrial heartlands of the twentieth century generated huge sulfur emissions, the main ingredient in twentieth-century acid rain.[24] Japan's emissions tended to wash out over the Pacific. The emissions of other countries created frictions between neighbors. The most important such cases involved acid rain in northern Europe, eastern North America, and East Asia.

In Europe, detectable international transport of air pollution annoyed sensitive observers from at least the 1860s. The Norwegian playwright Henrik Ibsen (1828–1906) complained about airborne British grime

[23]These paragraphs draw on Hashimoto 1989, Morris-Suzuki 1994, and Tsuru 1989 and 1993. A less sanguine assessment of Japan's recent environmental history is Hoshino 1992. See also Krishnan and Tull 1994; Broadbent 1998.

[24]Until about 1980, sulfur dioxide led nitrogen oxides 2:1 as a source of acid rain. Lately, in Europe and America, the ratio has been closer to 1:1. The other important acidifying compound was ammonia, the latter derived mainly from agriculture. Acid rain became more of a problem in Europe and America after smoke abatement policies reduced the particulate matter blasted out of smokestacks. Formerly ash and soot had neutralized much of the sulfur; once particulate emissions were checked, acid emissions became more serious. A sketch of the emission history of sulfur, nitrogen, and other acidifying compounds is given in Graedel et al. 1995.

crossing the North Sea.[25] But attention focused on the issue only in the 1960s, when Scandinavian scientists showed that acid rain derived from British coal combustion had strongly affected the rivers and lakes of southern Sweden and Norway. Further monitoring and research revealed enormous international flows of air pollution, mainly sulfur dioxide, broadly moving from west to east with the prevailing winds across Europe.[26] Iceland and Portugal received almost no imported pollution; Scandinavia and Poland plenty. The tall smokestacks required to avoid acute local pollution problems naturally increased transboundary pollution. Acidification affected soils and waters across Europe, but especially the broad belt from Birmingham to Bratislava, where coal and car use was concentrated. Even rural areas far from any industry showed ill effects, mainly damage to trees, lakes, and streams. The thresholds beyond which acid buildup causes biotic problems (so-called critical load) are quite variable: the limestone soils of Greece can neutralize 10 times as much acid as the sensitive soils of Scandinavia. Europe's industrial growth, energy use, and motorization from 1950 to 1980 poured enormous quantities of sulfur and nitrogen into the air, creating acid rain as never before.

Addressing problems of this scale and complexity required international cooperation. In 1985 and 1994 most European countries signed protocols requiring sulfur emission reductions, the second of which recognized the importance of varying critical loads. In fact, European sulfur emissions dropped 15 percent (1980–1995), driven at first by energy efficiency stimulated by the fuel prices hikes in the 1970s.[27] Controlling nitrogen oxides means controlling car emissions, on which Europe made less progress, despite an international protocol in 1988. Long-range pollution transport, and acid rain in particular, continued to burden Europe's ecosystems into the 1990s. The politics of air pollution control suffered from the fact that Europe comprised so many small countries, making a large proportion of air pollution international.[28]

Yet similar problems plagued eastern North America, where only two countries were involved. There, as in Scandinavia, lime-poor thin soils

[25]Brimblecombe et al. 1986.

[26]Russia received generous dollops of acid rain from the Ukraine, Belarus, and other former (upwind) Soviet republics (Cooper 1992).

[27]Murley 1995:387 gives country-by-country data on 1980–1992 sulfur emissions reductions. Austria and the Scandinavian countries lowered theirs by about 75%. Croatia, Portugal, and especially Greece (+27%) bucked the trend and increased sulfur emissions.

[28]Freemantle 1995 reviews Europe's acid rain accords.

and waters proved vulnerable to acidification. The industrial belt of the Great Lakes and Ohio River valley spewed forth abundant acidic emissions. In the 1970s and early 1980s, rain here typically was more acid than root beer, and occasionally more so than Diet Coke.[29] The ecological consequences, first detected in the 1950s, soon became conspicuous in the lakes and rivers of Quebec, the Maritime Provinces, New England, and New York. In the 1980s, half of Canada's sulfur deposition came from the United States, which shaped the politics of North American acidification. But the political situation was complicated by the position of northeastern states on the receiving end of midwestern sulfur emissions. Ontario's emissions wafted over Quebec and the Maritimes, as well as over the northeast United States. After 1975 the USA and Canada reduced sulfur emissions by 15 to 25 percent, roughly as much as Europe achieved after 1980.

The extraordinary industrialization of East Asia in the 1980s generated a new arena of transboundary air pollution. Here Japan found itself showered by acid rain from Korea and China. By 1996 in some parts of Japan, half of all acid deposition came from Chinese sources. Japan, like Canada and Norway, became a member of the unhappy downwind club, and began to sponsor more research and international dialogue on acidification. China, whose acidic fallout wafted across the Philippines, Taiwan, and Korea as well as Japan, joined the United States and the United Kingdom as net pollution exporters, generally serving as brakemen on international cooperation. In the 1990s, transboundary pollution in East Asia remained small in quantity and political weight by European standards, but was growing fast.[30]

The number and size of regions affected by transboundary pollution grew exponentially in the twentieth century. Although sharp impacts remained confined to Europe, East Asia, and eastern North America, faint impacts became hemispheric in proportion—and less faint.[31] In 1995 a space shuttle monitoring device noticed a sulfur plume wafting across the

[29]The pH of Hires root beer is 4.38; Diet Coke, 3.32. The average pH of rainfall in the industrial heartlands of North America in 1982 was 4.2. According to Ponting 1991:366, Wheeling, West Virginia, once recorded rain with a pH of 1.5, between vinegar (2.4) and battery acid (1.0).

[30]Kotamarthi and Carmichael 1990. China surpassed the USA as the world's biggest sulfur dioxide emitter in the 1990s, but in North China the acid rain problem was much reduced by the vast quantities of alkaline dust in the air (from the loess regions), which neutralizes atmospheric sulfate and nitrate. On long-range pollution transport and acidification I consulted Aamlid 1990, Ayers and Yeung 1996, Elder 1992, Hashimoto et al. 1994, Mason 1992, Rodhe 1989, and Stanners and Bourdeau 1995:ch. 4 and 31.

[31]See Rodhe et al. 1995. Rodhe 1989 says acid rain is regional, not yet global in nature, but in 1995 wrote of the "global scale transport" of acidifying compounds.

Atlantic from North America to Europe.[32] The atmospheric archives in Greenland ice caps show that sulfur and nitrogen emissions in the Northern Hemisphere as a whole circulate into the polar region. Greenland ice shows two surges in sulfur deposition, a modest one around 1875 to 1910 and a strong one around 1945 to 1975. Since 1950 the ice records a sharp jump in nitrogen deposition. Taking all acidifying compounds together, satellite monitoring in 1994 showed a giant region stretching from Britain to Central Asia in which highly acid rain regularly fell.[33] In the Southern Hemisphere, by comparison, acid rain and other air pollution remained local in scale with few exceptions.

Although after 1985 European countries achieved agreements on emissions controls, in most cases international air pollution flows proved harder to limit than local, urban pollution. Multiple centers of political power were involved, not merely nation-states, but provinces within countries, which sometimes worked at cross-purposes with national governments. This hampered coordination. Furthermore, distrustful neighbors, like Japan and China, found it difficult to cooperate. The actions that generate most long-range pollutants—coal combustion and driving—are difficult for most economies to sacrifice. For all these reasons, political will was rarely equal to the task of reduction in transboundary pollution.

Further Consequences of Air Pollution

Pervasive air pollution, on scales never previously reached, predictably enough had numerous consequences, few of them welcome. From the anthropocentric point of view, the most important ones involved human health.

---

[32]Freemantle 1995:11–2 reported that 2.5% of Europe's sulfur dioxide comes from North America.

[33]This rain had a pH of less than 4.6, roughly 10 times the acidity of ordinary rainwater, which is around pH 5.5 (Rodhe et al. 1995). Distilled water is pH 7.0, or neutral. Occasionally rain falls with a pH of less than 3 ($\approx$ 500 times the acidity of normal rain). On Greenland ice, see Laj et al. 1992, and the special issue of *Science of the Total Environment*, 1995:160–1. On Arctic air pollution generally, see Barrie 1986, Barrie et al. 1985, and Shaw 1995. Total sulfur deposition rates in Greenland ice tripled during 1870–1975. For comparison, over the eastern United States, anthropogenic sulfur deposition by 1990 was about 5–10 times natural rates (Husar and Husar 1990:416).

AIR POLLUTION AND HUMAN HEALTH. The health consequences of air pollution in the twentieth century were gargantuan, although hard to measure precisely. By 1992, on one World Bank estimate, particulates alone in the world's cities killed 300,000 to 700,000 people a year. (Car crashes killed about 880,000 annually).[34] In 1996, the Harvard School of Public Health put this figure at 568,000.[35] In 1997 the WHO estimated that all air pollution killed 400,000 people worldwide annually.[36] Taking the lower figure, and assuming that at-risk urban populations quadrupled since 1950, that the increasing lethality of air pollution in China, the Third World, and the Soviet bloc offset the air quality improvements in Japan, western Europe, and the United States, I reckon that air pollution killed about 20 million to 30 million people from 1950 to 1997. For the century as a whole the figure would be only a bit larger, because urban populations were smaller, although in the Western world air pollution was worse. All told, a "guesstimate" for air pollution's twentieth-century toll would be 25 million to 40 million, roughly the same as the combined casualties of World Wars I and II, and similar to the global death toll from the 1918–1919 influenza pandemic, the twentieth century's worst encounter with infectious disease.

That guesstimated, it must be said that death from air pollution and death from war, while ultimately perhaps equivalent, are from the social and economic point of view, quite different matters. In the twentieth century, war killed people mainly in the prime of life; air pollution killed the sick, the elderly, and the very young. If one esteems all individuals equally, then the toll from air pollution may be reckoned as equivalent to the toll from the world wars. But if one considers instead that the elderly have already made what contributions to society they are likely to make, and that the very young—having had little invested in them—are very easily replaced, the calculus changes.[37]

In the United States after the late 1970s, air pollution killed roughly 30,000 to 60,000 people per year, about 2 to 3 percent of all deaths (roughly the same percentage as in Poland or Czechoslovakia).[38] Among causes of death in the United States, it ranked with automobile accidents,

[34]The World Bank figure is cited in WRI 1996:22 and Hall 1995:77.

[35]Murray and Lopez 1996:28 (summary), vol. 9 (details).

[36]WHO home web page, 21 April 1997.

[37]I prefer the latter calculus intellectually, although I am offended by its heartlessness.

[38]WRI 1996:64–7. The Natural Resources Defense Council 1996 gives 64,000 deaths annually as the toll from particulates alone.

slightly ahead of guns, far behind tobacco. However, it killed mostly elderly people—those most prone to respiratory infection and cancers—whereas cars killed people of all ages and guns mostly the young. So in terms of total years of life lost, air pollution's significance in the United States was much less than auto accidents or guns.

In addition to killing tens of millions in the twentieth century, air pollution provoked or aggravated chronic illness among many more, perhaps hundreds of millions. Particulates, especially tiny ones, promote respiratory infections and probably cancer. Lead damages nervous systems and inhibits mental development. Some epidemiologists think the absorption of ambient lead by American children in the years 1950 to 1980 reduced average IQ scores. In Poland and Bangkok the health impacts of lead exposure were more severe.[39] Further health problems arose from sulfur dioxide, which causes respiratory problems, especially for people who already have bronchitis. In Japan, where a pollution compensation law in effect from 1973 to 1988 occasioned careful studies, 90,000 people qualified for payments on account of health problems attributed to air pollution, mainly sulfur dioxide, in the vicinity of the Yokkaichi petrochemical complex.

Disentangling the impact of air pollution on health from other causes of death and disease is no easy matter. Indoor air pollution, particularly in the poorer countries where biomass and coal served as domestic fuels, produced the same ailments and probably killed millions more. That said, it is well to remember that polluted water caused far more death and disease than did polluted air in the twentieth century.[40]

AIR POLLUTION, PLANTS, AND ANIMALS. The ecological consequences of twentieth-century air pollution were at least as difficult to assess as the human health impacts. Air pollution supplemented natural selection with "unnatural selection," favoring some creatures and harm-

[39]A high correlation between low IQ scores and high lead levels in the blood may of course indicate other things. Analysis of human hair in various places in the USA howed that late 20th century hair had much less lead in it than late 19th century hair, suggesting that American lead intake declined despite the upsurge in atmospheric lead. The reduction in lead content in pipes, cookware, and paints might be responsible. Weiss et al. 1972. Moore 1995:45 quotes a physician from Yale Medical School as saying that lead in Poland's air caused 10–15% of the population to be unfit for employment. On Bangkok children's IQ scores and lead exposure, see WRI 1996:47. Williams 1998 suggests that lead impaired mental development in tens of millions of children, including 17% in the USA and "90% in some African cities."

[40]WHO estimates in 1997 that 5 million are killed each year by water pollution (vs. 400,000 by air pollution). WHO home web page, 21 April 1997.

ing others. The sharpest impacts occurred in the extreme cases, such as Cubatão (Brazil) or around Sudbury (Ontario) and Norilsk (Siberia), where pollution created moonscapes. The broadest impacts derived from regional pollution, notably acid rain. In between were the coal cities, where air pollution affected the biota noticeably but not devastatingly.

The great coal cities, with their smoke and acid rain, stunted and killed local vegetation, as did the great smelters. Conifers disappeared from coal cities because of their sensitivity to acid. Poplars and willows did better. Several varieties of sycamore became favorites of urban planners in the late nineteenth and early twentieth centuries because of their stout resistance to air pollution. The legacy is visible in parks and along avenues from London to Tashkent. After 1970, in Britain and Japan at least, acid-sensitive trees have made a slow comeback.[41]

Coal cities took a small toll on animal life too. Lions in the London zoo used to die routinely of bronchitis, birds from necrosis of the lung, many other creatures from anthracosis and silicosis, all caused by particulates in the lungs. Prize beasts at London's Islington Cattle Show suffocated during the notorious fog of December 1873. The smoke of coal cities created a new selective pressure in evolution, most visible in changing moth populations in London and the English Midlands. In the course of the nineteenth and early twentieth centuries, darker moths came to predominate, presumably because they had better camouflage than whiter moths. But late in the twentieth century, when smoke abated, whiter moths recovered their numbers and the dark ones suffered.[42]

By far the greatest ecological impacts of air pollution derived from regional pollution. Sulfur emissions damaged trees from the late nineteenth century. By the late twentieth, acid rain and ozone damaged broad swaths of forest. In Scandinavia and central Europe, millions of trees after about 1970 showed signs of decline: growth slowed, leaf density dropped, and many died. Conifers and oaks suffered most. By the 1990s, about a quarter of Europe's trees showed damage.[43] In the northern Czech Republic,

[41]Gilbert 1991:33–40. For telling me about sycamores, I thank Greg Maggio. Poland's air pollution since the 1960s sharply affected the distribution of fungi, lichens, and leaf parasites in Krakow and Lodz and presumably elsewhere.

[42]Fitter 1946:179–184. See also the *New York Times*, 12 November 1996:C1, C6, which reports on the evolution of moth coloring in Lancashire and Detroit over the last 150 years. Air pollution probably affected the population dynamics of a few other invertebrates. Urban smoke probably reduced flying insect populations, making life harder for those birds that eat them.

[43]Damage here is defined as defoliation greater than 25% (Stanners and Bourdeau 1995:560).

whole forests died and acid deposition made soils too toxic for trees. The world's other acid rain centers, eastern North America and East Asia, suffered less. While most students of *Waldsterben,* as it is called in German lands, attribute this phenomenon to air pollution, chiefly acidic emissions and ozone, rival explanations existed.[44] Less visibly but more costly perhaps, air pollution also damaged crops.[45]

Acid rain unambiguously affected aquatic life. Where underlying rock did not neutralize excess acid, lakes and rivers in the course of the twentieth century often became inhospitable to crustaceans, snails, molluscs, frogs, and the more acid-sensitive fish such as salmon and trout. In Europe and eastern North America, hundreds of thousands of rivers and lakes lost all or part of their aquatic life on account of acid rain after 1960. In Norway, recipient of a goodly share of Britain's sulfur emissions, ill effects date from at least 1920.

AIR POLLUTION AND CULTURAL MONUMENTS. The acidic brew emitted by smokestacks and tailpipes corrodes stone, especially limestone and marble. Some of the great cultural monuments of the world consequently decayed at heightened rates in the twentieth century. Where susceptible stone and heavy pollution existed in close proximity, as in many cities of Europe, the cultural loss was most pronounced. Acidic London fogs gnawed away the stone facing of St. Paul's cathedral at a rate of about 8 millimeters per century. Athenian car exhaust damaged the ancient marble on the Acropolis more seriously in 25 years than did all the weathering of the previous 2,400 years. The magnificent horses of St. Mark, much tossed by the tides of history, finally succumbed to air pollution in 1974: Venetian authorities hid them away and replaced them with fiberglass imitations atop the cathedral. Michelangelo's *David* could not be trusted to Florentine air; a replica stands in his place at the Piazza

[44]Among these other explanations were climate change and disease; multiple causes were probably at work (Fuhrer 1990; Kandler and Innes 1995; Olson et al. 1992; Kuusela 1994:135–6). Forest death, writes Kuusela, reduced Europe's forest area by 0.5%, most of it in Poland, Czechoslovakia, and Germany. According to Holland and Petersen 1995:322–3, air pollution caused forest damage (not death) in 30–60% of forests in Austria, the Czech Republic, Germany, Poland, and Switzerland. Lonkiewicz et al. 1987 say that in 1970, 200,000 ha of Polish forests were severely damaged by air pollution; in 1985, it was 700,000 ha, or about 9% of Poland's forest area. In the former West Germany, von Maydell and Ollmann (1987) estimate that half of all forests were damaged by air pollution; in Austria, F. Tersch (1987) estimates that 20% of forests were damaged by pollution.

[45]In the USA, ozone reduced farm output by $1 billion to $7 billion annually; in Poland, acidification cut yields by 3–4% (Holland and Petersen 1995:323).

della Signoria. Less exalted monuments and buildings suffered too. In the early 1990s it cost Europe about $9 billion per year to replace corroded stone. With effective pollution control in parts of Europe, corrosion of building stone slowed somewhat after 1970 but did not cease.[46]

Air pollution's cultural corrosion, while fastest in Europe, emerged elsewhere after 1980. Refineries at Agra emitted sulfur dioxide that ate away at India's incomparable Taj Mahal. Mayan limestone monuments at Tikal, in Guatemala, showed the ill effects of Mexican oil combustion 100 kilometers away. The Sphinx, the Pyramids, and other great pharaonic monuments of Giza, survivors of 40 centuries of intermittent sandblasting by the Sahara, decayed visibly under the impact of Cairo's pollution. Coal dust damaged the fifth-century A.D. sandstone statuary in China's Yungang Grottoes, which house 50,000 statues, including Buddhas as big as three or four elephants.[47] Those now living will be the last generation to see these monuments and artworks in anything like their original forms.

AIR POLLUTION, ECONOMICS, AND POLITICS. Air pollution in the twentieth century had significant but mostly unmeasurable economic and political effects. The forests, crops, buildings, bridges, monuments, machinery, and human health damaged or destroyed cost untold billions. The World Bank in 1997 estimated that China's air pollution cost the country 8 percent of its GDP; the figure would be smaller in other countries.[48] Until 1970 or so, these damages, if recognized at all, were widely considered an acceptable cost of doing industrial business. After 1970, events showed this need not be the case, and that pollution control could be profitable in its own right. After 1980, air pollution control grew rapidly and became a multibillion dollar industry.

The shift around 1970 reflected one of the important popular political mobilizations of the twentieth century. It centered on the historic heartlands of pollution—North America, Europe, and Japan—but emerged elsewhere too, notably in India. This was all part of the general environ-

---

[46]U.N. Economic Commission for Europe 1992:53–81; Kucera and Fitz 1995; Matson and Miller 1991; Norwich 1991; Sikiotis and Kirkitsos 1995; Trudgill et al. 1990. See also the special issue of *Science of the Total Environment*, 1995, 167.

[47]Salmon et al. 1995. On this problem, see Deshpande et al. 1993 (India), and Keskinler et al. 1994 (Erzurum, Turkey). In Washington, D.C., many limestone and marble buildings and monuments show the corrosive effects of rain with a pH of 4.2.

[48]Reported in the survey "Development and the Environment," *Economist*, 21 March 1998:4.

mental movement (see Chapter 11). Beyond air pollution's role in catalyzing modern environmental politics, it played a small role in international affairs as well. Long-distance transport of air pollution provoked frictions between close allies, such as the United States and Canada, Britain and Norway, and the former East Germany and Poland. It exacerbated tensions between edgy neighbors such as China and Japan, and Norway and Russia. However, air pollution's role in international politics was not all dark. Numerous international agreements, protocols, accords, and treaties arose to address air pollution problems, beginning with small-scale bilateral agreements but after 1975 involving regional and even near-global accords. The complexity of the issues involved required the participation of scientists, which perhaps eased negotiations.[49]

Climate Change and Stratospheric Ozone

In the twentieth century, human activities increased the presence of greenhouse gases in the atmosphere and reduced the concentration of ozone in the stratosphere. These two changes have enormous potential consequences for the history of the twenty-first century. In the twentieth century their consequences were mild. But because their portent is so significant for human affairs, and because the origins of the changes lie in twentieth-century trends and behavior, I will treat both—briefly.[50]

CLIMATE CHANGE. The major greenhouse gases are carbon dioxide, methane, and ozone.[51] While climate is governed by many factors, among the crucial ones is the composition of the atmosphere. Without any greenhouse gases, our planet would be a chilly place, more like Mars, where temperature averages are around −23° Celsius (−9° Fahrenheit), like

---

[49]This argument—that scientists form an "epistemic community" devoted to interests other than those of diplomats, and to interests that mesh more easily—has been made for the accords on the Mediterranean Sea but may apply to international air quality agreements too (Haas 1990).

[50]Thorough reviews include de Gruijl 1995, IPCC 1996, and Turco 1997.

[51]Others include nitrous oxide and halocarbons (which include CFCs and HFCs—or hydrofluorocarbons—used in some instances as CFC substitutes). IPCC 1996 1:15–21 reviews the relative contributions, past and potential, of these gases to "radiative forcing" (as climate change driven by atmospheric change is called).

Earth's arctic winters. With more greenhouse gases, Earth would resemble Venus, where temperatures are well above the boiling point. For life adapted to Earth, changes in greenhouse gases are serious business.

For the thousand years before 1800, carbon dioxide levels in the atmosphere varied around 270 to 290 parts per million (ppm). Around 1800 an accelerating buildup began, reaching about 295 ppm by 1900, 310 to 315 ppm by 1950, and about 360 ppm in 1995. Two trends drove this buildup: fossil fuel combustion, which accounted for about three-fourths of the addition, and deforestation (now mainly in the tropics but in 1900 mostly in North America and temperate Asia), which accounted for almost all the rest. Meanwhile, methane, which like carbon dioxide is present naturally in the atmosphere, increased from about 700 to about 1,720 parts per billion. The main sources of methane buildup were agriculture, especially irrigated rice, livestock, garbage decomposition, coal mining, and fossil fuel use.

These minute but momentous changes in the atmosphere, in concert with some even tinier ones involving other greenhouse gases, made the atmosphere more efficient at trapping heat from the sun. At the same time, human actions injected lots of dust and soot into the atmosphere, which slightly lowered the amount of solar energy reaching the earth's surface. The net effect, since 1800 or so, was about 2 more watts per square meter of solar energy delivered to the earth's surface. This probably accounts for the modest warming the earth experienced in the twentieth century.[52]

The earth has warmed up recently, although no one knows for certain if human actions are the cause. Between 1890 and 1990, average surface temperatures increased by 0.3° to 0.6° Celsius. That happened in two surges, between 1910 and 1940, and then after 1975. From 1940 to 1975, average temperatures actually declined slightly. But nine of the ten hottest years on record occurred between 1987 and 1997, and the 1990s promised to be the hottest decade since the fourteenth century.[53] Changes of this magnitude and rapidity are well within the natural range of variation, al-

---

[52]IPCC 1996 1:5 puts it this way: "The balance of evidence suggests that there is a discernible human influence on global climate." The U.S. Global Change Research Program 1998:14 was less cautious: "New statistical tests indicate that much of the warming that has been observed can indeed be attributed to human activities."

[53]According to data gathered by the National Aeronautics and Space Administration (NASA), the National Climate Data Center, and the U.K. Meteorological Office (reported in *Science News* 17 January 1998, 153:38). Data from the National Oceanic and Atmospheric Administration (NOAA) support this view. See the *Washington Post,* 13 July 1998.

though rare within the last 2 million years, probably nonexistent within the last 10,000 years, and definitely absent within the last 600 years.[54]

Some places warmed more than others—and some have cooled. The greatest warming took place in the Northern Hemisphere's higher latitudes, above 40°, north of Philadelphia, Madrid, and Beijing. Here, for example, the growing season lengthened by more than a week between the late 1980s and 1997.[55] Antarctica too warmed up a lot, and its ice melted faster than before.[56] Whether or not the climate grew more unstable—bringing more droughts, hurricanes, heat waves, floods, El Niño events, and the like—remained uncertain. The Intergovernmental Panel on Climate Change (IPCC), the United Nations's authority on the issue, decided in 1995 that no conclusion was yet possible; by the late 1980s, insurance companies, careful students of the probabilities of extreme weather, suspected climate had indeed become more unstable.[57]

The consequences of the warming of the twentieth century remained small. Some creatures shifted their habitat range to avoid the heat. Some (but not all) glaciers and ice caps shrank, and sea level rose by 10 to 25 centimeters, up to one's ankle or midcalf. Quite plausibly but quite unprovably, this rise extended the damage done by storm surges to the coasts of Bangladesh and other low-lying countries. Malarial mosquitoes expanded their range, particularly into higher elevations in tropical Africa, raising the death toll from malaria in places such as Rwanda. In the larger picture of twentieth-century history, perhaps none of this amounts to much (although Bangladeshis and Rwandans might see matters differently). The much greater warming forecast for the twenty-first century and beyond, however, will be quite another thing.

In the late 1990s most climate modelers expected that in the twenty-first century average temperatures would rise perhaps 1° to 5° Celsius (2–10°F).[58] If their models were to prove correct, this warming implies vast changes in evaporation and precipitation, a more vigorous hydrological cycle making for both more droughts and more floods. The consequences for agriculture, while difficult to predict, would be sharp.[59]

[54]Mann et al. 1998.

[55]*Washington Post,* 17 April 1997.

[56]See Dennis 1996:165. In 1995 a chunk of the Larsen Ice Shelf, 200 m thick and about the size of Rhode Island or Luxembourg, plopped into the sea.

[57]Berz 1990.

[58]The range of forecasts for global warming by 2100 is considerable, from about 1°–5°C. For reviews of consequences of various scenarios, see IPCC 1996.

[59]Rosenzweig and Hillel 1995.

Human health would suffer from the expanded range of tropical diseases and their vectors. Species extinctions would accelerate. Coping with warming on this scale—assuming it happens—will be a major chapter in the history of the next century. For some low-lying island nations, such as the Maldives, it could also be the final chapter.[60]

OZONE AND CHLOROFLUOROCARBONS. The history of stratospheric ozone depletion also commands attention more for its future prospects than its past importance. Up in the stratosphere, sunlight and oxygen react to form ozone, which absorbs 99 percent of the ultraviolet (UV) radiation[61] entering the atmosphere. Before there was oxygen, there was no ozone layer and life on earth had to stay underwater. This indispensable sunscreen, consisting of ozone molecules circulating in the vastness of the stratosphere at concentrations of only a few parts per billion, took eons to form. This thin shield then protected life on earth admirably for about a billion years.

But in 1930–1931, Thomas Midgley invented Freon, the first of the chlorofluorocarbons (CFCs), which proved useful in refrigerants, solvents, and spray propellants, among other things. Freon replaced dangerously flammable and toxic gases previously used in refrigeration, and made air-conditioning practical. The virtue of CFCs and other like gases (as a group called halocarbons) is that they are very stable and react with almost nothing—until they drift into the stratosphere, where direct UV radiation breaks them up, releasing agents that in turn rupture ozone molecules.

Midgley, the same research chemist who figured out that lead would enhance engine performance, had more impact on the atmosphere than any other single organism in earth history. He was born in Beaver Falls, Pennsylvania, in 1889 into a family of inventors. (His grandfather invented the bandsaw.) He grew up in Dayton and Columbus, graduated from Cornell University, and went to work as a chemical engineer. Dur-

[60]If future generations burn all the known coal reserves, the atmosphere will have four times as much carbon dioxide as it had in 1850, average temperatures will exceed those of any time in the last 200 million years, and sea level will rise by perhaps 10–20 m, enough to drown many of the world's major cities. This would take a few centuries, because there is a lot of coal (Kasting 1998).

[61]Specifically, it absorbs UV-A very slightly and UV-B nearly completely. UV-B is biologically dangerous, and it is the reduced capacity of the ozone layer to absorb it that causes worry. Together with oxygen, ozone absorbs UV-C, the shortest-wavelength light, which is sufficiently lethal to use in sterilizations.

Thomas Midgely

ing World War I he worked on aerial torpedoes (robot bombs) and synthetic airplane fuel. In 1921, while working for General Motors Research Corporation, he found that tetraethyl lead reduced engine knocking, and in 1923 his employers marketed ethyl gasoline for the first time. Midgley's work made high-compression auto and aircraft engines possible.

At the request of General Motors's Frigidaire division, Midgley then tackled problems of refrigeration. In front of a large audience at the 1930 meeting of the American Chemical Society, he demonstrated the non-toxic and nonflammable properties of Freon by inhaling a lungful and then slowly exhaling over a lighted candle, which the Freon extinguished. Holder of over 100 patents, winner of all the major prizes in chemistry in the United States, Midgley was president of the American Chemical Society at the time of his peculiar death. He had contracted polio in 1940 and designed a system of ropes and pulleys to help him in and out of bed. In 1944 he died of strangulation, suspended above his bed, entangled in his network of ropes, killed by the combination of bad luck and his own ingenuity. Midgley was the Fritz Haber of the atmosphere.[62]

Only after the Depression and World War II did Midgley's invention catch on quickly. CFC releases remained small in the 1930s and 1940s, reaching perhaps 20,000 tons annually by 1950. But by 1970, emissions reached about 750,000 tons and an unnoticed assault on the ozone layer was in full career. In 1974, scientists Sherwood Rowland and Mario Molina suggested the theoretical possibility that halocarbons might thin the ozone layer; in 1985, J. C. Farman's observations confirmed that over Antarctica this had indeed happened.[63] Subsequent measurements showed miniholes over Chile and Australia. Meanwhile the ozone shield over the Northern Hemisphere thinned slightly (about 10%, 1960–1995), over the tropics not at all.[64]

The initial discoveries prompted an unusually quick political reaction, because the prospects of enhanced UV-B radiation were more than worrisome. It kills phytoplankton, the basis of oceanic food chains. It affects photosynthesis in green plants. In humans it causes cataracts and other eye ailments, suppresses immune system response, and in susceptible people causes skin cancer. The United States, Canada, and Scandinavian countries banned CFCs in aerosol sprays in the late 1970s. This had little overall effect: global CFC releases continued to climb. In response to Farman's findings, the United Nations Environment Programme

[62]Like Haber he also worked hard to extract something useful from seawater—not gold, but bromine for ethyl gasoline. His life is described in the *Dictionary of American Biography,* supplement 3 (New York: Scribner's, 1973):521–2.

[63]Farman's data aroused skepticism because they did not harmonize with satellite records. But the satellite data were wrong: the computer in which they were recorded had been instructed to discard observations beyond a certain range on the theory that these must be erroneous. A cautionary tale!

[64]Between 1979 and 1992, DNA-damaging UV radiation increased in northern hemisphere latitudes (45–55°N) by 4–9%. See U.S. Global Change Research Program 1998:37. Several gases other than CFCs can release agents that destroy ozone. It is actually chlorine and bromine that kill ozone. Bromine is released chiefly via an insecticide (methyl bromide).

(UNEP) organized the 1985 Vienna Convention on ozone depletion. It led to the 1987 Montreal Protocol, the 1990 (London), 1992 (Copenhagen), and 1995 (Vienna) amendments, altogether an extraordinary international response to an extraordinary problem. These agreements sharply curtailed production of CFCs. Chemical manufacturers, who had for years argued against the theory that CFCs might damage the ozone layer, initially complained about the protocol and amendments but quickly found substitutes. Some chemists who had used CFCs as solvents found that water or lemon juice worked just as well. After 1988 the rate of CFC releases dropped. The big users (USA, EU, Japan, and Russia) lowered CFC use by 75 to 100 percent (1986–1994). China and India used more than before, but worldwide CFC use declined roughly 80 percent. That was the good news.

Unfortunately, because CFCs are so stable, they last a long time in the atmosphere, and some of those released before the Montreal Protocol will still be destroying ozone in 2087. Hence the ozone layer will continue to thin for a decade or two at least, before slowly beginning to beef up again. The "ultraviolet century" in human history should prove to span from roughly 1970 to 2070.[65] It is fortunate for the biosphere, and especially for fair-skinned people and phytoplankton, that CFCs did not become commercially viable 10 or 50 years before they did; that the science of Molina, Rowland, and Farman came no later than it did; and that UNEP already existed to take a leading role in brokering international agreements.

In the first 25 years of the ultraviolet century (i.e., to date), perhaps 1 to 2 million excess cases of skin cancer derived from stratospheric ozone loss. That translated into about 10,000 to 20,000 early deaths, mainly among fair-skinned people in sunny lands such as Australia. The toll—to date[66]— was far smaller than that of respiratory ailments brought on by air pollution. No one knows the full effect of excess UV radiation on immune response, so the real impacts of CFCs erosion of the ozone layer on human health (let alone on the rest of the biosphere) remain entirely

[65]Projections, for what they are worth, suggest that ozone depletion early in the 21st century will be 5% on average, up to 10% at high latitudes. UV-B penetration to the earth's surface will be roughly 10% greater than at 1960, and skin cancers should increase in like proportion (Turco 1997:434–5). Predictions such as these involve many assumptions, mostly about future releases of CFCs.

[66]NASA in 1991 predicted 200,000 additional skin cancer deaths in the USA by 2040.

unclear. But stratospheric ozone depletion—another combination of bad luck and Midgley's ingenuity—will surely kill many thousands more before the close of the ultraviolet century.[67]

While climate change and ozone depletion had only modest effects by the end of the twentieth century, their futures were substantially set, although in ways no one can know with precision. Particulates are washed out of the air by rain, and most of the pollutants responsible for local and regional air pollution are similarly removed quickly by natural processes. Hence local and regional air pollution is easily and quickly reversible, provided one stops adding new pollutants. The global-scale issues are different. Were we to stop injecting ozone-rupturing chemicals into the air tomorrow, we would still need about a century before we enjoyed a healthy ozone layer again. With climate change, the turnaround time is longer still. Most of the carbon dioxide added to the atmosphere in the twentieth century will remain there for centuries. (The added methane will last only about 12 years). Carbon dioxide is not rinsed out of the air like soot; rather, it is very slowly absorbed by the oceans and by living things, at rates that cannot be significantly accelerated. Its impacts, great or small, cannot be reversed quickly. So the combustion history of the twentieth century has already partly determined conditions in the next several centuries to come.

Space Pollution

The scale of atmospheric pollution in the twentieth century ballooned from local to global—indeed, a little beyond global. Since the Soviet launch of *Sputnik* in 1957 the great powers (commercial as well as political) have strewn near-space with several thousand satellites, rocket boosters, and miscellaneous space junk, orbiting at speeds up to 30,000 kilometers per hour. At such velocities even paint flakes, of which there are millions zooming around, can damage satellites or space suits. Astronauts on space walks try to stay in the shadow of their space ships lest they blunder into the path of a rogue rivet. As yet, there are no regulations. Authorities tend to take the view that it is too expensive to do anything,

---

[67]De Gruijl 1995. On ozone politics, see French 1997.

and anyway space is so big that there's room for space junk. Any industrialist would have said the same of the atmosphere 150 years ago.[68]

Conclusion

The quantities and consequences of twentieth-century air pollution were vast. So were the changes in its scale and character. Taken together, our alterations of the atmosphere amount to several simultaneous, inadvertent, and uncontrolled experiments.

One of these experiments, urban air pollution, dates from earlier times, but reached levels in the twentieth century sufficient to kill many millions of people through respiratory disease. Another, acidification, altered regional ecologies widely in the Northern Hemisphere. A third, carbon dioxide loading, probably warmed the globe slightly and might yet warm it far more. A fourth, CFC emissions, frayed the ozone layer that makes terrestrial life possible.

All of these changes to the atmosphere were accidental effects of industrialization. The main driving force was fossil fuel use, but new technologies played a substantial role too (e.g., CFCs and tetraethyl lead). Behind this industrialization lay the attractions of higher living standards, or at least more consumption, for masses of people; of profit for hundreds of firms engaged in mining, metallurgy, electric power generation, and other polluting enterprises; and of political power for states, bureaucracies, and politicians. Geopolitical anxiety or ambition in particular helped spur Germany, Japan, Russia, China, Brazil, and other polities to industrialize as quickly as possible, almost regardless of consequences. Via air pollution, as well as via violence, the lives and health of millions were sacrificed on multiple altars of national interest.

The reductions of air pollution that took place from the 1940s onward were remarkable political events. They required alliances among medical researchers, engineers, citizen activists, legislators, and bureaucrats. In smoke abatement, at least in the United States, the United Kingdom, and Japan, women played a conspicuous role. In the international agree-

---

[68]On space junk, see Scheraga 1986; and the *Economist,* 29 March 1997:87–8. The U.S. Space Command, which tracks orbital debris, figures there were in 1997 some 8,649 man-made items in space larger than a grapefruit, and over 2 million smaller ones.

ments on acid rain and ozone after 1975, ministers, heads of state, and major corporations took part. Almost all air pollution, except for the global-level problems, proved quickly reversible when addressed at the source. In most of the world and on most issues, the political response to air pollution remained inadequate—in the sense that pollution was worse in the 1990s than in prior decades. But considering the inertia that governs the thinking and politics of most societies, the social and political response to changes in the atmosphere is extraordinary. Sometimes, as Samuel Johnson noted in another context, it is less remarkable that something is done badly than that it is done at all.

5

The Hydrosphere: The History of Water Use and Water Pollution

Water is good; it benefits all things and does not compete with them. It dwells in lowly places that all disdain.
                                        —Lao-tzu (6th century B.C.)

People need water just as surely as they need oxygen from the atmosphere and food grown in the soil. For most of human history we needed water only to drink. But in the last few thousand years we have relied on it to irrigate our crops, carry off our wastes, wash our bodies and our possessions, and more recently to power our mills and machines. Individuals and societies expended great effort to ensure reliable supplies of water, especially in the dry belt from Morocco to Central Asia. With the modern currents of industrialization, high energy use, and urbanization, societies acquired greater power with which to move and control water. But at the same time, they came to use more, waste more, and pollute more water more thoroughly. The health, wealth, and security of any and all societies depended upon getting sufficient supplies of sufficiently clean water to the right places at the right times, without doing too much damage in the process. Ordinary efforts to achieve wealth and security often complicated this task, by polluting water. Success guaranteed nothing: health, wealth, and security required many other things besides adequate water. But failure guaranteed ill health and a weakened economy. Water management, although not always seen as such, was a crucial technical and political challenge.

Water Basics

Earth is the water planet, the only place in our solar system where water exists as liquid. "Water, water everywhere, nor any drop to drink," wrote Coleridge in the "Rime of the Ancient Mariner," which could serve as a first approximation of the earth's condition.

Of the hydrosphere's 1.4 billion cubic kilometers, more than 97 percent is salt water in the oceans.[1] Happily for us, the solar desalination and purification machine pumps up about half a million cubic kilometers every year, which falls back to earth as rain and snow. This is the source of all the world's stock of fresh water. Most of that (69%) is currently locked up in ice caps and glaciers, almost all of it in Antarctica. Almost all the remainder (98%) lies underground in aquifers, mostly at inaccessible depths.[2] Only about one-quarter of 1 percent of the world's fresh water (approximately 90,000 km^3) is in lakes and rivers, where it is most easily accessible. Of this, about a quarter is in Lake Baikal in Siberia. Small amounts of fresh water are in the atmosphere, in permafrost, and in living things.

The world's renewable flow of fresh water is somewhat less than its total freshwater stock. Continents get more in rain than they surrender in evaporation, and the difference is the world's streamflow (about 40,000 km^3 every year). Two-thirds of this runs off in floods, so about 14,000 cubic kilometers per year are available for routine use. This comes to more than 2,000 cubic meters per person annually, which is plenty. But it is spread unevenly around the world. Twenty or thirty countries, mostly in Africa and southwestern Asia, have less than half this figure, and are, by the conventional measures of hydrologists, short of water. South America has ten times as much per person as Asia, and five times as much as Africa. The flow is also unevenly distributed throughout the year, so many

[1] The data here come from Shiklomanov 1990, 1993, and Postel 1992.
[2] Inaccessible with today's technology and energy prices. These, of course, will change.

places have either too much or too little water (for human designs) at any given time. In Cheerapunji (in Assam, northeast India), the second rainiest place in the world, the summer months bring about 9 meters of rain. For about six months every year, water supply is sufficient, indeed excessive. But for the remaining six months, it often is not.[3] The inconvenient distribution of fresh water, coupled with the expense of transporting it over long distances, has made water supply a major constraint on human affairs. Many societies have invested heavily in efforts to ease this constraint.

World Water Use and Supply

In the history of water use, some things change but one thing stays the same. In modern times, as in the past, we have used water mainly for irrigation. Most of our enduring societies and empires were based on the control of water, particularly river water. Egyptian, Mesopotamian, Indic, and Chinese civilizations all rested on irrigation, river transport, and use of river water to dilute and carry off noxious wastes. Skilled water management also underpinned Andean and Mesoamerican civilizations. We have used water for irrigation for 9,000 years and to power mills for 2,000. Now we also need it for industrial purposes, such as hydroelectric power generation, and for cooling and cleaning countless machines. And of course we continue to use water for drinking and for diluting wastes. Following hydrologists, I will divide water use into three main categories: irrigation, industrial, and municipal.

The clearest thing about the history of water is that people use a lot more now than they used to. In 1700, when the world had about 700 million people, total freshwater use amounted to perhaps 110 cubic kilometers, 90 percent of which went for irrigation, almost all in Asia. Table 5.1 sketches the quantitative history of water use since 1700.

If these figures are accurate, total freshwater use in 1990 was about 40 times greater than in 1700. In the twentieth century alone, water use

[3]Rao 1989. This has to do with deforestation and accelerated runoff in the vicinity of Cheerapunji. Prior to 1960, the water supply usually sufficed year-round. The world's rainiest place is a Hawaiian mountaintop.

TABLE 5.1 ESTIMATED GLOBAL FRESHWATER USE, 1700–1990s

| Year | Withdrawals (km³) | Withdrawals (per capita) | Uses[a] | | |
			Irrigation (%)	Industry (%)	Municipal (%)
1700	110	0.17	90	2	8
1800	243	0.27	90	3	7
1900	580	0.36	90	6	3
1950	1,360	0.54	83	13	4
1970	2,590	0.70	72	22	5
1990	4,130	0.78	66	24	8
2000[b]	5,190	0.87	64	25	9

Sources: Elaborated from L'vovich and White 1990 and Shiklomanov 1993.

[a]Percentages do not add up horizontally to 100% because of rounding. I have eliminated Shiklomanov's category of reservoirs, which is always tiny and before 1970 negligible.
[b]Projected.

spurted upward ninefold. Most of the increase probably derived from population growth, which in the same years was about fourfold. This means that per capita water use in the 1990s was slightly more than twice the level of 1900.[4] In the richer parts of the world, water use stabilized after the 1970s through improved efficiency, motivated in part by antipollution regulations. In the United States, total water use peaked around 1980 and declined by a tenth as of 1995, despite the simultaneous addition of some 40 million to the American population.[5]

While the world's thirst grew spectacularly in the twentieth century, water was still used in the 1990s for much the same purposes and in much the same places as in 1900. Although industrial and municipal water use grew, irrigation still took the lion's share. The distribution of

[4]We can't know for sure what proportion of this increase is accounted for by population growth without knowing exactly what the additional water use went for. The greater part was for irrigation, but it might have been for basic food production or for specialty crops that satisfied the fancy of wealthy consumers.
[5]Gleick 1993:396; U.S. Geological Survey data reported in the Economist, 14 November 1998:29.

water withdrawals among the continents appears even more stable, as shown by Table 5.2. Asia in the 1990s, as in 1900, used more water than all other continents combined. This comes as no surprise: Asia had more people than all other continents combined, and it had nearly a third of all the world's streamflow. The only notable changes in the continental distribution of water use were the rise of North America's consumption, mostly in the first part of the century, and the smaller rise of South American water use, almost entirely post-1950. Within each continent, however, the patterns of water use changed a good deal, not least because of the rise of cities.

TABLE 5.2 CONTINENTAL DISTRIBUTION OF WATER USE, 1900-1990

	Streamflow (% of world total)	Share of Global Freshwater Use (%)[a]		
		1900	1950	1990
Asia	32	71	63	60
Europe	7	12	13	13
North America	18	10	17	18
Africa	10	5	5	6
South America	26	2	2	4
Australia and Oceania	5	—	1	1
		≈100	≈100	≈100

Sources: Elaborated from L'vovich and White 1990 and Shiklomanov 1993.

[a]Percentages do not sum to 100% due to rounding.

Urban Water

Cities always faced the twin problems of procuring adequate drinking water and carrying off, or diluting, wastes. The simplest approach—dumping wastes in the nearest watercourse and drinking from it too—worked only where people were few and water plentiful. Early in human history, more complex approaches emerged, aiming to segregate drinking water from waste water. Failure meant disease and early death for urban dwellers. New knowledge of the transmission patterns of cholera (1850s) and typhoid (1880s) concentrated attention on urban water quality toward the end of the nineteenth century. The stories of Istanbul, an old city with scant fresh water, and Chicago, a new city with plenty, illustrate the issues of urban water supply and urban sewage.

Istanbul (once Constantinople and before that Byzantium) has a long history of sophisticated waterworks. Its situation, while strategic and spectacular, provided it with very limited fresh water, in the form of a stream that flows into the Golden Horn, Istanbul's historic harbor. The city's population easily polluted this water supply to the point where it was dangerous to drink. Shortages of clean water constrained the city's growth. Engineers in Roman and Byzantine times erected dams and aqueducts and dug giant cisterns to address the problem. When the Ottoman Turks captured the city in 1453 and made it their capital, they built still more aqueducts, most of them during a great sixteenth-century expansion of Istanbul. Sinan, perhaps the greatest architect of his day, designed much of Istanbul's water system. With this infrastructure, Istanbul collected water from distances of 20 to 30 kilometers away, allowing it to become one of the world's largest cities by 1600.[6] Its far greater growth in the twentieth century required much more water.

Consequently, while some of the Ottoman systems remained in use, Turkey, which emerged from the ashes of the Ottoman Empire in 1923, built several new dams and pipes, extending the reach of Istanbul to more than 120 kilometers. The new republic moved the capital to Ankara in the 1920s, which slowed the growth of Istanbul. But by the 1950s, rapid population growth throughout Turkey and rural exodus swelled the city by 10 percent per year. Most of the new arrivals built their own housing on the edges of the city, and lived without piped water or sewage. These settlements sprawled in every direction in the 1960s and 1970s, and eventually their residents acquired enough political weight to extract favors, such as connection to the city water and sewage systems, from the government. After 1980, Istanbul (nearing 10 million people) drew heavily on water from the Asian side of the Bosporus, carried beneath The Straits by pipe. Even this proved insufficient in the 1990s, when summertime often required strenuous water conservation.[7] As in many cities

---

[6]Orhonlu 1984:78–82; Çeçen 1992; Pinon and Yerasimos 1994.

[7]See Özis 1987. Sinan contemplated a diversion from Lake Sapanca, in Asia Minor, which the Turkish republic actually built in the 1980s.

5. Key Locations in Water Quality History

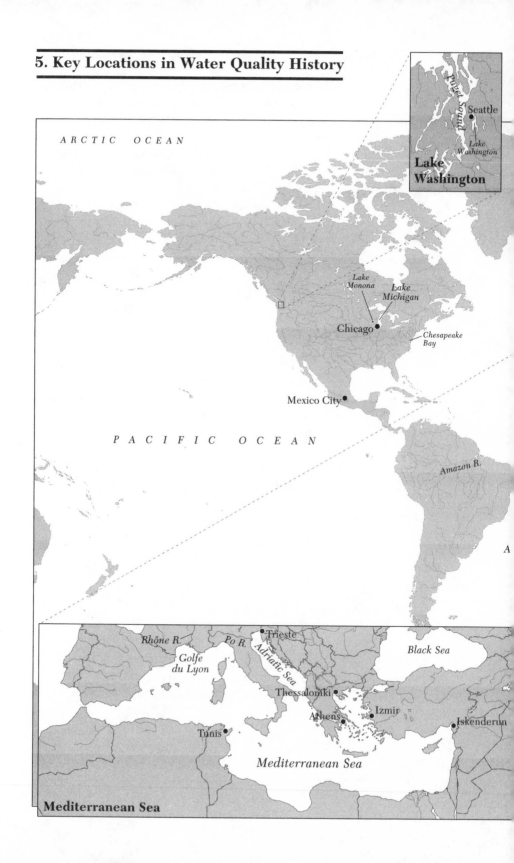

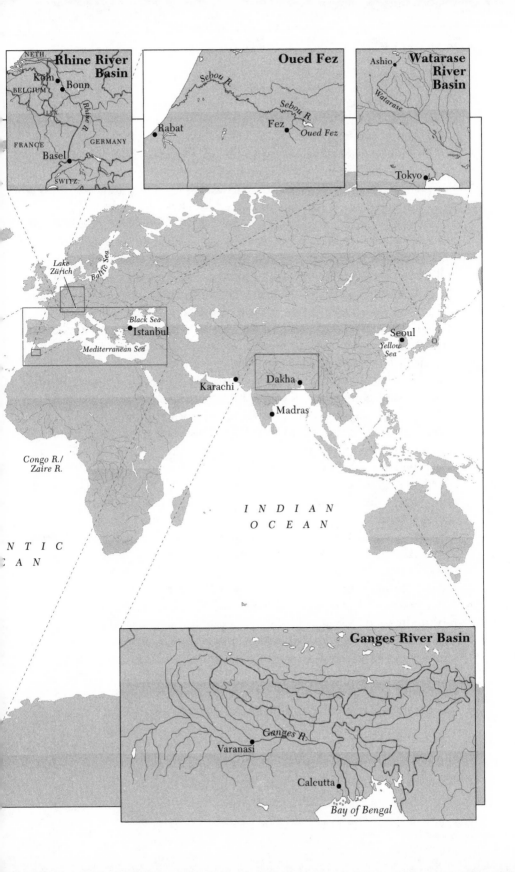

Rhine River Basin

NETH.
BELGIUM
Köln
Bonn
LUX.
FRANCE
GERMANY
Basel
SWITZ.
Rhine R.

Oued Fez

Sebou R.
Rabat
Sebou R.
Fez
Oued Fez

Watarase River Basin

Ashio
Watarase
Tokyo

Lake Zürich
Baltic Sea
Black Sea
Istanbul
Mediterranean Sea
Karachi
Dakha
Madras
Seoul
Yellow Sea

Congo R./
Zaire R.

INDIAN OCEAN

NTIC
AN

Ganges River Basin

Ganges R.
Varanasi
Calcutta
Bay of Bengal

around the world, water supply continued to vex authorities responsible for accommodating urban growth—and ordinary people who had to make do with less.

Chicago is a young city on one of the largest lakes in the world. But it too developed water problems with its rapid growth in the nineteenth century. Its population used the lakefront and the Chicago River (which flowed into Lake Michigan) to dump its wastes, contaminating the water supply. What 30,000 Chicagoans dumped into the river and lake in 1848 caused only modest problems, but when the city's population boomed after the Civil War, old arrangements had to change. City authorities built longer and longer pipes out into the lake to try to draw water unsullied by the city, but Chicago's rapid growth continually outstripped the pipes. Until 1900, Chicago had a deserved reputation for typhoid. In 1885–1886 alone, 90,000 people around Chicago (including 12% of the city's population) died of waterborne diseases. Typhoid sickened about 20,000 Chicagoans a year from 1891 to 1895. The epidemics provoked America's largest engineering project before the Panama Canal: the Chicago Metropolitan Sanitation District reversed the flow of the Chicago and Calumet Rivers, so that by 1900 they no longer emptied into Chicago's drinking water supply, but instead flowed toward the Illinois River and down to the Mississippi. Thus the sewage of Chicago, including the offal from the world's greatest stockyards, no longer menaced Chicagoans, but drifted away to Joliet, St. Louis, and New Orleans. Typhoid and other waterborne epidemics became only memories. The *New York Times* offered as headline news: "Water in Chicago River Now Resembles Liquid."[8]

What suited Chicagoans did not always suit their neighbors. Residents of the other Great Lakes states, as well as the Canadian province of Ontario, thought the Chicago River reversal lowered the level of the Great Lakes—it did lower Lakes Michigan and Huron by 6 or 8 inches. Great Lakes states, the province of Ontario, and the U.S. federal government repeatedly sued Chicago and Illinois, as did St. Louis and other communities on the receiving end of Chicago's sewage. Under legal restrictions after 1930, Chicago and vicinity increasingly turned to groundwater, but by 1959 they overused it, provoking a lawsuit by the aggrieved neighboring state of Wisconsin. After 1985, larger deliveries from Lake Michigan eased the situation. So did Chicago's world-class sewage treatment

---

[8]Changnon and Changnon 1996:104.

plants, which cleansed its discharged water to a high standard.[9] Getting enough water for growing cities was a problem that inspired ingenious solutions. Keeping that water clean, however, was another problem.

One of the great divides among humankind that has arisen since 1850 separates those societies that provide safe drinking water from those that do not. Prior to 1850, pathogen and biological pollution plagued almost all urban societies and some rural ones. Great changes began in northwestern Europe shortly before 1850. Partly to combat diseases, London and Paris built sewer systems. These drained directly into the Thames and Seine, making both rivers putrid and lethal. Britain's Houses of Parliament once required burlap saturated in chloride of lime hung over their windows to protect the nostrils of MPs from the Thames' stench. Following scientific discoveries about the contagion routes of cholera, typhoid, and other infections in the 1880s, western European and North American cities built thousands of filtration plants to purify domestic water supplies. Hundreds of cities adopted chlorination, which kills most microorganisms, after Chicago's experiments with it in 1908 to 1910.[10] Filtration and chlorination sharply lowered urban death rates. In the 1990s, waterborne diseases in the United States killed a few thousand unlucky people every year, a tiny fraction of their impact 150 years ago.[11]

By 1920 almost all big cities in the richer parts of the world provided citizens with safe drinking water. Treatment of sewage took a little longer. Big cities on small rivers, such as Moscow and Madrid, quickly overcame the waters' powers to assimilate organic wastes. Only primary sewage treatment—trickling through filters—took place until 1912–1915, when British engineers developed the sludge activation process.[12] In the 1920s and 1930s big cities in the Western world began to build sewage treatment plants. Washington, D.C., acquired its first in 1934. Moscow built

[9]Illinois Department of Energy and Natural Resources 1994, 2:75–84, 101–12; Stout and Ackermann 1987. Changnon 1994 is a complete guide to the Chicago River reversal; I thank Peter Campbell for this reference. See also Changnon and Changnon 1996.

[10]London pioneered filtration in a small way in 1829; the first U.S. city to do it was Poughkeepsie in 1879. Chlorination was first used on a small scale in London around 1800, but on a municipal level not until much later. On France, see Goubert 1989. On efforts in German cities, see Büschenfeld 1997.

[11]Outwater 1996:133–47. See Guayacochea de Onofrí 1987 on Mendoza, Argentina, where water and sewage treatment came very early in the 20th century.

[12]Primary sewage treatment consists of the use of filters to separate out solids. Secondary treatment also includes the use of bacteria to degrade labile organic matter and reduce biological oxygen demand in wastewater, as, for example, by the sludge activation process. Tertiary treatment is anything beyond that.

small treatment plants in the late 1930s and big ones in the 1960s. Tokyo's came after 1945. Technical refinements and expensive investments followed, especially after 1970, so that by the 1990s one could drink the effluent of many sewage treatment plants, which was often cleaner than the waters into which it flowed.[13]

This notable success in containing lethal biological pollution spread unevenly around the world in the twentieth century. At the other end of the spectrum from the United States and northwestern Europe, people in parts of India and China suffered increasing pathogenic loads because cities grew while sewage treatment made limited progress. Colonial cities in Africa and Asia often acquired water filtration and sewage disposal systems in the European quarters, but not throughout the cities, creating different health regimes that corresponded to distinctions of wealth and race in places such as Shanghai, Kampala, or Algiers. In Calcutta, where filtration and sewage systems existed from 1870 in richer neighborhoods, the water system degenerated after 1911, when the capital of British India was moved to New Delhi. In Madras the British left a sewer system built around 1940, but sewage treatment had to wait until around 1980. As late as 1980, half the world's urban population had no wastewater treatment whatsoever, and in China the proportion was 90 percent. In 1995, 89 percent of the residents of greater Manila had no connections to any sewer system, 82 percent in Dhaka, and 80 percent in Karachi. By contrast, Mexico City left only 20 percent and Seoul 14 percent without sewerage.[14]

The uneven history of the provision and treatment of urban water after 1880 was a case of escalating distinctions between the haves and have-nots. Those who had clean water and good sewerage got it because they were comparatively rich, and getting it made them healthier and richer still. Those who lacked it, lacked it mainly because they could not afford it, and lacking it made them sicker and poorer still. Economists use the term "increasing returns" to describe situations in which the more you have the more you get. The increasing returns generated by investment in clean water helped to create, and widen, the cleavages in wealth and health that characterize the world today.

---

[13]Windhoek (Namibia) cleaned its water so well by the 1970s that a third of it could be returned to the city's water supply (*Encyclopedia Britannica*, 15th ed., 1976, 14:753). On Moscow, see Goldman 1972:96–101.

[14]GEMS 1989:274; Headrick 1988:145–59; Smil 1984:100; Yeung 1997. For Madras, see Sundaramoorthy et al. 1991.

River Water

Most urban water was river water. For millennia, rivers carried off human wastes, and the big rivers sufficiently diluted it so that little harm came from the practice. Until quite recently there was not much the small populations of the Amazon or Congo (Zaire) river basins could do to pollute the enormous quantities of water in those rivers. But rivers passing through thickly settled landscapes (like the Ganges), rivers in the middle of industrial zones (like the Rhine) or mining zones (like the Watarase), and small rivers (like the Oued Fez) acquired toxic loads of biological and chemical wastes in modern times.

THE GANGES. The Ganges drains a quarter of India.[15] Its basin in 1900 contained about 100 million people, of whom perhaps 10 million dumped their wastes directly into the river. The Ganges' fetid condition gave rise to one of the world's first antipollution societies, in 1886. Mark Twain, who traveled the Ganges in 1896, found the water at Varanasi "nasty" on account of the "foul gush" of its sewers.[16] By 1990, 450 million people lived in the basin, and some 70 million discharged their wastes into the Ganges. Almost all the sewage, in 1990 as in 1900, went untreated. Its decay robbed the river water of oxygen, menacing fish populations with suffocation. The Ganges, because of population growth, suffered probably five to ten times more from biological pollution at the end of the century than at the beginning. The same must be true of hundreds of rivers around the world.

But the Ganges is unique in one respect. It acquired pollution for sacred as well as profane reasons. In Hindu belief, gods created the Ganges to give people a chance to wash away their sins. Hindus believe that death or cremation at Varanasi (Benares) ensures liberation of the soul, so Varanasi attracts millions of elderly and sickly Indians. In the 1980s, Varanasi's official crematoria burned 30 million bodies a year and deposited several million tons of human ash into the Ganges every month. Many additional bodies, partly cremated or not at all (fuelwood

---

[15]This section is based on Ahmed 1990, Basu 1992, Ghose and Sharma 1989, and Varady 1989.

[16]Twain 1899, 2:192–3.

Pious Hindus hope their mortal remains will be cremated at Varanasi on India's Ganges River, which act assures their souls of liberation from the cycle of reincarnation and suffering. Here fires and corpses are prepared for cremation in 1895. By the 1980s, Varanasi's official crematoria burned some 30 million bodies a year. The ash that is the residue of cremation was normally poured into the Ganges.

costs often being too much), were shoved into the river, as were roughly 60,000 animal carcasses. The Ganges was a bacteriological nightmare when the first systematic pollution studies took place in the 1960s; it then got worse. Government cleanup efforts, begun in the 1960s and coordinated into the Ganga Action Plan in 1985, had little discernible effect. In the Ganges, the important changes were the intensification of bacteriological pollution, not problems derived from industrial emissions, which the river's huge flow masked until about 1990.[17] Bathing in the Ganges may cleanse the soul but, now more than ever, not the body.

[17]Ghose and Sharma 1989:41–4. The tremendous silt load of the Ganges, second only to the Huang He, might also help by dragging heavy metals and other pollutants into the sediments of the river bottom. Meybeck and Helmer 1989:294 are less sanguine: they note that some 990 industries dumped untreated wastes into the Ganges in the early 1980s.

THE RHINE. In the industrialized world the situation was the opposite: chemical pollution, derived from technological change and economic growth—not from population growth—menaced rivers and lakes. The Industrial Revolution had a profound impact on the waters of the Western world, first in Great Britain. By the middle of the nineteenth century, factories dumped huge quantities of toxic effluents into British rivers. A Royal Commission in 1866 found that the water of the Calder, in northern England, made a tolerably good ink, which it demonstrated by using Calder water for a small part of the Commission's report. The Bradford Canal, fed by water that had passed through the grimy city of Bradford, was worse:

> ... [I]t was found practicable to set the Bradford Canal on fire, as this at times formed part of the amusement of boys in the neighborhood. They struck a match placed on the end of a stick, reached over and set the canal on fire, the flame rising six feet and running along the water for many yards, like a will-o-the wisp.[18]

Some ameliorative measures took effect in the nineteenth century, but by and large the river and lake water of the industrial world received an ever larger and more varied chemical cocktail at least until the 1960s. The River Irwell (U.K.), in 1869 "caked over with a thick scum of dirty froth," in 1950 could be a vivid orange in the morning and jet black by noon.[19] A 1972 description of the Sumida River in Tokyo sounds like an updated version of the British Royal Commission report:

> As a result of the pollution, the famous events which once took place on the river—swimming, regattas, firework displays—have vanished. The gases rising from the river corrode metals, blacken copper and silver ware and shorten the life of sewing machines and TV sets.[20]

Countless watercourses became variations on this theme. The story of the Rhine must stand for many rivers.

The Rhine flows about 1,300 kilometers from the Swiss Alps to the

---

[18]The quotation comes from the Royal Commission on Rivers Pollution, cited in Clapp 1994:74–5.

[19]Sherlock 1922:295 quoting Parliamentary papers in 1869; Sheail 1997:207 on 1950. By 1979 human action had raised the mineral content of rivers worldwide by 12% from c.1860 (Meybeck 1979:241).

[20]Quoted in Ponting 1991:364.

North Sea. Before 1765 it flowed unobstructed, and its waters were clean enough to accommodate sensitive fish such as salmon, which were so plentiful that servants complained about having to eat it too frequently. As cities and population grew, urban wastes affronted sensitive souls, such as the poet Samuel Coleridge, whose visit to Köln in 1828 inspired the poem "Cologne":

> In Köhln, a town of monks and bones,
> And pavements fang'd with murderous stones,
> And rags, and hags, and hideous wenches;
> I counted two and seventy stenches,
> All well defined, and several stinks!
> Ye Nymphs that reign o'er sewers and sinks,
> The river Rhine, it is well known,
> Doth wash your city of Cologne;
> But tell me, Nymphs, what power divine
> Shall henceforth wash the river Rhine?

No power, divine or not, did much to wash the filthy, frothy Rhine for a century and a half. After 1880, mounting chemical pollution added to the mix.

The proximity of coal and iron deposits in the Ruhr valley assured that the middle Rhine would in the nineteenth century become an industrial zone. By the 1890s its iron and steel production was very competitive worldwide, and production levels climbed. The Rhine's suitability for navigation—its flow is very steady throughout the year—attracted other industries, including the formidable German chemical industry. By 1914 the Rhine's pollution load was heavy, salmon were rare, and the mayfish, whose upstream migrations once occasioned public festivals, disappeared entirely from the lower Rhine. The last sturgeon was caught in 1931.

The Rhine got a brief respite with the wartime destruction and postwar stagnation of French and German industry (1944–1948), but recovery and *Wirtschaftswunder* (1950–1973) worsened the river's condition. By 1980, about 20 percent of the world's chemical production took place in the Rhine basin. Metallurgical and chemical plants oozed copper, cadmium, and mercury into the river, and urban wastewater contained zinc, nickel, and chromium. Between 1900 and 1977 the concentrations of heavy metals in Rhine sediments increased fivefold for chromium, twofold for nickel, sevenfold for copper, fourfold for zinc, twenty-sevenfold for cadmium, and fivefold for lead: Dutch hydrologists complained that German industry was metal-plating Holland. No one complained, but the same was happening to the sediments of the North Sea.[21] French potash mining in Alsace helped sextuple the salt content of the Rhine between 1880 and 1960, jeopardizing the Dutch flower business, which irrigated its orchids and gladioli with Rhine water. Nutrient loading with phosphorus and nitrogen (from detergents, sewage, and fertilizers) emerged as an additional problem after 1948, stimulating algae growth to the point that it clogged pumps and interfered with shipping. Algae's decay consumed oxygen, denying it to other species. New and toxic organic chemicals, such as DDT and PCBs, added to the brew, so that the Rhine between 1950 and 1975 was in its lower reaches almost devoid of fish, and fouler than ever before. Prudent sport fishermen threw back their catches in the 1980s, because Rhine fish often carried 400 times the concentrations of PCBs officially deemed safe to eat.[22] With high population, densely packed heavy industry and chemically dependent agriculture in its basin, the Rhine bore most of the pollution burdens a river can have.

Cleanup efforts began with sewage treatment after World War II. In 1964 Germany required biodegradable detergents. International accords among Germany, France, and Holland restricted many forms of pollution from the 1970s. Most heavy-metal concentrations in the river, but not the sediment, declined sharply after 1975. Fish populations, in decline since 1885 and especially since 1915, rose after 1976. More effective action followed the disastrous fire in a Sandoz chemical warehouse near Basel, Switzerland, in 1986. Firefighters sprayed water on the warehouse, washing pesticides, herbicides, and fungicides into the Rhine, killing virtually everything for 180 kilometers downstream. Although most of the aquatic biota recovered within two years, the affair concentrated the attention of ministers and captains of industry as never before. Regulations, incentives, and enforcement of all sorts followed. Fishermen caught salmon again in 1992.[23]

[21]Behre et al. 1985.

[22]PCBs are polychlorinated biphenyls, used as insulators, lubricants, and in many other applications since 1929; DDT is dichlorodiphenyltrichloroethane, first synthesized in 1874, and commercially introduced (as an insecticide) in 1942 by a Swiss chemical firm. On fishermen, see Reihelt 1986. In the USA, PCBs ingested from mothers' milk apparently lowered children's IQ scores (at age 11) up to 6 points (*Science News*, 14 September 1996:165, citing research of Joseph Jacobson and Sandra Jacobson).

[23]This account of the Rhine draws on Friedrich and Müller 1984, GEMS 1989:280, Habereer 1991, Lelek 1989, Malle 1996, Meybeck 1979, Reihelt 1986, van der Weijden and Middleburg 1989; and Van Urk 1984. In the Thames, salmon were caught in 1974, after a 140-year absence (Wood 1982:118).

THE WATARASE RIVER AND THE ASHIO MINE. Whereas the Rhine suffered from hundreds of industrial polluters, Japan's Watarase River suffered from one: the Ashio copper mine in Tochigi Prefecture (central Japan), active since 1610. During Tokugawa times (1603–1868) Ashio provided Japan with much of its copper, but had almost ceased production in 1877 when Furukawa Ichibei, a brilliant entrepreneur who had made and lost a fortune in silk, acquired the mine. He modernized and expanded operations, and in 1883 found a rich vein that made Ashio Asia's most lucrative copper mine. Japan's national policy of militarization needed Ashio, because Japan imported 95 percent of its steel in the 1890s and Ashio's exports earned foreign exchange that helped buy the steel. Copper was Japan's second or third largest export, and Ashio produced some 40 percent of it. It was Japan's single most important mine. So authorities stood firmly by Furukawa.

Furukawa's expansion and modernization brought far more serious pollution problems to the water and air around Ashio. By 1888, sulfuric acid rain from smelters had killed 5,000 hectares of forest and contaminated local waters. Floods became more common because hillsides lost their vegetation cover. Mine tailings seeped, or were dumped, into the nearby Watarase River, contaminating water used for irrigation in rice paddies. Local peasants became sickly and resentful. In the 1890s, death rates outstripped birth rates in the town of Ashio, which was home to about 30,000 people. Toxic waters killed off fish and fowl, depriving peasants of traditional supplements to their food supply. Everyone along the Watarase knew it was Furukawa's mine that jeopardized their rice, their health, and their lives.

Scholars, journalists, and the local member of the Diet, or parliament, Shozo Tanaka, took up the cause, demanding that the mind and smelter shut down. Thousands of peasants marched on Tokyo three times (1897–1898), clashing violently with police and attracting publicity, which obliged the government to require Furukawa to install antipollution devices. But the technology was primitive and ineffective, so acid rain and river pollution continued. A fourth march on Tokyo provoked strong government repression in 1900: the Ashio mine was too important to the state to allow objections by its neighbors. The peasant antipollution movement lost all momentum with the patriotism and repression that attended the Russo-Japanese War (1904–1905). Miners rioted in 1907, a landmark event in Japanese labor history that was in part motivated by pollution grievances.[24]

[24]Nimura 1997:21.

Some 450 Ashio households were banished to Hokkaido, Japan's northern island, ending popular protest. Thereafter, Ashio plagued its remaining neighbors in relative peace. Furukawa installed desulfurization equipment in 1955 and closed the mine in 1972. In a landmark judicial case, local farmers in 1974 won millions of dollars in compensation for a century of air and water pollution. The Watarase River basin had served its purpose as a sacrifice zone in the industrialization of Japan.[25]

OUED FEZ, MOROCCO. More typical of the world's rivers was the Oued Fez in Morocco.[26] It had neither copper mine, nor chemical industry, nor holy status. But it had a city and farms along its banks. The Oued Fez flows through the city of Fez (population c. 1 million in 1995) on its way to the larger Sebou River and the Atlantic. Water supply and sewage systems were built for Fez in the tenth century. That left the Oued Fez clean upstream of the city and filthy below it. In 1371, Lisanuddin Ibn Al-Khatib, an intellectual from Granada, feeling sullied after an encounter with the people of Fez, wrote: "I entered the city as their water did, and I left it as it did."

The impact of the city on the small river led rulers to build new cities upstream. The Merinids constructed a new upstream Fez in the thirteenth century; the French, who acquired control in 1912, did the same, yet further upstream. The pollution these efforts aimed to escape came from human wastes and from artisanal industries long established in Fez, such as leather tanning. From the 1960s, the Oued Fez also carried chemical fertilizer runoff from the Saïs plain, just upriver from the city. By 1990 the water just downstream of Fez carried five to ten times legal limits for several pollutants, and was perhaps 50 times as foul as in Ibn Al-Khatib's day. A small river flowing through a sizable city in a poor country, the Oued Fez was highly vulnerable to pollution and ill equipped to address it. In the twentieth century, there were a few thousand rivers like it.

[25]Hashimoto 1989; Miura 1975:259–86; Notehelfer 1975; Shoji and Sugai 1992; Tsuru 1989.
[26]What follows is based on Kettani 1993; the quotation is from p. 663.

Lakes and Eutrophication

Cleaning up rivers, like resolving local air pollution, can be easy. Lakes, where polluted water may linger for decades rather than days, are another matter. In the twentieth century, lakes in industrial areas acquired all manner of pollutants. In some cases, when downwind of smokestacks, they became acidic on account of acid rain (see Chapter 4). The most pervasive problem emerged in the 1930s: eutrophication.

Every ecosystem has limiting factors that constrain life. In most bodies of water either nitrogen or phosphorus plays that role. If somehow these limits are relaxed and unusual quantities of nitrogen or phosphorus become available (eutrophication), then aquatic plants and bacteria (especially blue-green algae) grow ebulliently. When they die, their decomposition consumes oxygen, which then becomes unavailable for other species. Vast changes in aquatic biotas often ensue. In extreme cases, all animal life suffocates for lack of oxygen. Algae population explosions can also render water unfit for drinking, swimming, navigation, and other uses. Warm water and stagnant water carry less oxygen to begin with than do cold and bubbly water, and thus are particularly at risk. Excess nitrogen and phosphorus usually came from urban sewage and, after the introduction of chemical fertilizers, from farm runoff.

Eutrophication can happen naturally, as lakes age. But after 1850, human action more and more often drove the process. Urban lakes led the way. Lake Mendota, in Madison, Wisconsin, featured algal blooms almost annually by 1850. Lake Zürich in Switzerland suffered from eutrophication by 1898, and regularly after 1930. The Italian Alpine lakes showed eutrophication beginning in 1946 and by the 1960s were occasionally covered by algal blooms. Small lakes near big cities eutrophied first, because of the impact of human wastes. After 1945, phosphate additives in detergents aggravated matters, and even large lakes (like Lake Erie) suffered from eutrophication. Seattle's Lake Washington illustrates the problem—and a solution.

Seattle's raw sewage caused eutrophication and small algal blooms in Lake Washington in the 1930s. The problem abated when the city diverted its sewage to Puget Sound in 1936. But in the late 1940s suburban growth revived the problem, and by 1955 algal blooms decorated the lake. Political tussles ensued, but by 1963 the suburbs too sent their sewage into Puget Sound. Lake Washington cleared up again. Puget Sound's size, regulations on phosphate additives, and improved sewage treatment prevented the Sound from suffering Lake Washington's fate.

With the proliferation of sewage treatment in urban areas of Europe and America after World War II, cities reduced their nutrient loading of lakes and rivers. But the tremendous increase in chemical fertilizers more than made up for it. Runoff from fields and animal feedlots became the main source of excess nutrients. Simple (if expensive) solutions such as treating urban wastewater did not exist, short of abandoning chemical fertilizers. Hence eutrophication of rural lakes and watercourses spread widely, first in North America and Europe, and then in the 1960s and 1970s around the world, wherever intensive fertilizer use held sway.[27]

In 1860 Chicagoans thought Lake Michigan was so large it could easily absorb the wastes their city might dump into it. Time proved them wrong. Similarly, most people living on the shores of the Black Sea or the Yellow Sea in 1900 gave no thought to problems that might arise from using the coastal seas as a sink for wastes. These seas appeared infinite for all intents and purposes. Time proved them wrong too. Today it is the deep oceans that are used as receptacles for all manner of wastes, on the theory that they are so huge they can safely dilute whatever human activity might inject into them. So far, that theory has held up.

Seas and Oceans

From the point of view of the deep seas and the oceans, the twentieth century was much like any other. Human impact scarcely extended beyond the inland seas and the coastal zones. These, however, are important, as they house most of the saltwater biota.[28]

Absent churning tides, enclosed seas easily suffered from eutrophication. The Baltic Sea acquired the most severe case, visible (and smellable) by the late 1950s. The urban wastes of Stockholm, Helsinki, Leningrad,

---

[27]Barica 1979; Bonomi et al. 1979; NRC 1992:188–91; ReVelle and ReVelle 1992:395–7; Schröder 1979. Details on Lake Washington are in Edmondson 1991.

[28]Gorman 1993:106–7; 90% of marine species reside in coastal waters on the continental shelves in what amounts to well under 1% of the oceans' space.

and (via the Vistula River) Warsaw added to the farm runoff from increasingly chemicalized agriculture, loading the Baltic with excess nutrients. Arms of the Mediterranean such as the Adriatic, recipient of the Po's nutrient-laced waters, bloomed with algae by the 1960s. So did the western Black Sea courtesy of the Danube. Europe's inland seas suffered first, because of the high urban populations and the early adoption of chemical fertilizers.

But enclosed and shallow seas elsewhere soon felt similar effects. After 1970, eutrophication affected Malaysian waters for the first time. Wherever human action delivered excess nutrients—Red Sea, Persian Gulf, Yellow Sea, or Sea of Japan—coastal fisheries were affected. In some cases, the heightened fertility of the seas translated into more food for more fish, increasing the catch. But wherever nutrient loading generated severe algal blooms, fish stocks—and catches—plummeted.[29]

Heavy metals flowed into the coastal seas and rained down upon them. An influx of marine heavy metals, apparent in sediments, took place in the Baltic after 1880 and on the southern California coast after 1940. Wherever metallurgy and chemical industries proliferated, heavy metals worked their way into the sea. The bays, estuaries, and enclosed seas of Europe, the USSR, and the United States received the heaviest doses, often sufficient to damage marine life. Cadmium and mercury accumulations sometimes made shellfish poisonous to human beings.[30] The most grievous case took place in southwestern Japan, in a fishing village called Minamata.

MINAMATA BAY. In 1910, Nippon Chisso built a chemical factory in Minamata, which gradually became a company town of 50,000 by 1950. From 1932 the Nippon Chisso factory manufactured acetaldehyde, which requires inorganic mercury as a catalyst. (Acetaldehyde is useful in the synthesis of acetic acid, used in printing, plastics, photo processing, and many other things). Nippon Chisso dumped mercury-laden waste into Minamata Bay. Bacteria converted the mercury into an organic compound, methyl mercury, which worked its way up the food chain in ever greater concentrations. Unexplained fish die-offs began in the late 1940s. In the 1950s, the factory accelerated production, and mercury dumping. Soon many Minamata cats went mad, danced as if drunk, vomited, and

---

[29]Elmgren 1989; Larssen et al. 1985; Linden 1990:8. *Ambio,* 1990, 19(3) is a special issue on Baltic eutrophication.

[30]Adler et al. 1993; Alderton 1985.

died; people called it "cat-dancing disease." By 1956, Minamata children began to develop brain damage: they had what would come to be called Minamata disease.[31] Fish were correctly suspected, and soon fishermen found it impossible to sell their catch. A prominent local doctor, Hosokawa Hajime, confirmed that Minamata disease was mercury poisoning, but his findings were kept secret under pressure from his employer, the Chisso company. In 1959 the local fishermen, unable to get the company to stop spewing mercury into the bay, attacked the factory. But the mercury kept flowing for another 10 years, while thousands of people developed symptoms, and more than 100 died.[32] The mayor, who consistently sided with Nippon Chisso, maintained in 1973—long after the connections between Nippon Chisso, mercury, fish, and death were clear—that "what is good for Chisso is good for Minamata." The aggrieved mounted a lawsuit. The Chisso company lost, and by 1977 had paid $100 million to Minamata victims and their families. For decades, no one from elsewhere in Japan would knowingly marry anyone from Minamata, on the theory that doing so might lead to deformed offspring. After 1984 the Japanese government dredged and, to its satisfaction at least, decontaminated the floor of Minamata Bay, which cost about $400 million. In 1997, authorities declared Minamata Bay free of mercury and removed the netting, installed in the 1970s, that had kept unsuspecting fish out of the contaminated waters.[33]

The Minamata episode was probably the twentieth century's (and any century's) worst case of contamination of the sea, but it was a simple one, involving only one nation, one factory, and one pollutant. A much larger and more typical case was the Mediterranean Sea, whose pollution history involved many nations and many pollutants.

THE MEDITERRANEAN. In 1798, Samuel Coleridge wrote, in "The Rime of the Ancient Mariner":

> The very deep did rot: O Christ!
> That this should ever be!

[31]Analysis of dried umbilical cords shows that the methyl mercury content in the diet of pregnant women peaked in the late 1930s and early 1940s, and again in the late 1950s through 1965. The Japanese kept umbilical cords as a cure for serious illness (Nishigaki and Harada 1975).

[32]By 1990 Minamata disease had claimed 987 dead, 2,239 still suffered from it, while another 2,903 sought official victim status (Ui 1992b:131, citing the Japanese Environmental Agency). Some of these victims came from elsewhere in Japan.

[33]Akio 1992; Kudo and Miyahara 1991; McKean 1981; Ui 1992b.

Yes, slimy things did crawl with legs
Upon the slimy sea.

He wrote these lines 15 years before he laid eyes on the Mediterranean Sea. When he did—he served as secretary to the Governor of British Malta for two years—the Mediterranean was good and slimy in only a few harbors. But 200 years later the deep did occasionally rot where algal blooms decayed. And from time to time slimy things did crawl or slither upon oily seas. The growth of modern industry in many Mediterranean countries, the emergence of chemicalized agriculture, and the rise in human and animal populations, sharply increased the basin's pollution load after about 1950. A great deal of this ended up in the sea itself.

The Mediterranean is the world's largest inland sea. In 1995 its catchment was home to about 200 million people divided among 18 countries.[34] It is a salty sea because evaporation is high and the freshwater inflow from rivers is low. At Gibraltar its heavier, salty water flows out to the Atlantic underneath an incoming current of lighter, less salty ocean water. On average it takes about 80 years for the Mediterranean to flush out fully. Pollutants linger longer than in the North Sea, where they stay about two years, but not so long as in the Black Sea, where they can subsist for about 140 years. Biologically the Mediterranean is at once rich and poor. It is rich in species diversity, home to about 10,000 animals and plants. But because its waters are normally thin in nutrients, its total biomass and biological productivity are extremely low. This is why, where it is not polluted, the water is so clear.

In the twentieth century the waters of the Mediterranean grew progressively less clear and more polluted. Marine pollution, of course, is not new. The ancient harbors of Ostia (near Rome), Piraeus, and Alexandria were strewn with wastes and garbage. Bays, estuaries, and inlets close to population centers—the Golden Horn, the Venetian lagoons, the Bay of Naples—were unsanitary long before the twentieth century. It is possible that the amount of pollutants dumped directly into the Mediterranean is now less than it was a century or two ago. But pollution also reaches the sea via rivers and through the air, and in greater quantity than ever before.

The main pollutants in the Mediterranean were and are much the same as elsewhere around the aquatic world. Microbes, synthetic organic compounds such as DDT or PCBs, oil, litter, and excess nutrients topped the

[34]The total population of countries bordering the Mediterranean in 1995 was around 400 million; that of the administrative districts with Mediterranean coasts more like 130 million.

list, with heavy metals and radionuclides less important. In the most general terms, by 1990 about a quarter of the Mediterranean's total land-derived pollution contaminated the northwestern coasts from Valencia to Genoa, and a third plagued the Adriatic. Earlier in the century these proportions were probably higher.[35] The main sources were and are the big cities, big rivers, and a few coastal industrial enclaves.

Microbial contamination from sewage existed in rough proportion to human population until the twentieth century, because sewage treatment scarcely existed. By the end of the twentieth century, about 30 percent of the raw sewage splashing into the Mediterranean received treatment, but the total quantity had tripled or quadrupled since 1900.[36] So the risks of gastrointestinal ailments, typhoid, or hepatitis to people bathing or eating seafood increased significantly. By the late 1980s, when the European Union (EU) had developed guidelines for permissible levels of microbial contamination, beach closings became routine from Spain to Greece. In any given summer in the 1990s, about 10 percent of Mediterranean European beaches failed EU standards, although they were not necessarily closed.

Oil, a negligible pollutant before 1900, became a major one with the energy transformation of the twentieth century. The emergence of the Persian Gulf oil fields after 1948, the existence of the Suez Canal, and the energy demand of European transport and industry assured that the Mediterranean would become one of the world's oil highways. The Mediterranean has yet to endure a major tanker spill, although it has suffered many modest ones. Offshore oil drilling, a major pollution source elsewhere, remained tiny in the Mediterranean. Most oil pollution came from routine operations, unregulated until the 1970s, such as cleaning out tanks and dumping bilge water. The quantities involved were minor before World War II because the oil trade remained small. During the war, routine shipments were suspended, but a large share of military shipments were sunk. Nonetheless, oil pollution only grew after the war, driven by the huge expansion of European demand for Middle Eastern oil. About a quarter of the world's oil shipments crossed the Mediterranean in 1990. A 1975 estimate calculated that half a million tons of oil leaked into the Mediterranean each year; another in 1980 to 1981 suggested about 820,000 tons. Typically one-third of it washed up on shore as tar, which plagued Mediterranean beaches more than any others in the

---

[35]De Walle et al. 1993a:59.
[36]Stanners and Bourdeau 1995:495.

world. Much of the rest floated atop the water as oil slicks, which at times thinly covered as much as 10 percent of the sea's surface. Around 1980 the Mediterranean absorbed a sixth of the world's oil pollution. Over half of this took place in routine loading and cleaning; Libyan waters were most affected.[37]

Industry contributed more to sullying the Mediterranean than oil. Many factories sprouted on the water's edge, taking advantage of the low costs of seaborne shipping. Others emerged on rivers that flow into the sea, for transport reasons or because industrial processes required fresh water for cooling or cleaning. Even factories far from the water polluted the Mediterranean via airborne deposition. Whatever the avenue, the sea received significant amounts of synthetic compounds and heavy metals from industry.

Industrialization proceeded spectacularly in the Mediterranean basin in the late twentieth century. Mediterranean countries accounted for about 5 percent of the world industrial production in 1929, about 3 percent in 1950, but 14 percent in 1985. A great surge came after 1960. For the next quarter century, industrial production in Mediterranean countries rose by about 6 to 7 percent annually, faster in Greece, Turkey, Spain, and North Africa, slower in France and Italy. This industrialization contributed mightily to the extraordinary economic improvement in Mediterranean Europe after 1950 and translated into welcome enhancements of nutrition, health, and life expectancy. It also, of course, brought greater pollution.

That pollution, naturally, concentrated where industry did: in Italy, France, and Spain. Despite the rapid growth of industry in North Africa, by 1990 it still accounted for only 9 percent of Mediterranean industry; the several countries ranging from Israel to Croatia accounted for another 10 percent. Italy generated two-thirds of the industrial production of the Mediterranean basin, Spain (mostly Barcelona) a tenth, and France (where little industry is in the Mediterranean catchment) only a twentieth.[38] The greatest pollution problems therefore arose northwest of the Mediterranean basin, around the mouths of rivers with industrialized basins, such as the Ebro, Rhône, and Po and around the centers of heavy industry, such as Barcelona, Genoa, and the northern Adriatic coast from the Po delta to Trieste. Industries poured the usual pollutants into the air,

[37]De Walle, et al. 1993b:6, 62–3; Le Lourd 1977; Stanners and Bourdeau 1995:118. There were in the 1980s about 50 oil refineries in the Mediterranean basin.

[38]Grenon and Batisse 1989:103–5. I have interpreted their data differently, considering all of Italy as Mediterranean, rather than, as they do, half. The unit of measure they use is industrial value added, in dollar terms.

rivers, and Mediterranean itself: PCBs, heavy metals such as mercury, lead, and arsenic. Table 5.3 gives a rough indication of the geographic origin of Mediterranean pollutants in 1985. The concentration of heavy industry along the coasts of the Golfe du Lyon and the northern Adriatic posed the most acute problems and accounted for the lion's share of the imbalances apparent in Table 5.3.

Table 5.3 North-South Environmental Comparison in the Mediterranean, 1985

Item	North	South
Population	73,000,000	50,000,000
Urbanization rate (%)	69	47
Cities with sewage treatment (%)	70	50
Wastes (million cubic meters)	2,295	544
Nitrogen influx (thousand tons)	128	48

Source: Grenon and Batisse 1989:245–6.

Note: In the scheme of Grenon and Batisse, North consists of Spain, France, Italy, former Yugoslavia, and Greece; South consists of Turkey, Syria, Egypt, Libya, Tunisia, Algeria, and Morocco.

Outside of these large hot spots, pollution problems developed more recently and more locally. In Greece, for example, two industrial clusters developed in the twentieth century, around Athens and around Thessaloníki. Between them, they contained all major Greek industry, except for some power plants. Neither city, as late as 1990, had sewage treatment plants. The pollution from almost all of Greece's metallurgical industry, from all of its refineries, paper mills, and shipyards, from three of its four fertilizer plants, and the sewage generated by half its human population were concentrated in the vicinity of Athens or Thessaloníki. The increasing severity of pollution, and perhaps a decline in the Greek public's tolerance for pollution, combined to produce remedial measures after about 1980. These took the form of incentives to relocate industries (some such incentives had existed since 1965) and some pollution control.[39]

[39] In 1982 striped mullet caught in the vicinity of Athens or Elefsis (an industrial area northwest of Athens), carried 15–18 times as much PCBs and DDT as did those caught 20 km away (Vassilopoulos and Nikopoulou-Tamvakli 1993; see also Katsoulis and Tsangaris 1994).

Numerous counterparts to these troubled Greek waters existed. The Gulf of Izmir, the Gulf of Iskenderun (once Alexandretta), the gulfs of Tunis and Trieste, and many others all developed major pollution problems. Istanbul's Golden Horn, which had suffered from biological pollution for many centuries, added growing concentrations of toxic metals to the mix after 1913.[40] Indeed, wherever urban and industrial centers grew up on bays, gulfs, or inlets exempted from the general counterclockwise currents of the Mediterranean, pollution accumulated.

Eutrophication in the Mediterranean derived less from industry than from agriculture and municipal sewage. From time to time algal blooms, often called red tides, occurred naturally in the Mediterranean, as elsewhere in enclosed waters. But they happened much more often in the twentieth century, because of urbanization and its untreated sewage, and because of the burgeoning use of chemical fertilizers. The most affected areas were the Golfe du Lyon, which suffered its most serious blooms after 1980; the Saronic Gulf around Athens, which experienced its first recorded red tide in 1978; and the northern Adriatic. Rivers delivered about three-fourths of the excess nutrients.[41]

The northern Adriatic is a shallow shelf with poor circulation. The water warms up every summer and is unusually susceptible to eutrophication. It receives the waters of the Po valley, where from early in the century farmers used chemical fertilizers. It absorbs the wastewater of several sizable cities and (less importantly in this context) industrial centers. Between 1872 and 1988 the northern Adriatic had 15 recorded eutrophication blooms. That of 1969 was the first large one, that of 1988 the biggest. Their frequency increased after 1969, which probably reflected increased nutrient loadings, but might reflect warmer water temperatures—perhaps both.[42] The increased frequency and severity of blooms, in the northern Adriatic and elsewhere, clouded the waters and sharply reduced the depths at which a particular sea grass (Poisidonia oceanica) grew. Beds of this seagrass play a crucial role as a nursery of many Mediterranean aquatic species, which will suffer (or somehow adapt) in the years to come.[43] In the meantime, algal blooms played havoc with fish populations, seabed life in general, and the tourist trade.

---

[40]Tuncer et al. 1993. Pollution data on the Gulf of Izmir c.1985 is in Türkiye Çevre Sorunları Vakfı 1991:216–20.

[41]Vassilopoulos and Nikopoulou-Tamvakli 1993:432 on the Saronic Gulf.

[42]Bethoux et al. 1990; Marchetti and Rinaldi 1989.

[43]Stanners and Bourdeau 1995:120–1.

Despite a century of intensifying marine pollution, the Mediterranean Sea in the 1990s was no cesspool. Long stretches of coastline in southern Turkey and North Africa, and smaller ones elsewhere, retained clean waters. On straight coastlines where currents scoured the shore unimpeded by capes or headlands, or out in the open sea, pollution, while far from negligible, had yet to make much difference. The Mediterranean was cleaner than the Baltic Sea, the Black Sea, or the Sea of Japan.[44] Three reasons explain this: its size, its lively mixing of deep and surface waters, and its currents helped to dilute the pollution load; the total pollution load, while growing, paled beside that of the Black Sea and some other unfortunate bodies of water; and Mediterranean societies after 1975 made some concerted efforts to reduce the Mediterranean's pollution.

MEDITERRANEAN ENVIRONMENTAL POLITICS AFTER 1975 As in much of the world, explicit environmental awareness and politics around the Mediterranean dates mainly from the 1970s. Most countries by 1975 had tiny bands of ecologically concerned citizens, such as the Corsicans who in 1973 demonstrated against an Italian chemical plant's pollution that disfigured their island shores.[45] By 1980 some countries had green parties. The general political and cultural milieu—if one can generalize—did not favor environmental movements. In none of this, however, were Mediterranean countries exceptional.

They were exceptional after 1975 for developing the Mediterranean Action Plan (MAP). Under the auspices of the U.N. Environment Programme (UNEP), all Mediterranean littoral countries except Albania convened in Barcelona and agreed to an ongoing process of environmental management for the entire basin. The plan supported (and supports) scientific research and integrated development planning. It produced several agreements and protocols to limit pollution. Enforcement normally left something to be desired. For instance, about 2,000 kilometers of coastline were "sacrificed" to development through lax enforcement or special dispensations.[46] But the plan, together with national regulations and EU restrictions, helped limit Mediterranean pollution from 1976. MAP helped in the construction of sewage treatment plants for Marseilles, Cairo, Alexandria, Aleppo, and several other cities large and small. At the end of the 1980s, work began on sewage works for

[44]See Albaigues et al. 1985 for the comparative judgment.
[45]Molinelli-Cancellieri 1995.
[46]De Walle, et al. 1993a:79.

Thessaloníki and Athens. While the sea 20 years later was more polluted than when MAP began, it surely would have been much more so without MAP.[47]

Any accords involving Greece and Turkey, Syria and Israel, and other pairs of sworn enemies must rank as high political achievement. In this case some of the credit must go to scientists who forged something of a pan-Mediterranean community. Scientific wisdom, normally quickly ignored when hard bargaining begins in international environmental politics, carried unusual weight for one reason: hundreds of billions of tourist dollars were at stake. No country could achieve clean beaches alone, given the circulation of the Mediterranean. Syria needed Israeli cooperation and Israel needed Egypt's. The quest for tourists, who contributed mightily to pollution, paradoxically helped stabilize, and in cases improve, the quality of coastal waters.

THE OCEANS. The high seas attract no tourists and have few defenders. Their capacity to dilute makes it very tempting to pollute them. After 1945 the high seas had to accommodate growing doses of metals, chemicals, oil, and nuclear radiation. Consider only plastics. They scarcely existed in 1950, but by 1992 accounted for 60 percent of beach litter worldwide. Thor Heyerdahl, the Norwegian adventurer and scientist, made two raft voyages across the Atlantic. In 1951–1952 he found no sign of human pollution. Eighteen years later in 1969 he saw oil slicks on 40 of his 57 days at sea, and plastics bobbing in the waves and swells all the way between Cape Verde and Barbados. Had he made a third raft voyage after another 18-year interval he likely would have found less oil but a lot more plastic. A huge increase in oceanic plastic debris took place after 1970, because the world—especially Europeans and North Americans—used far more plastic without making sufficient provision for its disposal.[48]

Efforts to limit pollution of the oceans suffered from two problems: the perception that their size negated any ill effects, and the oceans' international character. Individual nations restricted ocean dumping from the nineteenth century, and did so more energetically after 1970. The United States reduced its industrial waste dumping at sea from 6 million to 1 million tons between 1973 and 1983. Estuaries and bays, such as the Chesapeake, were partly restored. International waters required a more

---

[47]On MAP politics, see Antoine 1993 and Haas 1990.
[48]Earle 1995:254–55; Gorman 1993:34, 39, 114.

involved politics. The first international accord, on oil dumping at sea, came in 1954. More concerted effort dates from about 1972 but with scant results.[49] Sufficient incentives for international agreements mandating restraint did not yet exist. The oceans were big enough—like space—that even the twentieth century's garbage and pollution mattered only around the edges.

Conclusion

The biochemical changes to the earth's water—chiefly pollution and its abatement—followed the course of industrialization and urbanization, and affected almost every society. Where these trends waxed strongest, pollution became severe, especially after 1945 with the introduction and mass production of new organic chemicals. Pollution damaged lakes and rivers most severely, inland and coastal seas considerably, and the open oceans hardly at all. Before 1800, water pollution was a local matter, important only in the immediate vicinity of cities or certain industries like tanning or glassmaking. In the nineteenth century, in England and eventually in other industrial centers, it occasionally became a more regional affair. In the twentieth century water pollution often did, changing chemistry and ecosystems in areas as large as Lake Erie or the Baltic Sea.

Contaminated water killed tens of millions of people in the twentieth century, easily humanity's most costly pollution problem. The provision of safe water for modern cities, begun in the nineteenth century and carried forward in the twentieth, was decisive in shaping modern life. Without it, big cities would be far fewer—and far less healthy. But despite this success and our general hydrological ingenuity, a sufficient supply of clean water appears likely to be one of the sharper constraints on the human career in the next century.

The scale of pollution increasingly surpassed the thresholds at which waters could assimilate wastes. The favorite method of water pollution control for millennia—dilution—worked less and less often in the twentieth century. Newer methods worked only where they were tried, and even there not perfectly. Cleaning up the water of the Rhine to the point

[49]Prager 1993:87–131 reviews international efforts to limit marine pollution, as does Gorman 1993:69–92.

that fish returned was comparatively easy: the river gets new water all the time, and only a few nations, which after 1948 were fairly friendly to one another and fairly rich, were involved. Cleaning up the Mediterranean, quite incompletely achieved, proved harder. Decades rather than weeks were required for the system to flush itself out. More countries, some of them mutual enemies, many of them poor, had to cooperate. The oceans are difficult to pollute seriously, because of their size, but should it be done, they will be nearly impossible to clean up.

6

The Hydrosphere: Depletions, Dams, and Diversions

And thou beholdest the Earth blackened; then, when We send
down water upon it, it quivers, and swells, and puts forth herbs
of every joyous kind.

—Qur'an 22:5

When the well's dry we know the worth of water.
—Ben Franklin, *Poor Richard's Almanac*

In 1908 Winston Churchill, on a tour of Africa, stood near the northern
shores of Lake Victoria. He watched the waters of the world's second
largest lake flow over Owen Falls into the world's longest river, the Nile.
He later recorded his thoughts, provoked by the grandeur of the place:
"So much power running to waste . . . such a lever to control the natural
forces of Africa ungripped, cannot but vex and stimulate the imagination.
And what fun to make the immemorial Nile begin its journey by diving
into a turbine."[1]

---

[1]Quoted in Worthington 1983:101.

Imaginations remained vexed for some time. World wars and the Great Depression intervened, but in 1946, after suitable studies and plans, construction began on the Owen Falls Dam. In 1954, when Churchill was prime minister for the second time, the Nile waters at last dove into turbines. Lake Victoria became a reservoir, and Uganda and western Kenya got 150,000 kilowatts of electrical capacity.

Churchill's outlook reflected the dominant approach to water in the twentieth century. He saw it as a resource, and it irked him to see that resource unexploited. Its development, he thought, promised a better future, in this case for Uganda and the British Empire. And, of course, recasting nature could be fun, particularly for those with Churchill's boyish bent—and the confidence in the justice of one's cause that British imperialists in the age of Rudyard Kipling often possessed. A great deal has changed in the hydrosphere because of men who felt much the same way Churchill did. Lenin, Franklin Roosevelt, Nehru, Deng Xiaoping, and a host of lesser figures saw water in much the same way, and encouraged massive water projects in the USSR, the United States, India, and China. They did so because they all lived in an age in which states and societies regarded adjustments to nature's hydrology as a route to greater power or prosperity. And they had unprecedented technological means at their disposal. Since 1850, hydraulic engineers and their political masters have reconfigured the planet's plumbing. They did so to accommodate the needs of the evolving economy, but also for reasons of public health, geopolitics, pork-barrel politics, symbolic politics, and no doubt to satisfy their vanity and playfulness. To get water in the right amounts in the right places at the right times for whatever one wishes to do with it requires hydraulic engineering, one of humankind's oldest sciences. In the twentieth century, with methods ancient and modern, we built several million dams, tubewells, canals, aqueducts, and pipelines to divert water from the destinations gravity had in store for it. This chapter explains what physical changes humans wrought in the hydrosphere after 1900: how we came to divert much of the world's fresh water from its previous paths and rhythms in order to serve our manifold purposes.

Groundwater

One of the most important, and to many minds most troubling, changes in the hydrosphere fostered in the twentieth century was the intensive use of groundwater. People dug wells for millennia in order to irrigate fields and provide for drinking water. Considerable effort and ingenuity went into it. Ancient Chinese well diggers went as deep as 500 meters. But in the end, limits of muscle and wind power imposed constraints on the pace of groundwater exploitation.[2]

Cheap energy, a distinguishing feature of the twentieth century, helped make it feasible to pump groundwater on a massive scale. Although groundwater is abundant (at some depth below the surface) almost everywhere, it is only where surface water is scarce and energy cheap that huge quantities of water are brought to the surface. The Middle East and the American West are the best examples.

In the High Plains of the United States, most years lack the rainfall to raise the crops American farmers and markets prefer. From the late nineteenth century, wind-powered pumps helped farmers make a go of it from Texas to Montana. But the best High Plains windmills could draw water from no deeper than 10 meters, and could irrigate no more than about 3 hectares (7 acres) of wheat. With well-digging techniques improved by the oil industry and with cheap gasoline and natural gas, by the mid-1930s farmers could tap underground water without a huge investment. They began to pump up the Ogallala (or High Plains) Aquifer, a body of water equal in volume to Lake Huron (or maybe only Lake Ontario),[3] but stretching from the Texas Panhandle to western South Dakota. The Ogallala is actually a very slow underground river, dripping inches per day southeast through a gravelly bed at a depth of less than 100 meters. The water now there has seeped through the Ogallala's gravel for 10,000 to 25,000 years.

The searing drought of the 1930s sharpened the thirst of the High Plains, and postwar opportunity deepened it further. The newly accessible Ogallala seemed an answer to the prayers of Dust Bowl farmers: a re-

---

[2]Except in the case of *qanats,* dug in Iran, Morocco, and Afghanistan to tap higher elevation groundwater for agriculture at lower elevations.

[3]Opie 1993 says the Ogallala contains 3 billion acre-feet, the equivalent of Lake Ontario. But according to the conversion tables and lake-volume data in Gleick 1993:449 and 162, 3 billion acre-feet is much closer to Lake Huron's than Lake Ontario's volume.

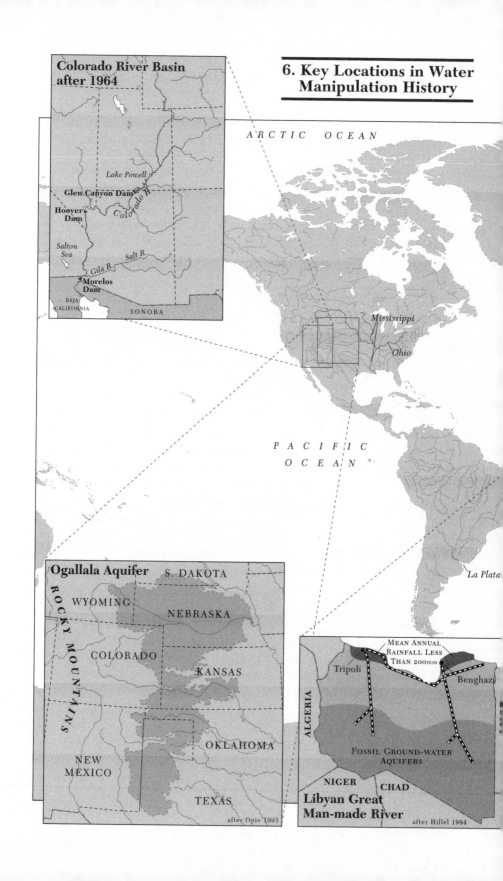

Colorado River Basin after 1964

ARCTIC OCEAN

Lake Powell

Glen Canyon Dam

Colorado R.

Hoover Dam

Salton Sea

Gila R.   Salt R.

Morelos Dam

BAJA CALIFORNIA   SONORA

Mississippi

Ohio

PACIFIC OCEAN

La Plata

Ogallala Aquifer   S. DAKOTA

ROCKY MOUNTAINS

WYOMING

NEBRASKA

COLORADO

KANSAS

NEW MEXICO

OKLAHOMA

TEXAS

after Opie 1993

MEAN ANNUAL RAINFALL LESS THAN 200mm

Tripoli

Benghazi

ALGERIA

FOSSIL GROUND-WATER AQUIFERS

NIGER   CHAD

Libyan Great Man-made River

after Hillel 1994

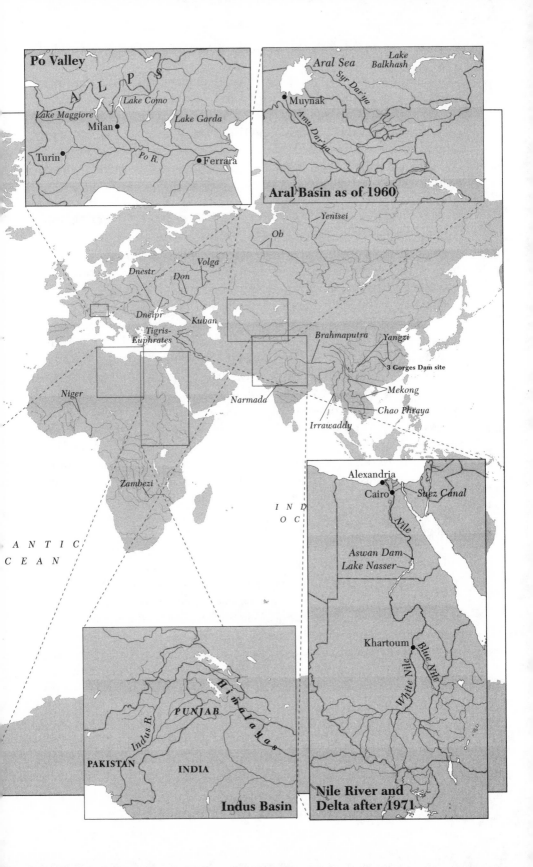

Po Valley

ALPS

Lake Maggiore

Lake Como

Lake Garda

Milan

Turin

Po R.

Ferrara

Aral Basin as of 1960

Aral Sea

Lake Balkhash

Syr Dar'ya

Amu Dar'ya

Muynak

Yenisei

Ob

Volga

Dnestr

Don

Dneipr

Kuban

Tigris-Euphrates

Brahmaputra

Yangzi

3 Gorges Dam site

Mekong

Narmada

Chao Phraya

Irrawaddy

Niger

Zambezi

I N D

O C

A N T I C

C E A N

Nile River and Delta after 1971

Alexandria

Cairo

Suez Canal

Nile

Aswan Dam

Lake Nasser

Khartoum

White Nile

Blue Nile

Indus Basin

Himalayas

PUNJAB

Indus R.

PAKISTAN

INDIA

liable source of water, equivalent to rain on demand. Use intensified after 1945, and quadrupled between 1950 and 1980, spurred on by droughts in the 1950s and 1970s. By the late 1970s, Ogallala water accounted for one-fifth of the irrigated area in the United States. A goodly chunk of the country's wheat, corn, alfalfa, and even cotton depended on it. Nearly 40 percent of the nation's cattle drank Ogallala water and ate grain produced with it. Farmers drained it at a rate of a little under 1 percent per year in the late 1970s, drawing water 10 times faster than the aquifer could recharge under the best conditions. Many good livings came from this. Clarence Gigot, a rancher and farmer who probably used more Ogallala water than anyone else, made himself a multimillionaire on the dry Sandhills of southwestern Kansas. He applied center-pivot irrigation to raise grain and cattle on land no one else wanted and made it pay.

The end of this water boom is in sight. In the southern reaches of the High Plains, where exploitation began first, farmers soon had to go deeper and deeper to get it, and many found the costs did not justify the results. In northern Texas, irrigation declined after 1974 and in the High Plains generally, irrigation contracted after 1983. While farmers' decisions to irrigate or not depend on many things, such as energy costs and crop prices, the drawdown of the Ogallala played a large role. Since the late 1970s states have come to agreements about who gets how much Ogallala water, and extraction rates have stabilized. But they have not declined. Across the High Plains, 150,000 pumps work day and night during the growing seasons. Farmers around Sublette, Kansas, figured in 1970 they had about 300 years' worth of water left. In 1980 they reckoned they had 70 years' supply, in 1990 less than 30. Half the accessible water was gone by 1993, and hydrologists and farmers agree the bonanza will end in 20 or 30 years, more if conservation triumphs, less if there is another drought. While it took many millennia to fill, the Ogallala's usefulness to humankind will almost surely last less than a century.[4]

The Arabian peninsula and Libya also combine considerable groundwater supplies with cheap energy. When oil markets in the 1970s sent billions of dollars their way, the Saudis invested some of that in schemes to exploit their aquifers. They now get 70 to 90 percent of their fresh water from underground. Although it takes a thousand tons of water (assuming not a drop is wasted) to raise a single ton of wheat, Saudi policy after 1975

---

[4]Schwarz et al. 1990:265–7 provide a brief account. The best detailed treatment is Opie 1993, from which most of this discussion is drawn.

was to grow wheat in the desert, at five times the international market price, in order to be self-sufficient in food. After 1984, Saudi Arabia exported wheat regularly. The Arabian aquifers scarcely recharge at all, so they too will not last. The Saudis hope that desalinization of seawater will become a practical alternative, and will relax the historical constraint of water supply in Arabia. It is a bold gamble.[5]

The Libyan scheme is water manipulation on the heroic scale. Libya is a big country with a small population. Southern Libya, in the heart of the Sahara desert, lies on top of vast amounts of fossil water. In the 1920s, when Libya was an Italian colony, Benito Mussolini hoped to match Britain's success with oil in Iraq by drilling in Libya. Disappointed Italian petroleum engineers found only aquifers. After Libyan independence in 1951, American oil men found more aquifers. But the water lay far from any center of population: 40 days by camel over shifting dunes. So the water stayed put. That changed after the 1969 revolution when Col. Muammar el-Qaddafi decided to make Libya self-sufficient in food and fiber. He convinced the American billionaire Armand Hammer, the head of Occidental Petroleum, to help build pipelines to deliver the Saharan water to the Libyan coasts.[6] A good share of Libya's oil revenues (about $25 billion) went into the Great Man-Made River, a system of two major pipelines buried under the sands, capable of delivering water equal to about 5 percent of the Nile's flow. The coastal regions around Tripoli and Benghazi can now raise crops on a scale quite impossible before the river began to flow in 1986. The water cost 4 to 10 times as much as the value of the crops it produced, and in conventional accounting was a massive money loser for Libya. After the collapse of oil prices in the early 1980s, Libya had trouble footing the bills. But Korean and American construction firms did well by the project, and Qaddafi found it useful in his efforts to secure the support of an occasionally restive population. This is the accounting that mattered, and explains Libya's persistence with the project, despite its economics and the troubles it caused with Egypt and Chad, which objected that the Great Man-Made River might poach their water.[7]

[5]Postel 1992:31–2.

[6]Coincidentally, Occidental Petroleum after 1981 owned the biggest meatpacking company on the High Plains, (IBP) in Holcomb, Kansas. Its operations use 600 million gallons of Ogallala water every year, making it one of the largest consumers of American fossil water (Opie 1993:154).

[7]Such poaching seems unlikely according to Hillel 1994:196–200, because the wells are too far from the borders to affect groundwater in Chad or Egypt (Allan 1994:75–6).

Less grandiose schemes to make use of groundwater emerged in Mexico, Europe, India, China, and elsewhere. While the water lasts, groundwater irrigation improves agricultural production. And its relative purity makes it attractive for urban water systems. But inevitably, sooner or later, the water runs short, and difficult adjustments must follow. In Uttar Pradesh, India's largest state, the government provided wells for 2,700 villages between 1970 and 1985. This made life notably easier until the water table dropped and 2,300 of the wells went dry. Despite massive investment in water infrastructure, in 1985 drinking water in most of India was harder to come by than in 1970. In Europe, where most big cities by 1990 were drawing down groundwater supplies, aquifer mining ran up against limits, especially in Greece and Spain.[8]

Many cities around the world relied on groundwater to slake their thirsts, with various results. Rapid urban growth, in Beijing and Mexico City among many other places, depleted local aquifers and obliged authorities to look further afield for imported water. Both cities were subsiding in the late 20th century as their groundwater disappeared, as was Bangkok, Houston, and many others. Tokyo too was sinking, by about 5 meters between 1920 and 1965, due to groundwater depletion, but after 1961 government controls on withdrawals stabilized the situation. Barcelona, while not subsiding, depleted its freshwater aquifers under the Llobregat River delta after 1965, allowing seawater to seep in. Some of the old industrial cities of Europe, such as Birmingham, Liverpool, and Paris, reduced their groundwater withdrawals after 1970 as industry declined. Fewer withdrawals led to problems as well: tunnels, basements, and other subterranean structures were built when industrial water use kept the water table low, and now with higher groundwater they flooded more often.[9]

New technologies and cheap energy in the twentieth century allowed aquifer mining on a grand scale. This made deserts bloom and cities grow, lubricating a few economic booms. But it was a short-term remedy for water shortage, and often clearly unsustainable, as the experience of northern Texas shows. In some cases the practice could—with luck— serve as a bridge to an age of cheap desalinization of seawater. The aquifer age will prove a passing phase, except in happy situations where pumping rates do not outstrip recharge.

[8]Jayal 1985:96 (on Asia); Stanners and Bourdeau 1995:66 (on Europe).
[9]GEMS 1989:152; Smil 1993:42–4; Stanners and Bourdeau 1995:67; Walker 1990:289–90.

Dams and Diversions

Dams and diversions, another ancient technology, provoked even greater changes in the hydrosphere. The earliest dam of which any trace or record remains diverted Nile water near Memphis, in ancient Egypt, 4,900 years ago. The early Han dynasty (second century B.C.) in China built earthen dams as high as 30 meters. Sri Lanka and Mesopotamia were other early cradles of dam technology. Earth and rock dams had their limits, which checked the potential of dam building for millennia.[10] But after 1850, applied science—civil engineering, hydraulics, and fluid mechanics—opened the way for larger and larger dams, first in Europe and then at the turn of the century, in the United States. Some states, like Italy and colonial India, built networks of modest-sized dams. Others, like Egypt, built on the heroic scale. A few, like the United States, the USSR, and post-colonial India, did both. Engineering and political considerations affected the choices. All dam builders sought changes in landscape, hydrology, economy, and society—and got them, although not always the ones they wanted.

Most dam construction in the nineteenth century, as previously, aimed to extend irrigation. Lesser goals included flood control and impoundment for reservoirs. By the turn of the century, dams were also built to create electricity. Dams generally had only one main purpose until the 1930s, when the United States pioneered the development of river-basin management and multipurpose dams. The Tennessee Valley Authority (TVA), the first such project, inspired imitation along the Volga in the USSR, in India, and elsewhere. The Boulder Dam (later renamed the Hoover Dam) on the Colorado River, the world's largest when it was built in the 1930s, inspired admiration almost everywhere.

The giant dams served larger political purposes wherever they were built. Communists, democrats, colonialists, and anticolonialists all saw some appeal in big dams. Governments liked the image they suggested: an energetic, determined state capable of taming rivers for the social good. Dams helped to legitimate governments and popularize leaders, something the United States needed more than ever in the Depression

[10]Garbrecht 1987; Rouse 1963.

The Boulder Dam (later renamed the Hoover Dam) on the Colorado River, here under construction in 1934, was the first giant multipurpose dam in the world, and helped remake the American Southwest to fit the demands and dreams of Depression-era America. It provided electricity, irrigation water, and flood control, changing an unruly river into a placid reservoir. Big dams proved popular around the world from the 1930s until about 1980.

years, and something Stalin, Nehru, Nasser, Nkrumah, and others all sought. Dam projects received great publicity, especially from about 1930 to 1970. Ambitious, modernizing states, especially colonial and newly independent ones with legitimacy problems, showed great fondness for dam building. So did Cold War hegemons, keen to display the virtues of their social and political systems. Their political utility helps explain why so many uneconomic and ecologically dubious dams exist.[11]

During the 1960s, more than one large dam (15 m or higher) was completed per day on average. The historic climax came in 1968. Although the pace tailed off, dam construction continued, so that by the 1990s about two-thirds of the globe's streamflow passed over or through dams of one sort or another.[12]

THE INDUS AND INDIA. The first great twentieth-century water project involved one of the world's great rivers, the Indus, the greatest colonial power of the day, Britain, its most important colony, India. The Indus and its major tributary, the Sutlej, rise in the western Himalaya and flow through some arid lands, in what is now eastern Pakistan and northwestern India, to the Arabian Sea. This region is known as the Punjab, or "land of the five rivers." The Indus is about 3,000 kilometers long, with a flow twice that of the Nile. People have used its waters for irrigation for 4,500 years. Today the Indus basin in Pakistan and India is home to the world's largest irrigation scheme.[13]

The modern scheme began in 1885, when the government of British India undertook to rebuild and extend some of the Moghul (16th–18th-century) waterworks in the western Punjab.[14] They did so, and much more. Engineering skill and endless toil turned the steppe and desert of the Punjab into wheat fields, creating farming settlements based on a network of irrigation canals—"canal colonies." By 1947 this scheme extended over an area of about 14 million hectares (the size of Greece or Alabama) and permitted the largest agricultural expansion in the history of British India. Punjabis from far and wide migrated to these new colonies, replacing the scattered cattle and camel herders who had pre-

---

[11]The connection between successful water manipulation and political legitimacy of course runs back through the history of hydraulic societies to the pharaohs. The link was probably tightest in China, where Confucian political theory makes explicit the role of prudent and productive water management in justifying political power.

[12]NRC 1992:200; Petts 1990a, 1990b.

[13]According to Neuvy 1991:173.

[14]The following is based mainly on Ali 1988; see also Agnihotri 1996 and Gilmartin 1994.

viously roamed the land and, to British dismay, had paid almost no tax. Thus a social transformation matched the ecological one. The British regarded it as a signal success because it created one of the more prosperous peasantries in Asia, raising two crops a year on the Punjab's sandy loam. By 1915 the transformed Punjab provided more tax revenue for the Crown than any other district in India, and created loyal subjects too. Punjabis volunteered in droves for service in World War I because veterans could expect irrigated land in return. The British India Army enjoyed strong support in the Punjab.

The British scheme shaped life and land in the Punjab long after the eclipse of the Raj. Indus irrigation helped prevent the success of the Indian National Congress, or an equivalent, in what became Pakistan, as residents of the canal colonies remained loyal to Britain until the eve of independence in 1947. This political effect did independent Pakistan no good. Army leaders and landholders inherited a powerful position upon independence, thanks in large part to the canal colonies, and they dominated the country ever after, successfully resisting land reform and personalizing politics. Pakistan expanded the irrigation network after 1947, and promoted export crops such as oilseed and long-staple cotton. In 1990, Pakistan had about 16 million hectares under irrigation (equivalent to Tunisia or the U.S. state of Georgia) and more irrigated area per person than almost any country in the world.[15] The old elites overcame all challenges to their power, despite their continual internecine quarrels.

The Indus irrigation system also brought international and environmental repercussions. With profitable irrigation in the Punjab, both in Pakistan and India, water became valuable enough to add to the friction between the two countries. A treaty in 1960 helped damp down high feelings. But it could do nothing to prevent salinization, a common curse of irrigation schemes. Salt had sapped the yields of Punjabi wheat from at least the 1860s, even before massive irrigation aggravated matters. By the 1960s the problem had grown acute, as repeated floodings left soils waterlogged and raised the groundwater level, carrying dissolved salts up with it, into the roots of crops, checking plant growth. Fortunately for Punjabi farmers, Pakistan mattered to the United States as a Cold War ally, which helped convince the Kennedy Administration to dispatch a technical mission to address these problems. It recommended pumping to lower the water table and protect crop roots from salt. Foreign aid

[15]In 1993, Pakistan irrigated 80% of its croplands, the fourth highest percentage in the world, after Egypt, Suriname, and Uzbekistan (WRI 1996:241).

money duly built thousands of tubewells.[16] But rising groundwater and salinization continued to haunt 80 percent of the Punjab canal colonies into the 1990s, when foreign financing for drainage schemes dried up, leaving Punjabi farmers high and dry and their fields increasingly saline.[17]

Independent India also took part in the rising tide of irrigation in the twentieth century. In the Ganges basin, the British had begun major irrigation works in the 1820s, again rebuilding Moghul systems. By 1947, irrigation nourished 22 million hectares of India, 32 million hectares in 1974, and by the 1990s, 45 million to 50 million hectares, equivalent to California or Spain. India accounted for about one-fifth of the global total, neck-and-neck with China as the world's leader in irrigated cropland. India also developed its hydroelectric potential, especially after 1975. India's first prime minister, Jawaharlal Nehru, called dams "temples of modern India." According to one reckoning, dam building accounted for 15 percent of planned state expenditure between 1947 and 1982.[18] Dams for irrigation and hydroelectric power were "the most prominent ingredient of the development effort following independence."[19]

This dam-building program buoyed India's food production and its industry in the second half of the century.[20] But the social and environmental costs of modern India's temples ran high. Dams and reservoirs displaced perhaps 20 million people between 1947 and 1992. In one instance, the Rihand project in Uttar Pradesh in the 1960s, ousted peasants received no advance warning and had to flee for their lives while rising waters drowned their homes.[21] India's tribal populations, often located in hilly areas suitable for hydroelectric development, lacked the political power to resist dam building and often found themselves refugees.

None of India's irrigation or hydroelectric projects after 1947 came in within budget or on time, and few lived up to their billing in terms of electric power, irrigation, or durability. Reservoirs silted up on average

---

[16]Michel 1967:455–6.
[17]GEMS 1989:150; Hillel 1991:146–7. The Punjab irrigation scheme also brought more malaria, cholera, and other waterborne (or mosquito-borne) infections to the region (Agnihotri 1996:54).
[18]Thukral 1992:9; Centre for Science and Environment 1982:59 gives 14% for 1947–1979.
[19]Gadgil and Guha 1995:78.
[20]India's two major grains, wheat and rice, increased in production sevenfold and threefold, respectively, from 1950 to 1990 (Mitchell 1995:196). New crop strains, fertilizers, pesticides, etc., also played a role. Hydroelectric capacity increased 12-fold during 1954–1979.
[21]Thukral 1992:13–14.

two to four times faster than planners promised. Waterlogged lands and salinization also plagued India, forcing the withdrawal between 1955 and 1985 of 13 million hectares from cultivation, more than a quarter of India's (1996) irrigated area.[22] Many dam reservoirs promoted malaria. Some obliterated forests in the Himalayan foothills or the Western Ghats.[23]

With problems like these, Indian dam projects aroused political opposition. The first notable peasant resistance came in the early 1920s. More followed, some of which succeeded in preventing planned dams; most did not. But in the 1980s and 1990s, popular resistance to dam building stalled some major projects. In 1989, 60,000 people rallied against the Narmada irrigation scheme, a gigantic project of 30 major dams and over 3,000 smaller ones. The Narmada River flows into the sea near Surat, some 350 kilometers north of Bombay. Its banks host an unusual number of holy sites, many of which stood to be flooded out. Despite the World Bank's withdrawal from the Narmada scheme, India's commitment to irrigation, and to prestige projects, remained. The Narmada scheme would displace at least another 100,000 people.[24]

THE SOVIET UNION AND THE ARAL SEA. The irrigation scheme for the Indus may be the world's largest and those for India the most contested, but Central Asia's scheme is the most dramatic in its consequences. In its quest for cotton, the USSR created the single greatest irrigation disaster of the twentieth century. Like India's dam-building difficulties, the case of the Aral Sea represents cavalier water manipulation by arrogant political and scientific elites, justified in the name of the people. The Aral Sea's demise is the climactic chapter in a long and checkered history of Soviet water manipulation.

Imperial Russia was despotic but not hydraulic. Central Asia's ancient irrigation systems declined, or were purposely destroyed in the nineteenth century, when Russia conquered the region.[25] The Bolsheviks in

---

[22]Jayal 1985:97. Waterlogging cost India 6 million ha, salinization 7 million ha, 1955–1985. The Centre for Science and Environment 1982:6–7 presents different data, suggesting that while 13 million ha have been lost to salinization and waterlogging, only 6 million of this total were due to recent irrigation. But another 10 million ha were in imminent danger.

[23]The Centre for Science and Environment 1982:56–69 reviews Indian dams and their environmental problems; Whitcombe 1995 considers colonial-era irrigation.

[24]Baviskar 1995; Drèze et al. 1997; Gadgil and Guha 1995:66–78.

[25]Moser 1894 surveys Central Asian irrigation. Central Asia probably had more irrigated area in the fourth century B.C. than in 1917. See Klige et al. 1996 for a review of historical estimates.

1917 inherited a country with little irrigation and little water management other than the provision of piped water to some major cities. One of Lenin's early decrees (1918) encouraged irrigation in Turkmenistan, but Soviet policy scarcely affected the hydrosphere until the 1930s.

Stalin and his successors believed that Soviet engineering could customize the hydrosphere to meet the economic and political needs of a rapidly industrializing country, struggling to build communism before its enemies destroyed it. In this climate, half-measures had no place, projects had to be big, targets heroic, and deadlines ambitious. Naturally, corners had to be cut. From Stalin's point of view, the availability of millions of cost-free laborers after the early 1930s, the political prisoners of the Soviet gulags, made giant projects all the more tempting.

A new era dawned with hydroelectric installations, the first major one of which spanned the Volga in 1937. More dams and irrigation schemes sprouted along the Volga, which became a series of large ponds, and on the Dnieper, the Don, and the Dniester. By the 1950s, diversions reduced the streamflow of all the large rivers in the southwestern USSR. As the Don and Ural Rivers shrank, engineers channeled Volga water through canals to top them up. That short-changed the Caspian Sea, already shrinking because of consumptive use of Volga waters.[26] Irrigation using the Don and Kuban Rivers deprived the Sea of Azov of its freshwater flow, making it far saltier and ruining its once magnificent sturgeon, bream, and perch fisheries.[27] By 1975 the USSR used eight times as much water as in 1913, most of it for irrigation.[28] The maximization of the USSR's economic potential required water, so these rivers were bent to the will of the state and its planners.

Technological capacity, ideological zeal, political ambition and much else combined in the 1950s to convince Soviet officials and engineers to tackle the great rivers of Central Asia, the Syr Dar'ya and the Amu Dar'ya. These rivers carried snowmelt from high mountains into the closed basin of the Aral Sea, then the world's fourth largest lake. For millennia they had provided water for Central Asian societies. In the early twentieth century, these rivers gave up slightly more of their water, as irrigation

[26]The Caspian Sea level declined 3 m, during 1930–1978, but rose 1.5 m in 1978–1994 (Micklin 1995:279).

[27]Kostin 1986; Shiklomanov 1990:38; Vendrov and Avakyan 1977; Pollution from agrochemicals surely helped kill these fisheries too.

[28]Gerasimov and Gindin 1977:61; Gleick 1993:70. In the First Five-Year Plan of 1929–1933 Stalin imported fascist Italy's irrigation and wetlands drainage expertise (Cecchini 1987).

slowly expanded. But by the 1950s Soviet planners had something grander in mind, the creation of a vast irrigated cotton belt that would make the USSR "cotton independent." The demise of the Aral Sea was a planned assassination. The president of the Turkmen Academy of Sciences, A. Babayev, voiced the conventional view in the late 1950s:

> I belong to those scientists who consider that drying up of the Aral is far more advantageous than preserving it. First, in its zone, good fertile land will be obtained. . . . Cultivation of [cotton] alone will pay for the existing Aral Sea, with all its fisheries, shipping, and other industries. Second . . . the disappearance of the Sea will not affect the region's landscapes.[29]

Such views pervaded the corridors of power. Skeptics were ignored, or worse. Irrigation in Soviet Central Asia spread to about 7 million hectares by 1990, an area the size of Ireland, and the USSR not only became cotton independent, it became the world's second-largest exporter of "white gold." It was among the world's lowest-quality cotton, marketable mainly to customers with few options in the Soviet sphere in eastern Europe.

The investment in cotton strangled the Aral Sea. Prior to 1960 its influx had averaged about 55 cubic kilometers a year, a flow comparable to the Po (Italy) Niger (West Africa), or Snake (United States) Rivers; it dropped sharply in 1960–1961, and continued to shrink with each passing year. By 1980 the Aral got only a fifth of its former water influx, and by the 1990s at most a tenth and occasionally nothing at all. The Aral Sea level began to fall, slowly in the 1960s but faster after about 1973. By the mid-1990s the Aral Sea stood more than 15 meters below its pre-1960 level, and covered less than half its former seabed. In 1990 it became two seas, as a land-bridge emerged in the north. Their total volume stood at about a third that of 1960. The salinity of Aral water tripled between 1960 and 1993.[30]

The Russians once called the Aral "the Blue Sea." Aral Dengiz in the Turkic tongues of Central Asia means "Sea of Islands." More and more islands have appeared, but soon it seems the Aral will be neither blue nor insular. It will be a salt pan the size of Ireland, dotted by a few brackish ponds. This will surely be the greatest hydrological change yet engineered by humankind.

[29]Quoted in Precoda 1991:111.

[30]Gleick 1993:6; Smith 1995:267 gives 30–37 grams per liter for 1993 salinity. The Aral Sea has experienced natural fluctuations in its 15,000-year existence, and probably some caused by changes in irrigation drafts—but nothing on the scale of the post-1960 changes (Klige et al. a1996).

In the meantime, some of the broader consequences for Central Asia became apparent in the 1980s. Dr. Babayev notwithstanding, the entire Aral Sea region was affected. The ability of the Aral Sea to moderate local climate shrank with the sea. Summer heat and winter cold grew more extreme, and the cotton belt's growing season shrank by about two weeks. With less evaporation from the Aral Sea, air became drier and the snowpacks that feed the Syr Darya and Amu Darya began to contract. The winds sweeping over the seabed picked up less moisture and more salt, lifted from the saline crusts left behind by the retreating waters. The Aral Sea had once received salt from the watersheds of its rivers; by the late 1980s the tables had turned, and the seabed exported airborne salt to the cotton belt. Aerial salinization lowered crop yields, ruined pastures, corroded powerlines and concrete structures, and caused eye ailments. In Kazakhstan, pastures some 200 kilometers distant from the Aral Sea began to acquire a blanket of salt.[31]

The fisheries of the Aral Sea yielded about 40,000 tons annually in the 1950s. They disappeared by 1990. The Muynak Cannery was maintained until the early 1990s by airlifting frozen fish from the Baltic and sending it across the Trans-Siberian Railroad from the Pacific, surely one of the great diseconomies of modern times. Twenty of 24 endemic fish species went extinct. Tens of thousands of jobs did too. Muynak's human population declined from 40,000 to 12,000 by 1995. The water table beneath the rivers' deltas dropped by 5 to 10 meters between 1970 and 1990, and the groundwater grew saltier as well. This did in alluvial forests, wetlands, and pasture, which gave way to salt-resistant vegetation. The local Kzyk-Ordinsk Cellulose and Carton Combine had to import its raw material from Siberia. By 1990, nearly half the mammal species present in 1960 had vanished, as had three-quarters of the bird species.

Aside from the impacts of the desiccation of the Aral Sea, the cotton belt suffered from the usual ailments of giant irrigation schemes and monocultures. About half the diverted water evaporated or seeped into the earth, serving no human purpose whatsoever. In Turkmenistan, seepage from the Karakum Canal (a 1,100-km-long artificial river) threatened to swamp its capital city, Ashkhabad. After 1970, engineers drilled scores of wells to pump the rising groundwater out of the city. Saliniza-

---

[31]That blanket was 50 mm thick in 1998 according to S. Ospanov (personal communication, May 1998). By the late 1990s, winds picked up 400,000 tons of dust and salt from the former seabed every day, and scattered it from the Himalaya to Belarus. By one calculation, the salt-sand plumes increased the particulate matter in earth's atmosphere by 5% (Lubin 1995:297).

tion damaged fields, affecting half the cotton area in Uzbekistan and four-fifths in Turkmenistan. Cotton pests thrived, causing farmers to drench their crop with pesticides, which contaminated drinking water. The human health effects of irrigation in Soviet Central Asia became severe by the 1980s. Kazakhstan officials discouraged breast feeding of infants because of dangerous pesticide residues in mothers' milk.

The redoubtable ambition of this Central Asian plumbing operation pales beside plans that the USSR shelved in 1986. In previous decades, visionary scientists and bureaucrats had hoped to divert water from northern Russian lakes into the Volga to address the difficulties arising from the sinking Caspian Sea level. Grander still, they hoped to reverse the northward flow of the great Siberian rivers, such as the Ob and Yenisei, so as to provide more irrigation water for Central Asia. Works began on the former project in 1984, but Mikhail Gorbachev quailed at the costs, and spiked both plans in 1986. The breakup of the USSR in 1991 made such schemes all the more improbable, as Russia no longer has much stake in the cotton export earnings of Uzbekistan—to the regret of many Central Asians who still saw Siberian water as their salvation.[32]

EGYPT, ASWAN, AND THE NILE. Egypt, as Herodotus noted, is an "acquired country, the gift of the Nile." For the last 10,000 years, that gift—water and silt, mainly from Ethiopia—made a long ribbon of the Egyptian desert habitable and gradually built the Nile Delta over the continental shelf. In the twentieth century, Egyptians spurned half the Ethiopian gift in trying to improve upon the other half.

The prominence of Egypt in history derives from its unique geography. The Nile was a two-way highway because northeast winds in most seasons allowed upstream sailing while the river's flow wafted traffic downstream. More important still was the annual flood, prompted by monsoon rains in the Ethiopian highlands. The late summer flood brought moisture and on average about a millimeter of fertile silt to the river's banks, floodplain, and delta every year, permitting cultivation of winter crops such as barley and wheat. Improvements upon the gift began some 5,000 years ago, with irrigation canals, later supplemented by mechanical devices

---

[32]This account is drawn from Feshbach 1995:55–6, Micklin and Williams 1995, Peterson 1993:111–8, Postel 1996, and Precoda 1991, and from remarks made by participants in a Social Science Research Council workshop in Tashkent in May 1998. On the Ob diversion, see Weiner 1999:414–28.

(*shadoofs*[33] and waterwheels). The flood immunized Egypt against the scourge of irrigation regimes—salinization.

The gift came with strings attached, however. If Ethiopia's monsoon rains were light, then the Nile did not rise—producing a "low Nile"—crops failed, and famine resulted. If the rains were especially heavy, the Nile rose too much and swept away the settlements along its banks. The rulers of modern Egypt changed the country's agriculture and economy in ways that made these ancient strings intolerable. So they changed the Nile too.

The first modern irrigation efforts began with Muhammad Ali (c.1769–1849), an Albanian adventurer in Ottoman service. As pasha of Egypt after 1805, he aimed to free himself from Ottoman control, to strengthen the country lest Britain or France take it, and to enrich himself and his followers. To these ends he proposed irrigation barrages, or low earthen dams, in the lower (i.e., northern) Nile, using stones from the Giza pyramids. He was dissuaded from this scheme, but beginning in 1842 he built barrages made of less precious stone. Finished in 1861, after Muhammad Ali's death, they adjusted the lower Nile's flow in an effort to suit the demands of cotton cultivation.

Cotton was a summer crop, harvested in August or September. Formerly, when barley and wheat were the mainstays of Egyptian agriculture, a heavy late-summer flood, while dangerous to life and property, could not ruin the springtime harvest. Only a low Nile, restricting cultivation, could bring famine to Egypt. But in the nineteenth century, with cotton cultivation and the growing importance of maize as a summer food crop, this changed. A surging flood could ruin the cotton and maize crops before the late-summer harvest, at a stroke destroying a good part of the country's food supply and almost all of its exports—bad for the hungry peasants; for the cotton kings whose incomes vanished; and for the state, whose revenues dried up. With the new crops and new economy, big floods became far more costly than before. Moreover, ordinary floods, which covered low-lying parts of the delta for months each year, now became a real constraint on the country's potential cotton production.

Cotton was the moneymaker which, Muhammad Ali had hoped, would allow Egypt to import the wherewithal for rapid modernization. The cotton grew—Egypt's production multiplied sixfold from 1855 to 1882—

---

[33]A *shadoof* consists of a long pole, working as a lever, with a bucket on the end.

but the scheme failed. Egypt defaulted on debts in 1876, providing the excuse, if not the genuine motive, for foreign intervention.

Britain occupied Egypt beginning in 1882. When Lord Cromer, consul general of Egypt from 1883 to 1907, convinced London authorities that this occupation was not to be temporary after all, Britain began where Muhammad Ali's descendants left off. Britain would tame the Nile to protect Egypt. The British acquisition of the Sudan in 1898 followed in part from anxiety over Egypt's water supply. With help from Cromer's family bank, Baring Brothers, a low Aswan Dam was erected in 1902. It was heightened in 1912 and 1934. This helped store water for the dry months, but it could not store sufficient water to safeguard against prolonged drought. Nor could it contain a big flood. In practice after 1934 it captured only the tail end, about a fifth, of the annual flood.[34]

A much larger dam could do more. As early as 1876 a British officer had proposed such a dam to the *khedive* (as the successors of Muhammad Ali were known) of Egypt. The idea vanished from official thinking but was resurrected by Adrian Daninos, a Greek-Egyptian engineer who hoped it might permit the electrification of Egypt. Daninos proposed a high dam at Aswan in 1912, but sparked no interest in high circles. He tried again in the late 1940s, but British hydrologists by then had settled on schemes to impound Nile water in Uganda, Sudan, and Ethiopia. But Daninos soon got his way.

In 1952 Colonel Gamal Abdel Nasser (1918–1970) and his fellow Free Officers—nationalist revolutionaries in the army—seized power in Egypt, determined to rid the country of residual British influence and the shame of poverty and weakness. A month after the coup, Daninos approached two army engineers of his acquaintance, members of the Free Officers' technical advisory group, with his idea of a high dam. They offered the idea to Nasser, who was fast becoming the strongman of Egypt. He seized upon it quickly. Nasser saw in a high dam at Aswan a symbol that would contribute to the heroic, vigorous image he sought for himself, his revolutionary regime, and for Arab nationalism; he also saw in it a reliable water supply for Egypt and sufficient hydroelectric power to transform Egypt into an industrial state. Linking himself to the pharaohs, Nasser said, "In antiquity we built pyramids for the dead. Now we build pyramids

---

[34]Engineers had to let big floods pass through the sluice gates of the old Aswan Dam (and the half dozen other dams between Khartoum and the Mediterranean) because too much silt would quickly fill its small reservoir, destroying the usefulness of the dam. See Waterbury 1979:88.

for the living." The dam would help him bring true independence and "everlasting prosperity" to Egypt.[35]

From the hydrological point of view, a high dam was misplaced at Aswan in southern Egypt, one of the highest evaporation zones on earth. A reservoir there would increase the surface area from which precious Nile water might evaporate. The proper site for water storage was far upstream, at high elevations, where cooler air would permit far less evaporation. The British hydrologists' scheme involved dams in Ethiopia and Uganda that would store water in existing lakes. This plan made impeccable hydrological sense. It made political sense to the British while Britain controlled Uganda and Sudan, and remained influential in Ethiopia and Egypt. It was even endorsed as policy in 1949 by the Egyptian cabinet.

But after 1952, different—Egyptian nationalist—sensibilities prevailed. Dams in foreign countries did not suit Nasser. He did not trust Britain or the emerging states of Sudan, Ethiopia, and Uganda with the lifeblood of Egypt.

Nasser, it turned out, had good reason: Britain's prime minister Anthony Eden (1897–1977) would soon be calling for his head. For many reasons Anglo-Egyptian relations deteriorated after 1952. Nasser concluded a deal with Czechoslovakia for Soviet weaponry early in 1955, and some in London and Washington feared he would soon be Moscow's puppet. In an effort to dissuade Nasser from accepting Soviet support for an Aswan Dam, Britain and the United States agreed to finance it in 1955. But this did not make Nasser cooperative, and in the summer of 1956 the United States and Britain announced they would not, after all, lend money for the dam. Within a week Nasser seized the Suez Canal, until then run by a British company, and proclaimed that its revenues would finance the construction at Aswan. This move led to the Franco-British-Israeli attack on Egypt in October 1956, known as the Suez crisis.

The Americans refused to back this attack, and Britain, France, and Israel could not pay for it alone. Amid the gnashing of ministerial teeth, London and Paris had to curtail the invasion and retreat. Nasser triumphed, becoming a great hero in Egypt, the Arab world, and the colonized world generally. Britain and France suffered keen humiliation and had to accelerate—or create—plans for the decolonization of their em-

---

[35]First quotation from Postel 1999:54; second from Nasser speech of 14 May 1964, quoted in Waterbury 1979:98.

pires. Meanwhile Soviet engineers drew up plans for the high dam in the late 1950s. Construction began in 1960 and ended in 1971, a year after Nasser's death. The Soviet Union—and receipts from Suez Canal traffic—paid the bill. The political consequences of the difficult birth of the Aswan High Dam reached around the world.[36]

Aswan's environmental consequences were only regional in scope, stretching from Sudan to the central Mediterranean. The Aswan High Dam can store about 150 cubic kilometers of water in Lake Nasser, equivalent to two or three years' worth of Nile flow, and 30 times more than the 1934 dam held. It stopped 98 percent of the silt that formerly had coated the inhabited part of Egypt.[37] It revolutionized Egyptian agriculture, allowing more systematic use of water, permitting two or three crops per year. It provided full flood control, safeguarding the cotton crop against even the heaviest of floods. Production of rice, maize, and cotton—all summer crops—flourished. The Nile below Aswan became a mammoth irrigation ditch, completely tamed. The high dam's turbines generated about a third of Egypt's electricity between 1977 and 1990.[38] In these respects, the Aswan High Dam fulfilled Nasser's expectations, although it did not make Egypt prosperous and independent. The dam improved markedly upon half of the gift of the Nile.

But it revoked the other half. The Ethiopian soil subsidy ceased to arrive after 1963. Without a topdressing of fertile silt, Egyptian agriculture had to turn heavily to chemical fertilizers, of which Egypt became one of the world's top users. Much of Aswan's electric power went to fertilizer factories. Salinization also emerged as a serious threat. Without the annual flood's flushing, soils kept more salts. Salt accumulation afflicted the northernmost delta, where seawater intruded as much as 50 kilometers inland; the Nile Valley, where reliable and free water led to overuse, waterlogging, higher water tables, and consequent salt buildup; and wherever groundwater was tapped. The Nile descends only 87 meters from Aswan to the Mediterranean Sea, a distance of some 1,200 kilometers. With such a gentle slope, drainage of irrigated fields was a costly problem

[36]On the international politics of Aswan and Suez, see Hahn 1991, Collins 1990, and Louis and Owen 1989.

[37]The irrigation channels of the delta trapped the rest of the silt, plus all that scoured from the Nile banks between Aswan and Cairo (Stanley 1996).

[38]The percentage of electricity ranged from as little as 18% in some years to 53% in others, depending on the volume of available water and the electricity production from other sources (Said 1993:241).

never satisfactorily addressed. Soviet engineers did no better with this problem in Egypt than they did in Central Asia. In a country with so slender a resource base as Egypt, and with a million more mouths to feed each year in the 1990s, menaces to agriculture were urgent matters.[39]

Perhaps more ominously, the Nile Delta began to shrink. The delta is home to 30 million people and accounts for two-thirds of Egypt's agricultural area. The delta was "born" about 7,500 years ago and by the early twentieth century covered about 24,000 square kilometers, roughly the area of Albania or Maryland. During the nineteenth century, the delta expanded into the Mediterranean by about 5 to 8 kilometers at some points. But with the 1902 dam, its advance stalled and in places reversed. Mediterranean currents swept away the deposition of prior ages slightly faster than new silt could make good the loss. After 1964 the sea routed the delta because silt delivery virtually ceased. The coastline retreated in places as fast as 70 to 90 meters per year, nudging people inland and stranding lighthouses offshore. Sediment instead collected in Lake Nasser, forming a new inland Nile delta, which by 1996 was a tenth the size of the original but less conveniently placed.[40]

The dam did more than withhold useful silt. It deprived the Mediterranean of the nutrients the Nile carried, destroying sardine and shrimp fisheries that had employed 30,000 Egyptians. The fisheries of Egypt's coastal lagoons also withered, with fewer nutrients and more pollutants. Without the flushing of the flood, the irrigation canals of Egypt became ideal habitat for water hyacinth, a beautiful but pernicious weed. The snails that carry schistosomiasis—a debilitating disease that attacks the liver, urinary tract, or intestines—love water hyacinth, need stagnant water, and consequently flourished in the new Egypt. Schistosomiasis infection rates increased 5- to 10-fold among rural Egyptians with the transition to perennial irrigation, and after 1975 approached 100 percent in many communities.[41] Thus the loss of silt and other effects of the Aswan High Dam imposed serious—and escalating—environmental and health costs on the Egyptian population.

[39]Kishk 1986; Stanley 1997; White 1988.

[40]Fanos 1995; Stanley 1996. It is interesting to note that the Three Gorges Dam, under construction on the Yangtze River, will reduce sediment flow to the Yangtze delta, where the sea currents, tides, and typhoons cause much faster shore erosion than in the Mediterranean. Fifty million people live in the Yangtze delta.

[41]Farley 1991; Hunter et al. 1993:43–4; Watts and El Katsha 1997. White 1988 has data on schistosomiasis showing that its incidence in Egypt declined sharply from the 1930s to the 1950s; then all progress stopped.

The dam also swamped and corroded the cultural heritage of the Nile Valley. Part of that heritage now lies beneath Lake Nasser. Elsewhere along the Nile, constant irrigation and poor drainage raised water tables, bringing moisture to the bases of countless monuments. This moisture crept into ancient stone and eventually evaporated, leaving salts behind. These salts recrystallized, cracking stone surfaces and destroying the art etched or painted on them millennia ago. Just as 25 years of air pollution did more damage than 2,500 years of weathering to the treasures of the Athenian Acropolis, so with 25 years of salt creep and the cultural legacy of pharaonic Egypt.[42]

The dam eliminated the costly consequences of irregular Nile floods, which helped the Egyptian population to double since the dam was built. This doubling, however, made the overall supply of Nile water, however distributed throughout the year, inadequate for Egyptian needs. The desert air evaporates a sixth or more of the Nile's annual flow from Lake Nasser, as hydrologists expected. Eventually this loss cost Egypt. In the 1970s in a moment of generosity, President Anwar Sadat suggested that Israel might make use of some Nile water. In the 1990s, Egypt had none to spare and lived in fear of the moment when Sudan or Ethiopia expanded its use of Nile water—and of climate change that might reduce the Nile's total flow.

The Aswan High Dam postponed a day of reckoning for Egypt. Nasser often pointed to Egypt's growing population as justification enough for the dam. But by the end of the century, water was short again, Egypt dependent again, and additional environmental costs were visited upon Egyptians, present and future. Muhammad Ali, Lord Cromer, and Nasser traded the only large, ecologically sustainable irrigation system in world history—one which sustained the lives of millions for five millennia and made Egypt the richest land in the Mediterranean from the Pharaohs to the Industrial Revolution—for this postponement. Such are the pressures of politics.

The impacts of the dam did not stop at Egypt's borders. To build it Nasser required an agreement with Sudan on water sharing. After difficult negotiations and a military coup in Khartoum, he got it in 1959. The new

[42]On Aswan High Dam effects, see Burns 1990, Howell and Allan 1994, Hvidt 1995, Kishk 1986, Mageed 1994, Neuvy 1991:187, and Stanley and Warne 1993. A book that remains useful after two decades is Waterbury 1979. A recent defense of the high dam is Said 1993; a fine account is Ayeb 1996.

Khartoum government faced a violent reaction from some 50,000 Nubians, a minority population in northern Sudan whose towns and villages stood to sink beneath Lake Nubia, as the Sudanese portion of the reservoir is known. Sudan had to use its army to force them to relocate. Twenty years later they still wanted their riverside homes and date-palm groves back again.[43]

In what may be its most durable consequence, the Aswan High Dam altered the waters and biota of the Mediterranean. After 1964, little Nile water—about 10 percent of former quantities—reached the sea, and since the Mediterranean receives little river water to begin with, the subtraction of the Nile's flow made a substantial difference to the salinity of the eastern Mediterranean. In this saltier sea, new species thrived. Ever since the Suez Canal opened in 1869, fish could swim between the Mediterranean and the Red Seas. Few that did so survived until the Mediterranean got salty enough to suit Red Sea creatures. But after Aswan, a mass migration began of fish, molluscs, and other creatures. They colonized the eastern Mediterranean, most thoroughly in Levantine waters, but all the way west to Sicily. These Lessepsian migrants (named for the builder of the Suez Canal, Ferdinand de Lesseps) in some cases proved useful commercial fish, especially for Israeli trawlers. The Indo-Pacific and Mediterranean fish faunas, separate for geological ages, were united in an irrevocable bioinvasion that will continue to reshuffle the food web of the Mediterranean for some time to come. The calculations behind the Suez Canal and the Aswan High Dam were responses to fleeting political circumstances; the biotic changes they brought will endure for millions of years.[44]

THE REVOLUTION OF THE WATERS: ITALY'S PO VALLEY. Struggles for power—political and electrical—revamped water, economy, and society in Italy too. The spectacular geopolitical and economic success of Britain and Germany in the nineteenth century inspired reforms, revolutions, and reconfigurations of nature throughout Europe. In southern

---

[43]White 1978. In Egypt, Lake Nasser flooded out 46 Nubian villages, against the wishes of most Nubians. Relocated Nubians are losing their ethnic identity, to the dismay of elders but to the satisfaction of Egyptian nationalists for whom a homogeneous population is desirable (*Washington Post*, 8 January 1998:A23).

[44]Some 10–12% of the eastern Mediterranean fish are now Lessepsian migrants. Few species went the other way, presumably because the Red Sea is too salty for Mediterranean fish (Ben Tuvia 1983; Por 1978, Por 1990).

Europe, as in Egypt, political elites sought to unleash the latent economic energies of their societies, foment population growth, and enrich the state by any number of strategies. A favorite was industrialization. It required inanimate energy, efficient infrastructure, additional food surplus for urban workers—as well as vast social change.

Northern Italy after 1890 featured all this in concentrated form.[45] The Po River basin covers a sixth of Italy and is home to a third of its population. It is for the most part flat, and for millennia posed drainage problems for those who would farm its fertile soils. Roman colonists, medieval monks, and Renaissance princes all tried their hands at taming the meandering waters of the basin. Despite their efforts, the lands between the Apennines and the Alps contained, in 1890, plenty of marshes, few roads, little industry—and, in the summer months, lots of malaria. But with political will, investment capital, and suitable technology, all this could change.

In the upper Po valley, landlords and governments from the eighteenth century sought to capitalize on the agricultural potential of wetlands and seasonal pasture. With massive investment and heroic labor, Lombards replumbed their landscape, building drainage and irrigation canals, and making much of Lombardy suitable for rice.[46] These efforts climaxed in the Cavour Canal, finished in 1866. After 1882 the government of newly unified Italy—as of 1861—generously subsidized further drainage, irrigation, and channelization. The help came at a convenient time, as Italian farmers after 1870 felt the effects of cheap American wheat and Burmese rice. Po Valley landowners quickly took advantage of state aid. Between 1882 and 1914 they confined the waters to chosen times and places, drained the marshes dry, regularized the shapes of their fields, adopted farm machinery and chemical fertilizers, and doubled or tripled agricultural output, mainly of wheat, maize, rice but of specialty crops like alfalfa, hemp, and sugar beet too. In so doing they almost eliminated the Po marshes and their fishing-hunting-horticultural way of life.[47] Malaria began to recede as well.

This was but the beginning of the "revolution of the waters."[48] By the

[45]The following section draws heavily on Sievert 1996.

[46]Bianchi 1989:462–4.

[47]By 1933 the Po valley had over a million hectares under irrigation, the Veneto another 400,000 or so. Italy had more irrigated area than any country in Europe (Bevilacqua 1989; Clark 1984:127–30). Details on fertilizers, machinery, and yields in Po valley agriculture before 1939 are found in Corona and Massullo 1989:375–426, and Robertson 1938.

[48]The phrase is adapted from Bevilacqua 1989.

1890s Milanese visionaries saw "white coal"—hydroelectric power—in the Alpine torrents that fed the Po. With enthusiastic state support, Piedmont and Lombardy built dams and power stations enough to underwrite the rapid industrialization of Milan and Turin. Italy lacked coal: Italian factories paid eight times as much for coal as did English ones. If Italy were to compete with the energy-intensive economies of the twentieth century, only hydroelectric power could drive the necessary transformation of Italian ecology and society. As one captain of finance, Giuseppe Colombo, put it,

> the transmission of electricity over long distances represents a fact of such extraordinary significance for Italy that even the most powerful imagination would have difficulty foreseeing all the possibilities. It is something that could alter completely the face of the nation, that could one day carry the nation to the ranks of the best endowed countries in terms of natural resources and industry. . . . When countries that had previously grown rich on coal run out it will then be the turn of nations with rich sources of flowing water.[49]

Italy had the water cascading down from the Alps, and tried to take its turn. Italy's first hydroelectric plant dates to 1885, its first big one to 1898. By 1905, Italy led all Europe in hydroelectric power use. By 1924, Italy produced 1.8 million kilowatts; over the next 15 years it grew 1,000-fold. In 1937, hydropower furnished Italy with almost all its electricity. Most came from the Alps, where the glaciated valleys lent themselves to dams. Scores of dams and artificial lakes appeared after 1890, inundating forest and pasture alike. The Alpine lakes of Como, Maggiore, and Garda became reservoirs. Power transmission lines festooned Lombardy. Milan was the second city in the world to electrify street lighting. The textile industry, which in the nineteenth century had migrated up into the Alpine hills to be closer to supplies of wood and water power, descended to the Po valley once electrification arrived (bringing pollution from chemical dyes with it). In the years 1901 to 1927, three-quarters of Italy's industrial workers toiled in the electrified triangle between Milan, Turin, and Genoa.[50] Rural electrification allowed farmers to pump water uphill and gave a fillip to marsh drainage efforts after 1920.[51] The ecological transformation of northern Italy fed upon itself.

[49]Quoted in Sievert 1996:89–90. Colombo was also a sometime finance minister of Italy.
[50]De Rosa 1989.
[51]Bevilacqua 1989:298–301.

Italy's emergence as a European and imperial power after 1890 rested on this electrification. Northern Italy created metallurgical, railroad, shipbuilding, aircraft, and other strategic industries before, during, and especially after World War I. Its armed forces, as late as 1896 humbled by defeat in Ethiopia, became a semi-industrialized military by the 1930s. Mussolini, dictator from 1922 to 1943, aimed for economic independence and military rearmament at the same time. Italian factories churned out ships, vehicles, munitions, and weaponry enough to allow Mussolini to fight first in Ethiopia (1935–1936) and then intervene in the Spanish Civil War in 1936–1938. Italy built and supplied a navy that by 1936 seriously concerned Britain in the Mediterranean and could dream of reincarnating that sea as Rome's *mare nostrum*. Without alpine hydropower, harnessed in the ongoing environmental transformation of northern Italy, Mussolini's geopolitics would have been impossible—instead of merely impractical.[52]

The pace of environmental change, while socially disruptive, caught the fancy of some Italians, notably the militantly avant-garde intellectuals and artists known as Futurists. The sculptor-painter Umberto Boccioni (1882–1916) gave voice to this view:

> Infinitely sublime is the juggernaut of man, pushed along by research and creativity, by the paving of roads, the filling of lakes, the submerging of islands, the building of dams—by leveling, squaring, drilling, breaking, erecting. This is the divine restlessness that will blast us into the future.[53]

It certainly helped blast Italy into fascism. The social change that accompanied the environmental reconfiguration of northern Italy was quick and wrenching. Industrialization fostered a class-conscious proletariat in Milan and Turin. The creation of rice and maize fields from the Po marshlands similarly spawned a growing class of landless rural laborers: capital-intensive enterprises took the place of family-based fishing, hunting, and horticultural exploitation of the Po floodplain. Po valley landowners, large and small, often saw their interests in conflict with those of the urban and rural workers. After 1919 these landed farmers provided strong support for Mussolini's fascist movement. So did the new industrial barons of the north, whom hydroelectricity had helped to make. The reconfiguration of

---

[52]See Sadkovich 1996 on Italy's navy. Mussolini never achieved autarky: Italy continued to import coal in the 1930s and during World War II, mainly from Germany and Poland.

[53]Fontana 1981:138.

environment and society went hand in hand. After 1890 changes came with disorienting speed and contributed a current to the river of political embitterment that climaxed in fascism.[54]

THE UNITED STATES AND THE COLORADO. After 1900 a gigantic water manipulation scheme transformed the American West. Three main river systems drain the West: the Colorado, the San Joaquin–Sacramento, and the Snake–Columbia. All three underwent radical redesign after 1900. I will tell the tale only once, for the Colorado.[55]

The Colorado River drains a basin about the same size as that of the Indus, but its flow is much smaller. Rain is rare in its basin, and most of it evaporates before reaching the river channels. However, the Colorado used to flood spectacularly. In springtime it carried many times its average flow, and carried off great chunks of southwestern soils: its silt load was 17 times that of the Mississippi. It was a wild and unruly river, but after 1900 the Colorado provoked a vehement response.

In 1900, growers in California's Imperial Valley opened the Alamo Canal, which diverted Colorado River water from Mexico to their fields. In 1905 they built another canal, but a major flood destroyed part of the waterworks and inundated the valley, creating the Salton Sea.[56] Before long, President Theodore Roosevelt and the Southern Pacific Railroad joined forces with California growers to control the Colorado and safeguard agriculture in the valley. The first high dam, the Roosevelt, appeared in 1911 on a tributary, the Salt River. Many more dams and irrigation canals followed, most notably the Boulder (now Hoover) Dam in 1935 at the Arizona-Nevada border on the Colorado itself. By 1964, 19 big dams controlled the Colorado system.

The post-1964 Colorado was a different river. It had a much more even flow, with smaller seasonal variations. Vegetation along its banks changed accordingly, as survival there no longer required adaptation to a dramatic flood regime. The physical bed of the river changed a great deal too, as dams checked deliveries of silt and the scouring effect of big floods. Physical and biological changes seemed so great that after 1983 U.S. authorities intentionally allowed huge releases of dammed water to simulate

[54]See, e.g., Corner 1975 on Ferrara.

[55]What follows is drawn from Graf 1985, Schwarz et al. 1990, and Worster 1985. On the Snake–Columbia, see White 1995. On the San Joaquin, see NRC 1989.

[56]The Salton Sea still exists, fed by drainage water from the Imperial Valley. It was born a freshwater lake in 1905, but is now roughly as saline as the oceans.

former floods, in hopes of recreating conditions of the old regime. The river's chemistry changed too. It came to carry much less silt and much more salt. Salinity tripled between 1917 and 1961, so that the paltry flow that passed through Mexico was nearly useless brine. By the 1980s the Colorado's waters irrigated fields equal in area to Connecticut or Lebanon, including much of the Imperial Valley's fields of plenty. The waters generated a good chunk of the electricity used in the American Southwest. Roughly 15 million people depended directly on the new-regime Colorado.

Recasting the river suited some people but not others. Seven American states and Mexico share the Colorado. The United States and Mexico wrangled over water quantity and quality from early in the century. Mexican farmers used the dregs of the Colorado for irrigation on the fields of Baja California and Sonora, especially after 1950 when Mexico built the Morelos Dam on the lower Colorado. Salinity became the major issue, on account of return of irrigation water to the river. A 1973 agreement obliged the United States to reduce the Colorado's salinity through desalination works. In 1979 the United States rerouted some of the return irrigation water, achieving the same effect. But continued irrigation withdrawals and returns brought the problem back. The riparian American states locked horns over Colorado water from 1905, generating a spate of lawsuits unsurpassed by any river anywhere. Mark Twain allegedly said that whiskey is for drinking and water for fighting over: Colorado water, at least in this century, was for litigating over.

As in India, dam building attracted political opposition in the American Southwest. The opponents were rarely humble folk who stood to be displaced by reservoirs, as they often were in India. Rather, they objected fundamentally to the transformation of the river, particularly its most scenic stretches, which often were the steep ones best suited to hydroelectric development. Gathering force in the 1950s, opponents organized by conservationist outfits such as the Sierra Club and the Audubon Society prevented dams in the Grand Canyon and in Dinosaur National Park (Utah). They lost in the case of Glen Canyon Dam, the second largest on the Colorado, completed in 1965–1964. But no significant new projects followed.

The taming of the Colorado system reconfigured the American Southwest as thoroughly as the Indus scheme did the Punjab or the Soviet scheme did Central Asia. It made agriculture possible and air-conditioning practical via cheap electricity, and encouraged rapid development and population influx. The modern prominence of agriculture and industry in California and the Southwest, and the rise of the United States as a giant on the Pacific Rim, depended on this harnessing of the Colorado—and equivalent events on the Columbia.

The international character of the Nile, Indus basin, Syr Dar'ya, and Colorado points up another issue which late in the century came to concern states and people around the world: water security. With growing populations, growing economies, and insufficient water conservation, the demands for water use grew continually (see Table 5.1). In parts of the world where supply was short and shared among nations, conflict regularly emerged. Most of the big rivers of the world are international ones, and many provide the only, or almost the only, significant source of water supply in dry regions. Negotiations rarely led to mutually satisfactory arrangements—certainly not for Mexico—but they prevented outright war. Disputes over water from the Nile, Jordan, Tigris–Euphrates, Niger, Mekong, Brahmaputra, Zambezi, and la Plata—to name a few—will test the skill of negotiators in the future.[57]

The experience of the Indian subcontinent, Central Asia, the American Southwest, and the Mediterranean with large-scale water manipulation was a mixed one: it allowed great increases in food and electricity supply, and created a welter of serious environmental problems at the same time. Almost every country in the world practiced irrigation on some scale and experienced similarly mixed results. Dams and irrigation held great appeal for those who made decisions. The benefits were tangible and immediate, and a tantalizing share could easily be captured by the state, big landowners, and powerful industries. The costs could often be shunted onto the poor, the powerless, foreigners—or the future. For these reasons, the world's total irrigated area spread from 50 million to nearly 250 million hectares between 1900 and 1995 (see Table 6.1).

As the table shows, the fastest expansion took place between 1950 and 1980. China more than doubled its irrigated area between 1950 and 1976, one of the distinctions of Mao's rule.[58] By 1980 the world's best dam sites had been used, but the search for good ones continued. At the end of the century, Brazil, Quebec, Venezuela, and Nepal were building or planning

[57]Postel 1992. On the Jordan, see Lowi 1993; the Nile, Waterbury 1979; the Niger, Grove 1985.
[58]Smil 1993:45.

TABLE 6.1 GLOBAL IRRIGATED AREA, 1900–1990	
Date	Global Total (million ha)
1800	8
. . .	. . .
1900	48
. . .	. . .
1950	94
1960	137
1970	168
1980	211
1990	235
1995	255

Sources: Gleick 1993:265; Postel 1999:41.

further gigantic irrigation or hydroelectric schemes, and China was building the mother of all dams.

China in the 1990s revived a long-held ambition to dam the Yangtze, Asia's greatest river. The Three Gorges Dam, suggested as early as 1919 by the nationalist Sun Yat-sen (1866–1925), existed only on paper for 70 years. Then, in the wake of the uprising that climaxed at Tiananmen Square in the spring of 1989, venerable leaders such as Li Peng decided to show China and the world that the Communist Party remained firmly in charge. Reviving the Three Gorges project—and imprisoning its critics—suited this agenda because, with a few exceptions, foreign governments, banks, and environmental groups did not support it. A heroic-scale project, firmly in the tradition of Chinese Marxism, would help burnish the image of the Party.

The resuscitation of the Three Gorges Dam proved controversial and divisive among the Chinese elite, but if finished according to plan it will be the largest water project in world history. It will create a lake as long as Lake Michigan and displace 1 million to 2 million people. It will radically alter conditions for fish, water birds, and the river dolphin that inhabit the Yangtze waters. It will trap the sediment of one of the world's most silt-laden rivers, depriving the Yangtze delta—China's richest soil—of its customary replenishment. And it will drown a slice of China's cultural patrimony and some of its most admired scenery, inspiration to centuries of poets and artists. In return for these costs and risks, China hopes to improve navigation on the Yangtze, to control the floods of an often dangerous river, and to add about 10 percent to its hydroelectric capacity.[59] The Three Gorges Dam will change the Yangtze and the East China Sea as much as the Aswan Dam altered the Nile and the Mediterranean Sea.

Water manipulation served the political interests of the powerful, but it also eased the lot of millions. For those not flooded out by dams, not beset by waterborne diseases, and not affected by salinization, dams and diversions often proved a great help. Irrigated fields accounted for 16 percent of the world's total cultivated area in 1990, and for some 30 percent of total food production. Hydroelectricity in 1995 supplied about 7 percent of the world's total commercial energy and 20 percent of its electricity. For these reasons, large-scale water manipulation in the twentieth century helped ease the lot of humankind.

Against this slate of successes in food and energy production stands a dismal environmental record. More than half the century's diverted water was wasted: it either evaporated or infiltrated before reaching a crop or a turbine. Dam-building techniques developed in the United States or the USSR were transferred unreflectively to zones of high evaporation, maximizing water loss. That same technique was also exported to lands highly susceptible to erosion, such as Algeria and China, where siltation forced the early abandonment of reservoirs, in one case in China even before the dam was finished. By 1980 salinization corroded agriculture on about a quarter of the irrigated land in India, Pakistan, the United States, and Egypt.[60] In the 1990s, salinization seriously affected about 10 percent of the world's irrigated lands. By 1996 it ruined land as fast as engineers could irrigate new land, so that the world's total irrigation area remained roughly constant.[61] Waterlogging and nutrient leaching aggravated problems. Throughout the century, irrigation almost always amounted to a short-term maximization strategy: to save money and effort on drainage and conservation measures, farmers and engineers mortgaged the future.

[59]Gleick 1999:84–91.
[60]GEMS 1989:150.
[61]Meyer 1996:77.

At N'Debougou along the Niger River in Mali, French colonial authorities hoped to create irrigated rice fields. Six men working this American-made ditcher could dig a mile-long irrigation channel in one day. The power of such machines brought the promise and perils of irrigation to dry lands around the world after 1950. In Africa the last decades of colonialism saw many efforts to refashion landscapes to suit the needs of economic development programs. This photograph is from about 1950.

This grand global plumbing project not only jeopardized future farming; it also destroyed the livelihoods, and sometimes the lives, of masses of people in the twentieth century. Dams displaced millions without compensation, perhaps 40 million over the course of the century, three-quarters of them in India and China. Reservoirs and canals helped spread diseases whose pathogens or whose insect vectors breed in water. The list includes malaria, schistosomiasis, cholera, typhoid, and many other notable killers: inadvertently, water manipulation certainly killed millions in the twentieth century.[62] Recasting the world's rivers ranks among the signal environmental changes of the twentieth century.

[62]See Neuvy 1991:187–9; Obeng 1977. Some disease data appear in Gleick 1993:170–224. Displacement figures are from Gleick 1999:78 and Gutman 1994. Adams 1992:22–3 describes dams and irrigation in Africa as "remarkably unsuccessful" and large dams as "extremely unsuccessful."

Taming Floods and Draining Wetlands

Diversions and dams account for the most important physical changes in the world's cycling of fresh water. The major point of such actions was to get more water to the right places at the right times. But people have also sought to get water out of places where it interfered with agriculture or other useful activity. In the twentieth century this happened most notably in river floodplains and in wetlands.

The point of river channelization is to tame floods, ease navigation, and farm fertile bottomlands. It is an expensive business because it involves building a channel for wayward rivers, confining waters to that channel, and keeping them out of floodplains. Only prosperous societies have undertaken it on any scale. Imperial China tried hard to constrict its rivers, but in modern times Europe and North America led the way. The Rhine's route was shortened and straightened beginning soon after 1800. In Illinois, where floodplains were managed more by beavers than people until after 1800, a quarter of the state's river length was channelized by 1990, including half the Illinois River. About 6 to 7 percent of U.S. rivers flow between man-made banks today.[63] The greatest efforts went into straitjacketing the Mississippi.

The Mississippi is the third-longest, the sixth-muddiest, and by volume the eighth-largest river in the world. Its basin covers 41 percent of the lower 48 U.S. states. When it flooded, people noticed. The first levees appeared on the lower Mississippi in the eighteenth century, and others popped up throughout the nineteenth century in communities that had enough money and ambition to try to deflect floodwaters onto their neighbors. The federal government sponsored levees after 1895. But a serious concerted challenge to Big Muddy's flood regime came only after 1927. That spring and summer saw one of the biggest floods in the river's known history. From the confluence of the Ohio River and the Mississippi (at Cairo, Illinois), on south, the water overflowed banks and levees in 170 counties, killed several hundred people, and at points formed a shallow lake 160 kilometers wide. New Orleans was saved only by dynamiting levees upstream to let the water spread out over rural Louisiana and Mis-

---

[63]NRC 1992:169, 194. Denmark has channelized about 90% of its river length, and Holland is not far behind (Stanners and Bourdeau 1995:81–2).

sissippi. Herbert Hoover, then Secretary of Commerce, successfully organized a huge relief effort, which helped earn him enough support to win the Presidency in 1928.

After the flood, the U.S. Army Corps of Engineers tackled the Mississippi systematically. Flood Control Acts in 1928 and 1936 authorized levees on almost the entire lower half of the river, connecting local ones into one great system of levees, dams, and reservoirs that would in theory confine the river to a single channel. The Corps straightened the river to ease navigation, shortening it by 229 kilometers between 1932 and 1955. The Corps's defenses worked fairly well against floods in 1951, 1965, 1969, and 1973. By 1990 the Mississippi had 26 dams (the Missouri had 60) and several thousand kilometers of dikes and levees. But levees have their limits, which a 1993 flood overtopped, drowning half a million hectares in nine states and costing $12 billion in damage. Channelization meant that fewer floods inflicted serious damage to people and property, but because it invited settlement and investment in floodplains, it also meant that the biggest floods caused more damage than they otherwise might. The 1993 flood prompted major reconsideration of channelization in the 1990s. Nonetheless so many people and so much investment is fixed in the floodplains of the Mississippi and its tributaries that it is hard to imagine a major departure from the pattern of the recent past.[64]

Channelization affected entire alluvial ecosystems. It cut the main branch of the Mississippi off from its previous banks, oxbows, and floodplains, presenting aquatic wildlife with novel conditions. Many species lost their spawning grounds, and the river's fish catch dropped dramatically. The freshwater mussels, which used to filter river water and reduce the consequences of pollution, declined too. Several mussel species went extinct. The southernmost levees channeled silt out into the Gulf of Mexico, where it fell (and falls) over the continental shelf, starving the Mississippi Delta of silt. The delta and the bayou country began to sink and shrink.[65]

[64]Daniel 1977; Galat and Frazier 1995; Outwater 1996:128–30; Phillips 1994; Turner and Rabalais 1991.

[65]NRC 1992:177. In Europe, animal biomass declined in channelized rivers by 65–75% (Stanners and Bourdeau 1995:82). On the whole, channelization cost too much for poorer countries. But colonial regimes attempted it here and there, and postcolonial ones did too when they could attract funding. See Adams 1992:128–54 on Africa.

In the late spring of 1927, the Mississippi overflowed its banks and the levees designed to constrain it, flooding 170 counties in the southern United States. After the flood, the U.S. Army Corps of Engineers redoubled its efforts to shackle the river to its bed. Here enterprising barbers in Melville, Louisiana, take the tools of their trade around to their stranded customers.

The Mississippi flood of 1927 drove half a million people from their homes and drowned several hundred. Here a farm family takes refuge in their attic, while barn-yard fowl roost on the roofline. A family of pigs is marooned atop a levee. This picture was taken May 18, 1927, but the location is unknown.

WETLANDS DRAINAGE. One of the reasons Mississippi floods were hard to manage was the draining of wetlands, which buffered floods, in the American heartland. Draining wetlands is probably almost as old as agriculture. Ancient civilizations engaged in it, and medieval Europeans became very skilled at it. Until the 1960s, almost no one thought a wetland more useful than drained land. Hence a spectacular assault took place on the world's wetlands, wherever sufficient money, labor, and technology were assembled.

The fringes of the North Sea were one such place. Diking and drainage began in ancient and medieval times, and expanded especially in Holland in the sixteenth to seventeenth centuries. The famous Dutch windmills, built to pump water from former peat bogs, date from this era. But technology and skills of the twentieth century allowed more ambitious work than ever before, culminating after a disastrous storm flood in 1953 in the closing off of several estuaries of the Rhine, followed by extensive reclamation. Half the Dutch population now lives below sea level, thanks to reclamation work. The English Fens, where ambitious drainage began in 1630, also underwent a transformation in the twentieth century. Diesel and electric pumps replaced steam engines and windmills, and a lot more of The Fens dried out. In Holland and England together, an area equal in size to Luxembourg was reclaimed, most of it after 1900, half of it from the North Sea, half from inland bogs and wetlands. It ranks among the best farmland in Europe. Irrigation is now practiced in The Fens.[66]

In the nineteenth century, one of the bigger wetlands areas in the world dotted North America, from Manitoba and the Dakotas to Ontario and Ohio. Most of this expanse was seasonally waterlogged, often malarial, and impossible to farm. The novelist Charles Dickens visited southern Illinois in 1842 and found

> [a] dismal swamp, on which the half-built houses rot away: cleared here and there for the space of a few yards; and teeming, then, with rank unwholesome vegetation, in whose baleful shade the wretched wanderers who are tempted hither, droop, die, and lay their bones; . . . a hotbed of disease, an ugly sepulchre, a grave uncheered by any gleam of promise: a place without one single quality, in earth or air or water, to recommend it. . . .[67]

[66]De Jong 1987:83–5; Lambert 1971; Williams 1990a.
[67]Dickens, *American Notes,* quoted in Prince 1997:121.

Locals set to work trying to drain the swamps and raise more wholesome vegetation. Digging and dredging drainage ditches was slow work, sometimes involving ditch plows that required up to 68 oxen.[68] After 1870, prairie farmers resorted to tile drains, ceramic tubes that carried water underground to the nearest stream. By 1880 over a thousand tile factories in Illinois, Indiana, and Ohio were converting clay prairie soils into drain tile, and farmers were converting wet prairie and year-round swamps into rich grainland. Reclamation was especially rapid in the years 1900 to 1920 and 1940 to 1970. By 1970, American farmers had drained about 17 million hectares (roughly equivalent to the state of Georgia), providing the United States with some of its best farmland, creating the cornbelt and obliterating wildlife.[69]

Elsewhere in the United States, farmers and the Army Corps of Engineers were nearly as active. In the South, especially in the Mississippi River bottomlands in Arkansas, Mississippi, and Louisiana, an area equivalent to Belgium was drained after 1930. Part of this took place in the context of systematizing the levees on the river (see above). But after 1960, high prices for soybeans and rice, which do well in the frost-free bottomlands, prompted further drainage. The great Central Valley of California was converted from wetland to farmland and pasture after 1870. The Florida Everglades, in 1880 about 1.6 million hectares (about the size of New Jersey), declined by half by 1970. According to a recent estimate, wetlands in 1780 covered about 100 million hectares, or 15 percent of the lower 48 states. By the 1980s, 53 million hectares remained.[70] Most of this drainage occurred in the twentieth century, almost all since 1865. As usual, agriculture's gain was wildlife's loss: about a third of the endangered species in the United States make their homes in wetlands.[71]

[68]Outwater 1996:126.

[69]Prince 1997; Williams 1990a. A fair bit of American (and Canadian) wetland destruction took place by accident: as a result of the fur trade. Once there were perhaps 200 million beaver in the USA, and therefore about 20 million beaver dams. Trapping in the 17th–19th centuries drastically reduced beaver numbers and wetlands. Today there are 7 million to 12 million beaver. Outwater estimates there were once 300,000 square miles (770,000 km²) of beaver-maintained wetlands in the USA, an area the size of Texas (Outwater 1996:17–21, 31–2). See also Naiman et al. 1986.

[70]Data from T. E. Dahl, *Wetland Losses in the United States: 1780s–1980s,* Report to Congress, U.S. Department of Interior, cited in Gleick 1993:295–6. Alaska adds another 69 million ha, almost none of which have been drained. See also Vileisis 1997.

[71]ReVelle and ReVelle 1992:55.

Another area where frontier agriculture and modernizing states took on wetlands was the mangrove coasts of South and Southeast Asia.[72] From the nineteenth century, ambitious colonial regimes in India, Burma, and Indochina sought to settle peasant farmers in the expansive deltas of the Ganges–Brahmaputra, the Irrawaddy, the Mekong, and several other rivers. After independence, the successor regimes eagerly continued. Thailand encouraged similar development of the Chao Phraya River. The point was to grow more rice, either for the world market or for domestic self-sufficiency. In the Sundarbans—the Ganges–Brahmaputra coastal delta—and the lower Irrawaddy, much the same scale of transformation occurred between 1880 and 1980: in each place about 800,000 hectares of wetlands went into other uses, mostly wet rice cultivation. Population quintupled in both deltas. And tremendous, unknown changes to the deltas' biotas resulted. Indonesia after 1970 attempted much the same program on a far larger scale in the coastal mangroves of Kalimantan and Sumatra. On one calculation, wetlands loss in six South and Southeast Asian countries came to about half of 13 million hectares between 1900 and 1980. The Philippines eliminated two-thirds of its coastal mangroves between 1920 and 1980.[73] After 1980, vigorous global demand for shrimp prompted the replacement of mangrove wetlands by intensive—and highly polluting—shrimp farming in the Philippines, Vietnam, Thailand, and elsewhere in Southeast Asia. All this meant a good deal more rice and shrimp and a lot less mangrove, a change no less welcome to most people than the transition from wetlands to farmlands in the North American heartland. The countless creatures that lost their habitats in these transformations, and the people who once made their livings there, might of course see matters differently.

In the world as a whole over the twentieth century, people drained about 15 percent of perhaps 10 million square kilometers of wetlands, the area of Canada. The United States drained half its wetlands, Europe 60 to 90 percent, New Zealand more than 90 percent, the great majority in the twentieth century.[74] Large-scale drainage continued in Africa, where in the 1970s the Sudanese government undertook a scheme, delayed by civil war after 1983, to build the Jonglei Canal through the huge Sudd swamp, depriving it of some its water so as to reduce evaporation loss and

[72]Richards 1990a; Richards and Flint 1990.

[73]J. F. Richards's data in Gleick 1993:293. Zamora data on the Philippines cited in Tolba and El-Kholy 1992:170. On Indonesia, see Ruddle 1987.

[74]Meyer 1996:72; Tolba and El-Kholy 1992:169–73.

improve the water supply to the Nile downstream. The plan aimed to improve the lot of the Arab population, who supported the government, while visiting unknown ecological effects upon the rebellious Dinka and Nuer peoples of the south. In 1998, half the world's remaining wetlands lay in Siberia, Alaska, and northern Canada. They did not reward drainage efforts. More than a quarter lay in South America, including the world's single largest wetlands, the Pantanal of western Brazil—a candidate for drainage in the 1990s. This great frenzy of wetlands drainage in the twentieth century stands with river-taming among the major environmental changes of our time. Like several of the others, it meant more room for people, for crops, and for livestock, and less room for creatures less useful to humankind.

Coastlines

Perhaps the ultimate indication of our urge to push the waters where we want them is the history of customizing coastlines. This requires keeping the ocean at bay, an act that requires confidence as well as expertise. The twentieth century left an impressive record of coastal change—in places. Coasts are among the world's naturally most dynamic environments. Sea level changes, land subsides or emerges, mountains send down their silt, while tides and storms shift it about. Much of the twentieth-century effort to alter coastlines came as simple reaction to unwelcome natural trends or events. The human impact, while strong in the North Sea and Japan, was modest in Brazil or Mozambique. On those long coasts, nature's own lively rhythms held sway.

The most ambitious recasters of coasts were the Dutch. In 1916 a storm surge spilled some of the North Sea onto the low-lying land surrounding the bay called the Zuider Zee. This provoked the Dutch parliament to act on a plan developed in 1890 to seal off and drain the bay. Construction of the enclosing dam began in the 1920s and finished in 1932. Work continued over the next 60 years, adding 13 to 14 percent more land to Holland, creating a new freshwater lake, and shortening the coastline by 300 kilometers.[75]

---

[75]De Jong 1987:83–4; Tolba and El-Kholy 1992:112; van Lier 1991.

Similar but smaller efforts around the world made land out of sea in dozens of countries, especially those, like Holland, where people were many, land short, and money abundant. Japan, Hong Kong, Singapore, Bahrain, and Saudi Arabia showed particular interest in coastal modification after 1970. Japan's reclamation was on such a scale that in the 1980s, 40 percent of Japanese industry stood on man-made land. Tokyo Bay, where reclamation began in the 1870s, shrank by a fifth between 1960 and 1980 to accommodate the Japanese economic miracle. Singapore expanded its national territory by 10 percent through reclamation. The worldwide total is hard to estimate, but a good guess is that reclamation added to the world's land area by 100,000 to 500,000 square kilometers in the twentieth century, equivalent to something between Iceland and Spain.[76]

Conclusion

In the twentieth-century, humankind has altered the hydrosphere as never before. We used and diverted water on a scale no previous age could contemplate. By one calculation, at the century's end we directly consumed 18 percent of the total available freshwater runoff of the globe, and appropriated 54 percent of it in one manner or another.[77] In places, successful water diversions eased constraints on economic development and human well-being. Irrigation made a big difference in the human condition: without it we would either have eaten less, eaten differently, or farmed a third again as much of the earth. In dry lands it created constraints almost as fast as it eased them, and may in cases now seem a mistaken course. Groundwater use often became groundwater mining, permitting local and regional population and economic expansions while the water lasted. The straitjacketing of rivers with dams and channels recast habitats in order to make rivers more convenient for human use. Draining wetlands obliterated habitats in order to make land for human use. The physical changes to the hydrological cycle were vast in their consequences, for wildlife, for people and societies, and insofar as we

[76]Walker 1984.
[77]Postel et al. 1996. The definitions of "use" and "appropriation" are lengthy and involve many assumptions; one must take the figures very cautiously.

have constrained the future in order to liberate ourselves from the past, for posterity as well.

In the distant past, only societies that could concentrate vast armies of laborers could make big changes to the hydrosphere. In the twentieth century, societies with powerful technologies and sufficient wealth could do it. Rich countries customized their shares of the hydrosphere much more thoroughly than did poor ones. Colonial powers often could concentrate armies of laborers to execute large-scale public works, as in Punjab. After the waning of colonialism, ambitious rulers often seized upon water projects as useful for their domestic and international ends, as in modern India or Egypt, using energy-intensive technology when available, and armies of laborers when needed. The height of dam building and wetlands drainage occurred in the decades of the Cold War, when the United States and the Soviet Union saw themselves locked in an economic and public relations struggle for which water projects, both within their borders and without, seemed most useful. As was so often the case in twentieth-century environmental history, political agendas helped propel the replumbing of the hydrosphere.

7

The Biosphere: Eat and Be Eaten

It is far from easy to determine whether she [Nature] has proved
to man a kind parent or a merciless stepmother.
—Pliny the Elder, *Natural History*

A now obscure Austrian geologist, Eduard Suess, coined the term "biosphere." Two titans of twentieth-century science, the Russian geochemist
Vladimir Ivanovich Vernadsky (1863–1945) and the French Jesuit pale-
ontologist Pierre Teilhard de Chardin (1881–1955), popularized the con-
cept. Here I use the term as they did, to mean the space inhabited by
living things, from the ocean depths to the mountaintops. It includes
bubbling seafloor vents teeming with bacteria, Himalayan glaciers home
to the occasional beetle, and everything in between. It thus intersects
parts of the hydrosphere, lithosphere, and atmosphere. It is the sum of
all habitats, home to all the biota.

That biota evolved without reference to humankind for about 3.5 billion years. Humanity, at least in our primitive form, arrived 4 million
years ago. For most of the last 4 million years, most of the biota contin-

ued to evolve without our influence, because we stayed in small sections of the biosphere, mostly in Africa. We were few in number and technologically little superior to other primates, merely one species among many. Then, perhaps half a million years ago, we domesticated fire, allowing us far greater influence over plants and animals. This helped us colonize much of Africa and Eurasia. About 30,000 to 40,000 years ago we developed new tools, perhaps better communication (the date of the origins of human language remains murky), and more formidable social organization. With this departure we became a rogue primate, genuinely dangerous to many other forms of life, and disproportionately influential in coevolution. Our capacity to alter biotas grew further with the domestication of plants and animals some 10,000 years ago. That permitted quicker expansion of our numbers, greater division of labor, and faster technological change—which in turn led to further domestication, in a positive feedback loop that accounts for the general direction and character of subsequent human history.

By the twentieth century, our numbers, our high-energy technologies, and our refined division of labor with its exchange economy made us capable of total transformation of any and all ecosystems. Some remained little affected, such as seafloor vents. But in most of the biosphere, coevolution gave way to a process of "unnatural" selection whereby chances for survival and reproduction were apportioned largely according to compatibility with human action.[1] In this new regime those creatures symbiotic with us prospered greatly. These included those that suited our needs and adapted to domestication (cattle, rice, and eucalypts), and those that found suitable niches in our changing, churning biosphere (rats, crabgrass, and the tuberculosis bacillus). Creatures we found useful but incapable of domestication (bison and blue whales) and those that could not adjust to a human-dominated biosphere (gorillas and the smallpox virus) faced extinction or at best survived on sufferance. In the twentieth century we became what most cultures long imagined us to be: lords of the biosphere.

The theme of this chapter and the next is the progress of this selection for symbiosis with humanity. Their refrains are the massive scale of human impact on the biosphere in the twentieth century; the prominence of science, technology, and transport in this impact; and its often

---

[1] I use "unnatural" in the traditional sense. Some people consider that, since humanity is part of nature, everything humans do is therefore natural.

inadvertent character. What Marx said of human history became true more broadly: men make their own biosphere, but they do not make it just as they please.

Microbiota: The First Lords of the Biosphere

For 2 billion years microbes reigned as lords of the biosphere. They helped shape climate, geology, and all life. Since they outnumber all other life forms handily, even insects, one might say life on earth was democratic. It is less so now, with the success of *Homo sapiens.* We engineered a biological coup d'état, culminating since 1880 with an attempted extermination of some of the previous masters. But, as with most coups, the aftermath was full of surprises and the new regime unstable. The twentieth century was a tumultuous one in the balance between people and microbes, partly because of conscious human efforts to attack disease and pests, and partly as a side effect of large-scale social and ecological changes.

Microbial life comes in a bewildering variety of forms. Some of it, from the human point of view, is most useful: you couldn't digest a meal without the help of millions of bacteria in your alimentary canal. A small portion of it has long interfered with human designs and desires by causing disease, among us or among the plants and animals that nourish and work for us. It is these microbes that I will focus on here.[2]

People have struggled to prevent and cure diseases for millennia, but before 1880 these efforts only occasionally affected the balances between society and microbes.[3] After the 1880s, with the work of Louis Pasteur (1822–1895), Robert Koch (1843–1910), and many others who deciphered the role of microbes and insects in disease transmission, doctors and public health officials knew enough to effect basic shifts. The more people learned about microbes, the easier it became to kill them.

---

[2]I do so because these microbes are of most immediate concern to human affairs. They are not, perhaps, the most important to us; the latter group probably consists of those microbes that allow proteins to be synthesized from nitrogen, without which we could not have genes or bodies.

[3]England is among the rare examples of successful disease reduction. Smallpox, tuberculosis, and a few other infectious diseases retreated well before the arrival of bacteriology and useful drugs (Mercer 1990).

Armed with emerging knowledge of microbiology and inspired by ideas of progress, medical officials attacked disease on many fronts. This began in western Europe and the United States where scientific knowledge in the nineteenth century was expanding quickly, and the gospel of progress, which held that disease control was not merely possible but an obligation, held sway. Two trends posed sharp challenges. Rapid industrialization and urbanization created huddled masses of often malnourished people in European and American cities, among whom disease spread easily. In addition, especially after the burst of imperialism that began around 1880, military campaigns and occupations in Asia and Africa put European soldiers within range of malaria, yellow fever, and other formidable infections. These challenges made the war on disease seem especially urgent and helped recruit funds and manpower. There were three main responses: environmental manipulation, meaning here sanitation, insect, tick, and rodent control, and other measures to make human environments less propitious for pathogens; discovery and manufacture of antibiotics; and vaccination and immunology. With these developments, humanity between 1880 and 1960 stole a march on pathogens, sharply reducing the human disease burden, encouraging rapid population growth, and fundamentally changing the human and microbial condition.[4]

ENVIRONMENTAL DISEASE CONTROL AFTER 1830. Public health measures designed to adjust human ecology and behavior in ways that would inhibit the spread of disease became widespread and systematic after 1850, first in northwestern Europe. The pioneers were doctors and social reformers; the sponsors were governments, armies, and philanthropic organizations. The main achievements were cleaner water (as discussed in Chapter 5), better housing, and insect control.

The cholera epidemics that terrorized European and North American cities after 1832 prompted the organization of public health authorities. Cholera had long existed in India, but improved transport and greater human migration, especially of soldiers, allowed it to break out in the early nineteenth century, killing millions around the world. In Europe, cholera and other scourges provoked official reaction. Slowly authorities began to improve urban sanitation and check the ravages of infectious diseases, especially cholera, typhus, and tuberculosis. Simultaneously, Muslim authorities, notably Muhammad Ali (the dam builder on the Nile

[4]Hays 1998, a source I discovered only after I had written these pages, covers much of the same ground in much greater detail.

described in Chapter 6) in Egypt, organized efforts to combat cholera, which had become a frequent companion to pilgrims to Mecca after 1831. By 1890 the *hajj* came under medical regulation in Egypt. By 1920 those diseases controllable by improved housing (tuberculosis), by cleaner water (cholera and dysentery), or by port quarantines had receded considerably, not only in Europe and America, but in Bengal, Argentina, Japan—everywhere that public health measures took effect.[5]

The program of late nineteenth century imperialism posed stiff challenges for environmental control of disease organisms. Armed with the tools and wealth created in the crowded, unhealthy industrial cities, western Europeans, and later Americans and Japanese, created colonial empires that spanned the globe by 1900. The process of building such empires required moving soldiers all over the world, and the outcome— more mines, plantations, and taxes—sent streams of refugees and laborers this way and that. Migrations and invasions often put people in microbial harm's way.

New disease regimes came with patterns of large-scale, long-distance labor migration in the nineteenth century. In their time, slave trades, particularly the Atlantic trade (1500–1850), had helped shape the global distribution of diseases. But with the suppression of slave trades and slavery, after 1850 tens of millions of indentured laborers moved throughout the plantations and mining camps of Africa, Asia, the Pacific, and the Caribbean. Like slaves before them, they carried old infections with them and encountered new ones where their journeys ended. This swirl of human and microbial migration put many at greater risk.[6] Peasants moving from Mozambique to South African mines, Melanesians working in the sugarcane fields of Australia, and Indians toiling in Trinidad all faced alien and adverse disease environments. Imperialism also knitted cities around the world into tighter contact, and gave a fillip to urbanization, especially of port cities such as Alexandria or Durban. All these developments promoted epidemics by exposing millions of people to diseases new to them, and by creating social conditions more propitious for the spread of infection. Many Asians and Africans paid for it with their lives. So did European soldiers and settlers, from West Africa to Indochina; their plight provoked an effective medical response.

[5]In the USA the lethality of waterborne diseases declined sharply in the 1910s and became trivial by 1940. Breath-borne killers, less susceptible to environmental disease control, became trivial only around 1960.

[6]Northrup 1995:120–4.

From the 1880s, researchers began to unravel the transmission cycles of malaria, yellow fever, and other killers that interfered with imperialism. After scientists established the central role of mosquitoes as disease vectors, they quickly devised measures such as mosquito netting and wetlands drainage to limit the circulation of infection. Dutch, French, and British doctors made tropical Asia and Africa far safer for Europeans after 1890.[7] American counterparts performed the equivalent for Cuba and the Philippines after 1898, when the United States acquired these lands from Spain, and then for Panama's Canal Zone, created in 1904. Efforts to safeguard local populations in the tropics sparked less interest. But with the vast expansion of imperialism, especially in Africa, European authorities intervened in health matters where their interests were threatened, and sometimes that meant fighting disease among locals as well as among Europeans. A significant reduction in the toll of tropical diseases took place after 1900 (after 1925 in tropical Africa), mainly among whites and mainly through successful interruption of disease transmission cycles.[8]

Colonial authorities were not alone in trying to improve health. Private philanthropy also played a major role in altering the relationships among humans and their pathogens in the early twentieth century. For example, beginning in 1908 the Rockefeller Institute's Sanitary Commission bankrolled and organized campaigns to eliminate hookworm in the American South, where it had for generations caused iron anemia among a large proportion of southerners. Oil tycoon John D. Rockefeller (1839–1937), a Baptist, took particular interest in the health and vitality of the millions of Baptists in the South. In the 1920s, Rockefeller's men joined with the United Fruit Company to carry the struggle to Central America, where hookworm weakened its labor force. Rockefeller's medical philanthropy soon extended to Mexico, where yellow fever, a scourge in the coastal oil fields in which Rockefeller's Standard Oil held an interest, was virtually eradicated between 1920 and 1923. This program and others like it, improved the health of millions in Brazil, West Africa, Ceylon, China, and elsewhere and played no small role in making plantations and colonialism pay.[9]

A striking feature of this shift in the history of pathogens, in temperate

---

[7]Curtin 1989.

[8]This reduction was from levels possibly boosted by colonial disruption in the 19th century. See Lyons 1992 and Kunitz 1994 on Central Africa and Oceania, respectively.

[9]Jennings 1988:28–32. On hookworm, see Ettling 1981; on Ceylon and Rockefeller medicine after 1916, see Hewa 1992. Chomsky 1996 deals with health and United Fruit Company labor.

and tropical zones alike, is the large role played by military doctors.[10] Big agglomerations of men crowded together in alien environments had always invited disease. Infections killed far more soldiers than did combat until the twentieth century. European armies after 1880 managed to lower the toll taken by tropical diseases especially, making imperialism in the tropics feasible. But the Japanese army proved the most systematic, protecting its forces with multiple vaccinations in the successful war against Russia in 1904–1905—the first war in which battle deaths outnumbered disease deaths.[11] Indeed the protracted mass slaughter of World War I required effective military medicine: only after 1905 could doctors keep mass armies healthy enough that they could butcher one another en masse.

ANTIBIOTICS AFTER 1940. The success of public health measures based on environmental manipulation continued throughout the twentieth century. After 1940, antibiotics added another weapon in the human assault on pathogens and pests. People accidentally employed antibiotics from the time they first stored grain in earthenware pots. Chinese intentionally used antibiotics, in the form of moldy soybean curd, against infections 2,500 years ago. But no one had a clear idea of what microbes could do to one another until after 1877 when Louis Pasteur and J.-F. Joubert noticed that certain bacteria killed off anthrax bacteria.[12] Subsequently, some soil bacteria served as antibiotics.[13]

But the systematic development of antibiotic drugs dates from 1928, when Alexander Fleming (1881–1955) departed his lab in a London hospital for a weekend. Upon his return he noticed his staphylococcus bacteria did not grow in the presence of a green mold. Instead of discarding his petri dish as contaminated, Fleming studied it, published his observations in 1929, and named the decisive agent penicillin. It was devilishly hard to produce in useful quantities, and Fleming eventually gave up on

[10]A theme developed in McNeill 1976:235–91; disputed by Cooter 1993.

[11]The first colonial war in which this was true was the French conquest of Morocco, 1907–1970. Miège 1989:211.

[12]Dobson and Carper 1996. Bacteria are simple cells very different from those that make up human flesh, and so are often vulnerable to antibiotics that do us no harm. Viruses are not cells, but penetrate our cells and are invulnerable to antibiotics. Protozoa (single-celled animals) and parasitic worms are the other main agents of human disease. These are sometimes vulnerable to chemical attack, but often not as their structure and metabolism may be too close to our own.

[13]Presumably these bacteria had always played this role among those peoples (e.g., Masai) who bathed themselves in dust and dirt.

it. But by 1940, Oxford scientists had taken up penicillin research, and one, Australian Howard Florey (1898–1968), went to the United States, where he helped launch the golden age of medicine. With his colleagues, Florey figured out how to make penicillin efficiently, at first culturing it in beer vats in Peoria, Illinois. He soon interested U.S. drug manufacturers. The U.S. Army saw the possibilities of penicillin and by 1943 employed it widely, radically reducing infections in open wounds. Soon penicillin and other antibiotics, many of them derived from soil microbes, proved effective against a wide range of bacterial infections, including pneumonia, diphtheria, syphillis, gangrene, spinal meningitis, tuberculosis, and some dysenteries. By 1990 some 25,000 antibiotics existed, curtailing microbial careers and improving human and animal health.

VACCINES AFTER 1897. Antibiotics offered no protection against viruses, but vaccines could. Like antibiotics, vaccines derived from ancient practice, in this case inoculation, but had modest consequences until the twentieth century. Smallpox inoculation existed for centuries in the Middle East and China, and in Europe from 1721, but its principles remained mysterious until the early 1880s, when Pasteur, Koch, and other microbiologists figured out how immune systems work.

A hunt for useful vaccines soon commenced, often involving heroic risks in experimentation.[14] A typhoid vaccine appeared in 1897 and became routine in the British army in 1915. A partially effective tuberculosis vaccination followed in 1921. The tetanus vaccine developed from early work in the 1890s into an efficient safeguard by the 1930s. An effective diphtheria vaccine arrived in 1923. With the invention of the electron microscope in the 1930s, viruses could be easily studied and immunology took giant leaps forward. Vaccine against yellow fever came in 1937, influenza in 1945, polio in 1954, and measles in 1962. These and other scourges disappeared as serious risks in life wherever vaccination programs took hold. In the United States this happened chiefly between 1945 and 1963, just in time to safeguard the largest generation in American history.

The combination of environmental disease control, antibiotics, and vaccination effected a partial epidemiological transition, in which infectious disease declined as a cause of death, to be replaced by noncommunicable diseases such as cancer and heart disease. These typically befall older people, so the transition helped add about 20 years to worldwide human

[14]The early history of vaccination is recounted in Moulin 1992.

life expectancy between 1920 and 1990.[15] The epidemiological transition made life longer, healthier, richer, and more predictable wherever it took place.

SOCIAL CONDITIONS AND HUMAN-MICROBIAL RELATIONS. Social conditions determined the degree to which public health measures worked. The containment of infection worked best in well-ordered societies. Soviet history illustrates this well. Under the stress of war and revolution, the feeble public health system in Russia broke down in the period 1915 to 1922.[16] Millions died in cholera, typhus, and typhoid epidemics. While typhus raged in 1919, Lenin told the Bolshevik Party Congress that either socialism will defeat the louse or the louse will defeat socialism. The outcome remained in doubt until after 1923, when the USSR organized its public health service and began systematic campaigns of vaccination and environmental disease control. By 1930, typhus, dysentery, malaria, and other killers had retreated as the Soviet Union self-consciously sought to create a "hygienic nation."[17] The Soviet emphasis on preventive medicine increasingly inhibited the circulation of disease in the USSR until the 1970s.

The extraordinary success in public health after 1880 affected human politics. Formerly, disease-experienced and immunologically toughened populations had enjoyed great competitive advantages in intersocietal relations, a fact shown most clearly in the conquest of the Americas, in which population declines of 80 to 90 percent occurred among Amerindians over the 150 years following Columbus's voyage of 1492. After 1880, those societies able to administer public health programs enjoyed competitive advantages. They could keep their armed forces alive to fight and their labor force healthy to work. Of course, well-organized societies always enjoy an edge over the badly organized. The effect of the medical changes since 1880 was to sharpen that edge and widen the gaps in wealth and power among societies.

The combined effect of environmental disease control, antibiotics, and

[15]See RIVM/UNEP 1997:96 on the evolution of life expectancy and of health. In the USA, cardiovascular disease outstripped infectious disease as a killer by 1920; cancer eclipsed infectious diseases by 1945. In poor countries the transition only began late in the 20th century.

[16]Tsarist Russian authorities resisted the bacteriological revolution, driving many doctors toward political revolution. Not until after 1917 did bacteriology benefit Russia (Hutchinson 1985).

[17]The phrase is from Feshbach and Friendly 1992:37. See also Johnson 1988; Solomon and Hutchinson 1990.

vaccination seemed to promise complete victory in the human struggle against infectious disease. In 1948 the U.S. Secretary of State, George Marshall, foresaw its imminent elimination. In 1967 the U.S. Surgeon General told Congress it was time to "close the book on infectious diseases."[18] The World Health Organization (WHO), a U.N. branch created in 1948, targeted several diseases for eradication. It did close the book on smallpox after a 10-year campaign. Smallpox, a human nemesis for at least 5,000 years and the killer of some 300 million in the twentieth century, finally met its end in 1977, in Somalia. It claimed its final victims in 1978 in Birmingham, England, when the virus escaped into the air ducts of a research laboratory, killing one scientist and driving the lab director to suicide. The WHO declared the world smallpox-free in 1980, but samples of the virus remain in freezers in laboratories in Atlanta and the Siberian city of Koltsovo. The anti-smallpox campaign represents perhaps the first deliberate extinction of a species and a signal achievement of the WHO.[19] Other infections, if not eradicated, were marginalized. But the golden age did not last.

MICROBIAL RESISTANCE AFTER 1946. In 1945, Fleming warned that penicillin's use would soon produce resistant staphylococci. Indeed, by 1946 London hospitals hosted resistant strains, which spread wherever penicillin was used. Other antibiotics normally killed off these strains until the 1970s, when multiple-drug-resistant (MDR) bacteria emerged. Thereafter, incurable strains of tuberculosis, malaria, and numerous other infections threatened human health.

Evolution made the emergence of MDR bacteria inevitable, but human weakness hastened the day.[20] The use of antibiotics selected strongly for resistant strains, and improper use made this happen sooner. Fleming thought oral antibiotics dangerous because anyone could take them easily, whereas intravenous antibiotics required medical personnel. In the United States, nonprescription antibiotics could be bought at any drugstore until the mid-1950s, and in much of the world this remained true in the 1990s. In any case, doctors succumbed often enough

[18]Bloom and Murray 1992:1055 put this statement in 1969, as does Porter 1997:491. Tenner 1996:58 and Garrett 1994:33 say 1967.

[19]See Fenner 1993 and Oldstone 1998:27–44 on smallpox history and eradication. The Soviet Union promoted the anti-smallpox campaign to the WHO in 1958. WHO by 1996 had nearly eliminated polio, guinea worm disease, and river blindness—all usually nonfatal infections. Measles was also a candidate for annihilation.

[20]A survey of MDR pathogens appeared in the *Economist,* 31 May 1997:73–4.

to impatient patients and prescribed antibiotics recklessly. Others did too. The Indonesian Ministry of Religion in 1981 provided tetracycline to 100,000 pilgrims making the *hajj* to Mecca, hoping to ward off cholera but in the process hastening the evolution of tetracycline-resistant bacteria. The U.S. livestock industry after the early 1950s fed mass quantities of antibiotics to American cattle and pigs to keep the animals healthy and help them grow faster in the epidemiologically hazardous feedlots of postwar America.[21]

All these practices represent a microbial "tragedy of the commons." It suited doctors, patients, and cattlemen to use antibiotics recklessly, because they were cheap and easy to use and the benefits were quick, real, and personal. The costs came in the future, they were shared among all society, and they were inevitable: restraint by any individual would at best only marginally delay matters.

RESURGENT CONTAGIONS. Microbial infections eventually struck back. Incurable strains of tuberculosis appeared first in South Africa in 1977, and by 1985 powered the first statistical rise in tuberculosis in the United States since the mid-nineteenth century. MDR tuberculosis thrived in hospitals from the 1980s, and colonized other settings, such as prisons and homeless shelters, where overcrowding and other conditions assisted its spread. In the United States in the 1990s it killed some 70 percent of those who contracted it. About 50 million people around the world harbored MDR tuberculosis in 1997, with the highest infection rates in Southeast Asia. The emergence of HIV (human immunodeficiency virus), which compromises immune systems, added to tuberculosis' renaissance. It was responsible for about half of the post-1985 tuberculosis epidemic in the United States and about a quarter of it in South Africa. Worldwide, tuberculosis in the early 1990s killed about 2.5 million people a year, almost all in poor countries. By 1995 the toll surpassed 3 million a year, and continued to rise.[22]

Other infections proved just as resilient. In 1955 the WHO hatched plans to eradicate malaria from the earth. In 1992 it gave up. The WHO

[21]This practice was made illegal in Canada and western Europe in the 1970s. In the USA, 30 times more antibiotics are used on animals than on people (Levy 1992). On reckless prescriptions: A study by Ralph Gonzales (reported in the *Washington Post,* 17 September 1997:A2) indicated that in 1992 a fifth of all antibiotics prescribed in the USA were for viral infections, mostly colds.

[22]Ewald 1994:65; Raviglione et al. 1995. In 1995 about 2 billion people carried the bacillus— one-third of humanity (Dobson and Carper 1996).

efforts inadvertently selected rigorously for resistant malaria and anopheles mosquitoes, malaria's vector. The WHO initially followed successful precedent, deploying DDT and other insecticides that after 1945 had slashed infection rates. But mosquitoes evolved resistance to DDT, and malaria resurged. India in 1977 had roughly 60 times more malaria cases than in 1960. Drugs, mainly chloroquine, still provided some hope for breaking the transmission cycle—if social conditions permitted their systematic use. But malaria evolved new strains, requiring new drugs. Then in the 1980s, MDR malaria appeared (amid chaotic social conditions) along the Thai-Cambodian border. Chloroquine-resistant malaria appeared in East Africa, Amazonia, and Southeast Asia, among populations who could scarcely afford more expensive substitutes. Resurgent malaria killed about 2 million people per year in the 1990s, half of them in Africa, making it second in lethality to tuberculosis among infectious diseases. It afflicted some 250 million to 300 million.[23]

Pneumonia, one variety of dysentery (shigella), and a few other diseases also developed MDR strains by the late 1980s. Cholera evolved an MDR form by 1992, apparently incubated in algal blooms off the coasts of Bangladesh.[24]

The evolution of resistant infections and disease vectors put an end to a golden age and locked pharmaceutical researchers into an endless arms race with virulent bacteria.[25] It stalled and may yet reverse the epidemiological transition that so agreeably limited the sway of infection in human affairs in the twentieth century.[26]

[23]Perhaps tied for second place with measles (see Murray and Lopez 1996).

[24]Epstein et al. 1994.

[25]Bacteria evolve quickly as their generations can be as short as 20 minutes and the evolution of drug-resistance can take only weeks to months. They can also swap genes for drug resistance casually, without the cumbersome procedures of sex. It now takes 5–10 years to research, patent, and market a drug (though it took half as long in the 1960s). This means medical researchers face a stern challenge, made more difficult by the inconsistent budgets accorded to public health services, which allow useful programs to decay before their work is done. The effects remain to be seen, but one likely outcome is diseases that plague richer populations will attract more research and more breakthroughs, while those diseases that predominantly affect poorer populations will not—as long as pharmaceutical research is conducted primarily by drug companies. The main beneficiaries of the evolution of MDR pathogens were drug companies and their research staffs, for whom it served the same purpose as planned obsolescence in automobiles. See the Society for General Microbiology 1995; Levy 1992; Garrett 1994.

[26]A bleak future is forecast by Garrett 1994, a cheerful one by Murray and Lopez 1996. No predictions are worth much given the uncertainties involved.

SOCIAL CONDITIONS AND RESURGENT CONTAGION. In many societies the deterioration of public health systems compounded the evolution of resistant microbes and mosquitoes. This happened in the Soviet Union after 1970. The morale and economy of the USSR frayed during the Brezhnev years, and public health programs suffered. Vaccination regimes lapsed, trust in the medical establishment eroded, and soon a substantial population without important immunities existed, setting the stage for resurgent epidemics with the progressive breakdown of order after 1989. These have helped lower average life expectancy from over 70 (in the mid-1960s) to about 65, and 58 for men in 1995.[27] Similar deterioration happened on a smaller scale in the cities of the USA, where after 1975 public health programs declined to the point where, for example, successful treatment rates for tuberculosis in Chicago and New York were far lower than in Mozambique and Malawi.[28]

The basic principles of immunology, environmental disease control, and antibiotics where applied made a great difference, but in many parts of the world, political and social conditions delayed the application of these principles before 1950. Revolution, invasion, and civil war in China between 1910 and 1949 inhibited public health programs. But after 1950 the relative stability of Mao's rule allowed great improvements. Communist China, like the USSR before it, and later Castro's Cuba, made health one of the highest priorities, one of the targets for mass mobilizations, one of the justifications for the new regime. Mao exhorted peasants to hunt down and kill snails to reduce schistosomiasis, and urged the equivalent of class warfare—species warfare—against mosquitoes and ticks. Mao also targeted syphilis, regarded as a moral stain upon the new China. Under Mao, China was not prosperous nor entirely orderly, but it was orderly enough to change some human-microbial relations and effect a revolution in health.[29]

In the world at large, Cold War stability and the great economic boom after 1945 permitted a fuller harvest of the fruits of microbiology, making human lives far healthier, happier—and numerous—than before. The

[27]Feshbach and Friendly 1992: app. 4; Population Reference Bureau, 1996, *World Population Data Sheet*. Bridges and Bridges 1996:178 say the decline was sharper still.

[28]Budget cuts played some role in this disparity, but not much. Curing TB is a long-term process, and the typical TB patient in Chicago was less likely to stay with it than were Mozambicans (Bloom and Murray 1992:1059–60).

[29]Lucas 1982. On pre-1949 difficulties in carrying out public health measures, see Yip 1995:105–14.

years 1950 to 1990 were unusually calm, orderly, and prosperous in world history, a rare moment propitious for the complex administrative requirements of successful public health programs. Neither vaccination, antibiotics, nor environmental disease control can work their miracles amid war and unrest: human containment of microbial pathogens remains precarious, strongly contingent upon public order and a stable international system.

ACCIDENTAL SHIFTS IN HUMAN-MICROBIAL RELATIONS.[30] The victories against infections were all the more precarious because they succeeded despite large-scale changes that created conditions favorable to the spread of diseases. Several fundamental features of twentieth-century history were involved, notably the expansion of irrigation, acceleration of transport, human disruption of tropical ecosystems, changing relations between humans and animals, and the spread of large cities.

Irrigation and Disease. Irrigation in the twentieth century expanded nearly fivefold, covering by 1990 an area the size of Sudan or four times the size of Texas (see Chapter 6). This proved a bonanza for certain disease vectors, notably snails and mosquitos. Almost all tropical diseases are waterborne in a loose sense, and several have extended their sway with irrigation.[31]

Egypt illustrates the epidemiological perils of irrigation.[32] Irrigation expanded greatly after 1902 in Egypt, and waterborne disease, especially snail-borne schistosomiasis, with it. Improvements and expansions in irrigation took place after the heightening of the Aswan Dam in 1934. In 1942, when German armies were invading from the west, mosquitoes bearing falciparum malaria, the most lethal kind of malaria, invaded from the south, flourished in the new irrigation regime, and killed about 130,000 Egyptians. Rockefeller Foundation forces eradicated the offending mosquito in 1944–5. Subsequently, Egyptian and international

---

[30]Many of these shifts altered the relations between pathogens and animals too. For example, the 1987–1992 viral epidemic among seals and dolphins may represent a new infection cooked up in algal blooms in polluted estuaries. But these matters, while indirectly important to humanity, are beyond the scope of this book.

[31]Malaria, schistosomiasis, onchoceriasis (river blindness), filariasis, and Japanese encephalitis to name a few (Brinkmann 1994:304–6). Brinkmann suggests that economic development, as recently pursued in the tropics through irrigation, mining, and road building, eases the transmission of tropical disease. See also Kunitz 1994:11.

[32]See Gallagher 1990.

authorities made great strides toward control of infectious disease, but the extent of stagnant irrigation water limited their success. The same was true almost wherever irrigation expanded, excepting the richer, better-organized societies.[33]

Transport Systems. Faster transport inadvertently offered efficient highways for disease transmission, helping to sustain and spread infections. Before 1850 crossing the Atlantic or Pacific Oceans would take weeks, which inhibited the spread of some vectors and pathogens; by 1910 it took several days, and by 1960 only a few hours. People unknowingly transmitted diseases still incubating within them; and mosquitoes, ticks, and other vectors survived more readily with briefer transit. For example, particularly efficient malarial mosquitoes *(Anopheles gambiae)* entered Brazil from West Africa by airplane in 1930. Malaria carried by other mosquitoes had plagued the Americas since the sixteenth century. But since they bit other animals more often than people, they did not spread malaria among humans as rapidly as did the newcomers, which relished human blood. This vector invasion brought 20,000 deaths, the worst malarial outbreak in Brazilian history, prompting a Rockefeller Institute insecticide campaign that in 1938–1939 succeeded in eradicating *A. gambiae's* toehold in the Americas.[34] Far more people and goods moved around the globe late in the century, accelerating microbial traffic. Early in the century it took a world war to put tens of millions of people on the move, providing diseases with opportunity for rapid spread. Troop movements occasioned by the end of World War I made the 1918–1919 influenza a worldwide pandemic, killing some 30 million people in a few months. But by the 1990s, routine travel potentially offered infections even better chances to spread: about 500 million people flew from one country to another each year.[35] Public health services, with numerous small exceptions, met this unprecedented challenge successfully.[36] Had they not, recent history would look quite different.

Disturbance of the Tropics. People have inhabited tropical zones for many centuries, but until lately their numbers remained small, their economies weak, their technologies modest. Ferocious disease regimes—

[33]Ibid. 34–5; Hunter et al. 1993:43–4.
[34]Curtin 1993:346–7.
[35]Chen 1994:323.
[36]Shope and Evans 1993.

to which local people evolved partial resistance—kept strangers at bay. But in the late nineteenth century, more and more tropical regions became linked to the wider world; goods and people circulated as never before, effecting a general microbial exchange of the sort promoted by the colonial empires. That exchange accelerated in the twentieth century, as the comparative isolation of places such as New Guinea, Amazonia, and Central Africa decayed.

In East and Central Africa, for instance, from 1880 to 1930 the winds of political change reshuffled wealth and power among local, Turco-Egyptian, Afro-Arab, and European merchants, kings, pashas, and governors. The slave, rubber, and ivory trades flourished. African plantations and a whole new political economy emerged, with increased human migration and instability among wildlife populations. When the incentives to farm increased, forest and bush cover shrank back in some places. But it extended elsewhere when famine and rinderpest (a cattle disease) reduced the human and cattle populations, and ivory hunters killed off most of the elephants, which had formerly kept woody plants in check. These disturbances upset previous relationships between the people, animals, tsetse flies (*Glossina*), and trypanosomes, the parasite that causes trypanosomiasis, or human sleeping sickness. The changes undercut land management practices, learned presumably from bitter experience, by which Africans had minimized the toll of trypanosomiasis.

In northeast Tanzania, for example, German colonial control upset social networks that before 1890 had allowed many people to survive drought and famine. The Germans undermined the position of rich patrons, who no longer fed the poor (in exchange for subservience) in time of drought. Hungry people had to flee, which left only a remnant population too small to maintain the forest-burning practices that hitherto had kept the tsetse fly at bay. Trypanosomiasis flourished, killing cattle and people. From 1890 to 1940 or so, colonial policy effected social changes with inadvertent ecological consequences, which in turn affected society, making the people of northeast Tanzania poorer and sicker.[37]

Through roughly similar chains of events, sleeping sickness, long endemic in many pockets of tropical Africa, moved into new territories and populations. Brutal epidemics killed a quarter million people in Uganda (1900–1905) and more elsewhere, contributing mightily to a general population decline in Central Africa from the 1880s to at least 1925. Many Africans developed a lasting sense that colonialism was a

[37]Giblin 1992.

form of biological warfare, an idea that resurfaced in a new guise when AIDS began to ravage Central and East Africa in the 1980s. Colonial authorities found in sleeping sickness justification for enhanced control of African life, as the disease seemed eradicable only if Africans changed their ways of cultivation, seasonal migration, and livestock management. In British East Africa, some Africans had to surrender their lands to settlers in the name of tsetse control. The European "scramble for Africa" after 1885 disrupted African ecologies, economies, and health and proceeded all the faster for it: political and ecological trends fed on one another.[38]

Subsequent disruptions of tropical ecology have loosed other infections on humankind. Hemorrhagic dengue fever first appeared in Southeast Asia in the 1940s. Lassa fever was first identified in Nigeria in 1969, the extremely lethal Marburg virus turned up in Central Africa in 1967, while Ebola was recorded in Zaire in 1976. AIDS, which killed about 2.3 million people around the world in 1997, ranking fifth among global causes of death, apparently also derived from the rise of human activity in tropical forests. It probably came from central African chimpanzees and jumped to human hosts shortly before 1959. Thereafter it circulated slowly in central Africa until the late 1970s, when it began to break out, assisted by widespread warfare in Angola and by refugee movements and labor migration in southern Africa generally. By the early 1980s, the AIDS virus had appeared in the United States and quickly became a global menace. Its epicenter remained in Africa, however. Two-thirds of the 47 million people who carried the virus in 1998 were Africans, as were most of the 14 million killed by AIDS between 1978 and 1998.[39]

[38]Ford 1971; Lyons 1992; Hoppe 1997; Giblin 1990 and 1992; Maddox et al. 1996; Headrick 1994:67–94, 273–384. On the Congo-Océan railway, built by the French in the 1920s, infectious disease killed 10–30% of the laborers annually, about 20,000 men in all. Italian miners and Chinese laborers were imported to help fill out the ranks. The impact of trypanosomes on tropical African history is very powerful: it inhibited stock raising, use of the plow, and protein supply, while favoring shifting cultivation and low population density—and preserving wildlife (Headrick 1994:68). Waller 1990 disputes the role of colonialism in producing sleeping sickness epidemics in at least one part of Masailand.

[39]See the many papers in Morse 1993 and the useful overview by Murphy 1994. The AIDS figure comes from WHO, reported in the *Economist*, 4 July 1998:79. The AIDS date is that of the first known infected blood sample, preserved in Kinshasa (*Science News* 153:85, reporting the work of Juofu Zhu; and the *Economist*, 6 February 1999:86, reporting the research of Beatrice Hahn). Changes in ecology altered disease patterns outside the tropics too. The resurgence of Lyme disease in the northeastern USA after 1974 followed gradual reforestation (from 1920), an upsurge in deer populations (mainly after 1960), and suburbanization (after

Resurgent malaria, while largely a matter of drug resistance, also gathered force from human perturbations in the tropics. In Amazonia after 1970, Brazil's government built a road network that brought armies of loggers and miners, as well as peasant farmers, into the rain forest. These immigrants usually had no resistance to malaria, were too poor to afford antimalarial drugs, and provided anopheles mosquitoes with new targets. Irrigation schemes, in Brazil and throughout the warm latitudes, gave mosquitoes more and better breeding grounds, promoting all mosquito-borne diseases.

People and Animals. Many human diseases derive from animal infections. For several millennia, human-animal contact shaped human disease history. In the twentieth century, new corridors of contact opened up. The human surge into the tropics put people into firmer contact with more species. Beyond this, there were far more domestic animals than ever before; and far more "synanthropes"—animals that live with us but are not domesticates, such as rats and seagulls. And of course there were more people. Although in 1990 a smaller proportion of humanity lived in close contact with livestock than 100 or 200 years ago, more animals of more species—and hence more potential pathogens—lived in contact with more people. With these shifts in human-animal relations the opportunities for species-jumping infections improved.

One example of such a shift was our inadvertent aid program for rodents. They eat about one-fifth of the world's grain harvest. By storing and transporting food in massive amounts, by building our urban warrens, and by our attack on foxes and other rodent eaters, we made the world safer for rats, mice, and their cousins. In so doing we made a world more suitable for rodent viruses, of which several are communicable to humans. More people and more rodents lived cheek by jowl than ever before after 1950 because of these changes. Several of the major emergent viruses after 1970 derived from rodent populations.[40]

1945). All of this put people in contact with deer ticks on a scale unknown for at least 150 years in those parts (Spielman 1994).

Inadvertent shifts in the human-microbe relationship need not bring on epidemics; they can result in the disappearance of disease, as for example with the elimination of malaria from northern Europe in the 19th century. This was achieved by sheltering cattle in barnyards thus providing a preferred food supply for mosquitoes. This broke the malarial transmission cycle, for the parasite that causes malaria does not flourish in cattle as it does in people.

[40]Hantavirus, Argentine hemorrhagic fever, Lassa fever (in its Liberian and Sierra Leonean variant) are examples.

It is also possible that the prevalence of influenza in recent centuries derives from the close quarters kept by ever more ducks, pigs, and people, mainly in China. The vast majority of influenza epidemics originated in China, although that of 1918 incubated on an army base in Kansas. Pigs serve as the "mixing vessel" in which avian and human flu viruses exchange genes, occasionally yielding a new strain with the right qualities to pull off a pandemic.[41]

Human use of animals often wrought havoc with the animals—sometimes with human consequences. In 1889 the Italian army, campaigning in Somalia, imported cattle bearing the rinderpest virus. It was new to Africa and highly contagious. The density and mobility of susceptible animals in East Africa led in the 1890s to the worst epizootic (outbreak of an animal disease) in recorded history. Millions of cattle died. So did millions of wild buffalo, antelope, giraffe, and other ruminants. South of the Zambeze River, perhaps 90 percent of grazing animals succumbed. Similar proportions died among domestic cattle throughout East and southeastern Africa. The basis of the pastoral economies of East and southern Africa evaporated, leading to famine, violence, desperate migrations, as well as to a spate of religious revivals and sudden conversions to Christianity and Islam. Among the Masai, for instance, perhaps two-thirds of the people died. Changes in the ways in which we lived amid animals powerfully affected the human and animal disease experience in the twentieth century. So did the way we lived with one another.[42]

Urbanization. Lastly, urbanization and population growth generally have altered the circumstances that govern human-microbial relations. Many infectious diseases required a minimum host population in order to keep circulating. They needed a concentration of nonimmune bloodstreams to sustain cycles of infection, which in practice usually meant a concentration of newborns, something cities could provide. In village settings an infection more often would burn itself out, exhausting the supply of susceptible hosts. Hence, cities hosted infectious disease far more persistently than did the countryside. In the twentieth century, cities grew in number and size—a fact made possible by antibiotics, vaccinations, and environmental disease control. But at the same time cities provided conditions suitable for many pathogens, and some more than

[41]Beveridge 1993; Murphy and Nathanson 1994.

[42]On rinderpest, see Dobson and May 1986, Ford 1971:138–40, Iliffe 1995:208–11, Ranger 1992, and Spinage 1962.

others, favoring tuberculosis and typhoid, for example.[43] The fourfold growth of human population in the twentieth century provided plenty of extra bodies in which pathogens could feed. Indeed, from their point of view, the twin processes of urbanization and population growth set the table nicely: but a breakdown in the social order and public health apparatus may still be required before they can begin to feast.

Of the 30 or more infections that emerged after 1975, almost none were genuinely new: they were merely new to humans, or even to particular populations of humans.[44] The human invasion of tropical ecosystems, the growing contact with animals, advances in transportation—all these broadened the variety of microbial traffic that passed through human bodies.

Major ecological shifts have always created new pathways for microbial traffic. Edward Jenner, who pioneered the use of cowpox inoculation as protection against smallpox, wrote in 1798: "The deviation of man from the state in which he was originally placed by nature seems to have proved to him a prolific source of diseases."[45] In the twentieth century such deviations were especially numerous and sudden. Fertilizers and pesticides created a brave new world for soil bacteria. Pollution of estuaries and waterways forced new evolutionary directions for bacteria and viruses that live there. Global warming after 1980 expanded the range of mosquitoes and other disease vectors. All these added to the effects of irrigation, transport, disruption of the tropics, new corridors of human-animal contact, and urbanization.

Taken together these shifts posed a fearsome challenge. Human-microbial relations altered fundamentally, and very favorably from the human point of view, in the twentieth century. But the new regime was provisional and precarious, hostage to public order, and subject to further shifts. The grand ecological systems of human society, animals, vegetation, vectors, and microbes will continue their enormously complex coevolution well beyond any conscious control: it would be strange if no surprises resulted.[46]

---

[43]Guayaquil (Ecuador) at the beginning of the 20th century is an example of this process. It grew fast during the cacao export boom and became notorious for its disease regime. Highland immigrants without useful immunities provided infections with a constant flow of new victims; sanitation efforts lagged well behind requirements (Pineo 1996).

[44]On the figure of 30 emergent infections, see WHO 1996.

[45]Quoted in Wills 1996:29.

[46]One potential source of disruption is the recently discovered bacteria populations of deep-sea vents. They live happily in searing heat and amid a chemical cocktail, but some of them might find conditions among us more to their liking. Cliff et al. 1998 review the disease implications of contemporary environmental change.

Land Use and Agriculture

The modern expansion of agriculture served as the main engine behind the drastic changes to the earth's vegetation in the twentieth century. Here I will offer an overview of vegetation and land-use history, followed by a closer look at the revolutions in agriculture. These affected the human condition nearly as fundamentally as did our containment of pathogenic microbes.

The main trend in twentieth-century vegetation was the increasing human management and appropriation of it. At the end of the twentieth century, about a third of the world's area under vegetation supported domesticated plants—crops and pasture grasses—roughly twice as much as in 1900. Some 35 to 40 percent of the earth's terrestrial biological production was used for human needs.[47] This represented an acceleration of a trend as old as agriculture, and one that had no major interruptions after the 1346–1352 pandemic, known in European history as the Black Death.

The broad lines of land-use and land-cover history, regrettably, are all one can get.[48] About 30 percent of the globe's surface (133 million square kilometers) is land not covered by ice or sand. Of that, no more than a quarter (26 million to 30 million square kilometers) permits cultivation. Of that, only about a quarter—roughly the area of Australia—was cultivated by 1900; by 1995 that area was more nearly the size of Russia or South America. Table 7.1 shows how global land cover has evolved.[49]

CROPLANDS. Inconsistently but inexorably, the world's cropland expanded for 10,000 years. The global trend paralleled that of population

---

[47]RIVM/UNEP 1997:75; Vitousek et al. 1986.

[48]Multiple definitions exist for all land-cover terms (e.g., forest, cropland, pasture), so the statistics, even if accurate, are incommensurable. Since 1980 or so, satellite imagery has made study of vegetation changes easier, but numerous complexities remain. Houghton 1994 discusses discrepancies between ecologists and agronomists in their views of land use and land cover.

[49]An alternative effort to classify the world's lands, regrettably not historical, is Hannah et al. 1994. The authors find that 27% of the world's vegetated surface remains undisturbed by human influence in the 1990s, the largest areas being in Amazonia and the miombo woodland/savanna of southern Africa. "Undisturbed" means fewer than 10 people/km² in their lexicon.

because enduring changes in yield were rare, modest, and slow. By 1700 roughly 2 to 3 percent of the world's land surface was cropland.

After 1700, with burgeoning European overseas colonization, the pace of cropland expansion quickened. Settlement frontiers took shape in North America, South America, South Africa, Russia, and Siberia. Agriculture colonized new lands both on the perimeter of China and in the

TABLE 7.1 APPROXIMATE EVOLUTION OF GLOBAL VEGETATION COVER TO 1990

| Date | Types of Land Cover (million km²) | | | |
	Forest and Woodland	Grassland	Pasture	Cropland
8000 B.C.	65	63	0	0
. . .	. . .	. . .	. . .	. . .
1700 A.D.	62	63	5	2.7
. . .	. . .	. . .	. . .	. . .
1850	60	60	8	5.4
. . .	. . .	. . .	. . .	. . .
1890	58	55	13	7.5
1900	58	54	14	8.0
1910	57	52	15	8.6
1920	57	51	16	9.1
1930	56	49	19	10.0
1940	55	47	21	10.8
1950	54	45	23	11.7
1960	53	41	27	12.8
1970	51	38	30	13.9
1980	51	35	33	15.0
1990	48	36	34	15.2

Sources: Graetz 1994; RIVM 1997:table A22; Richards 1990b; WRI 1996; WRI 1997.

Note: this table vastly simplifies land-cover categories and hides distinctions within them, as well as ignoring the (controversial) process of desertification. Totals are inconsistent because of rounding, and because the total landmass not covered by ice has changed slightly.

interior, especially on mountainsides. From 1830 to about 1930, settlement frontiers moved rapidly in the Americas and Russia, somewhat less so in northern India, driven as always by demographic growth, but now also by an increasingly integrated world market in grain. By 1930, world cropland amounted to four times the 1700 total. Most of the cropland expansion (1700–1930) came at the expense of grassland in the Americas and the Eurasian steppe.

In the twentieth century, population growth and the rapid rise in international grain markets continued to drive this frontier conversion to cropland. In North America the process ended with the settlement of the Peace River valley in Alberta in the 1930s, the last large-scale plow-up in Canadian history. Political ambitions sometimes provided the driving force, as in the last great effort at extensification of agriculture in temperate latitudes: Khrushchev's Virgin Lands scheme (1954–1960), in which broad swaths of Russian and Kazakh steppe were broken to the plow. After 1960, new settlement and cultivation of former steppe and temperate forest lands—one of the great trends of modern history—ceased.

But humankind acquired two more cards to play in its struggle to feed itself. Irrigation beyond the immediate confines of riverbeds (Chapter 6) and chemical fertilizer (Chapter 2) raised crop yields abruptly, shattering the old equations between population and cropland. These two practices permitted Europe after 1920, North America after 1930, and Japan after 1960 to forgo net cropland expansion. By the 1960s in temperate zones, efforts to expand food production focused almost solely on obtaining more harvest per acre rather than on farming more acres. The Soviet Union's formal commitment to "chemicalization" of agriculture in 1966 completed this transition.[50]

After the end of the Virgin Lands scheme, the active frontier zones of the late twentieth century lay in the tropics, mainly in West Africa after 1950, the interior of South America after 1960, and Indonesia after 1970. Cropland expanded elsewhere—India, Southeast Asia, and Iran—but at slower rates. By 1960 tropical agricultural expansion overtook that of temperate zones.[51] Nigeria's Igbo yam farmers were the latter-day equiv-

---

[50]Ioffe and Nefedova 1997:71–5.

[51]Expansion is measured in crop area added annually (Houghton 1994). In 1960 both the tropics and the temperate zones added about 4 million ha per year of farmland. After 1970, temperate zones added 1 million ha or less per year, while the tropics added 6 million to 10 million ha/yr. If one excludes the impact of Soviet Virgin Land programs in the 1950s, the tropics have led temperate zones in agricultural expansion since 1940.

alents of Saskatchewan sodbusters. These new settlement frontiers often ate into tropical forests, with consequences for biodiversity, the global carbon cycle, and human-microbial relations much greater than those of prior settlement frontiers in temperate forests or grasslands. They may also represent the final stage of a 10,000-year trend.

In the 1980s the growth in world croplands slowed. In Europe and North America, cropland continued to contract. The trend spread to Russia and Kazakhstan, where by 1997 a quarter of the grassland plowed up under the Virgin Lands program was abandoned. According to official figures, Africa's cropland area declined in the years 1972 to 1989.[52] Around the world, farmers gave up on degraded, eroded, and desertified land at faster rates. On a smaller scale, cities spilled out onto farmland wherever urbanization or suburbanization proceeded. In China between 1978 and 1992, urban and industrial pursuits replaced farming on about 6 percent of the country's cropland. For these and other reasons, around the world lands went out of agriculture in the mid-1990s almost as fast as others went in.[53] Whether this post-1985 trend is the wave of the future or a temporary aberration remains to be seen.

The main reason that human population could quadruple while cropland only doubled in the twentieth century is that farming became more productive. Several elements combined in this, most notably chemical fertilizers and pesticides, irrigation, agricultural machinery, and plant breeding. Fertilizer and irrigation were discussed earlier. Here I take up the mechanization of agriculture and the plant breeding of the Green Revolution, each of which, like fertilizer and irrigation, and linked with them, brought momentous changes to agroecosystems, to other ecosystems, and to society at large.

In 1900, farming around the world generally involved the same basic procedures as a thousand years before. Farmers still used animal or human muscle for nearly all farmwork; their fertilizers were dung, crop residues, and other organic matter gathered locally; they bought very little in the way of inputs; they controlled pests by crop rotation and fal-

---

[52]Biswas 1994 has the figures, showing an 8.4% loss of croplands during these years.

[53]From 1945 to 1990, abandoned farmland averaged about 2 million ha per year; in the mid-1990s it was more like 5 million to 10 million ha (Gardner 1997:49; and Xu and Peel 1991:258). Smil 1993:57 puts the Chinese cropland loss to settlement and infrastructure at 35 million ha between 1957 and 1990. See Douglas 1994 on urban area expansion. Biswas 1994 puts farmland loss at 1.5 million ha/yr from salinization, and 7 million to 8 million ha/yr to erosion and urbanization.

lowing; they rarely specialized in a single crop; and they generally produced no more than 1 to 2 tons per hectare, whether rice, wheat, maize, manioc, or millet. Something like 70 to 90 percent of people worked in this low-technology, labor-intensive agrarian world.[54]

By the 1990s farmers in Europe, North America, Japan, Australia, New Zealand, and elsewhere accounted for less than 10 percent of the population but had radically transformed agriculture and agroecosystems. They used huge amounts of fossil fuels; they controlled pests with chemicals; they often specialized in a single crop; they bought many of their inputs from factories; and they produced 4 or more tons per hectare. This amounts to a modern agroindustrial revolution. It rested on labor-saving devices—machines—mostly developed in North America where land was cheap and labor costly; and on land-saving techniques— crop breeding, fertilizer, pesticide—developed in many places. The most dramatic change happened after 1945, but the beginnings came much earlier.

MECHANIZATION. Farm mechanization began with horse-drawn threshers and reapers, introduced in the 1830s and made common in the United States by labor shortage during the Civil War (1861–1865). Steam-powered threshers, used in Britain and the United States as early as the 1850s, found few customers. Steam engines were too big to move smoothly over fields. Gasoline-powered tractors arrived in 1892. The United States led the way in tractor adoption because of high American labor costs and large farm size. The spread of gas stations, repair shops, and mechanics provided farmers with the needed support system. The American conversion to the tractor took place between 1920 and 1955. It attracted imitation in the USSR, where something of an official tractor fetish developed in the 1930s—some enthusiastic parents named their children for tractors. Government devotion to tractors was connected to the demands of collectivization, with its bigger fields and expanded scale of production.[55] Table 7.2 gives an idea of the history of tractor adoption

[54]Exceptions existed, as among the grain farmers of the North American High Plains or among dairy farmers of New Zealand's North Island, who by 1900 had begun to specialize, to sell most of their output, and in other ways tended in the direction that would prove to become the future of agriculture in the West. Yet even U.S. farmers in 1900 spent only 45% of farm income on outside inputs in 1900; by 1990 they spent over 80% (Solbrig and Solbrig 1994:224). On the variety of pre-1900 agriculture, see Vasey 1992.

[55]Fitzgerald 1996 details the importation of American tractors and combines into the USSR from 1929.

TABLE 7.2 TRACTORS IN THE USA, USSR, AND WORLD, 1920-1990

| Year | Number (millions) | | |
	USA	USSR	World
1920	0.25	0	0.3
1930	1.0	0.05	1.1
1940	1.6	0.5	3
1950	3.4	0.6	6
1960	4.7	1.3	10
1970	4.6	2.0	16
1980	4.8	2.4	24
1990	4.6	2.7	26

Sources: Stanton 1998; Vasey 1992; Volin 1970.

in the United States and the Soviet Union. Tractors worked their magic on farms in the United Kingdom and Europe only after 1950, and have never found much room to operate in Japan.[56] Their use in Brazil expanded after 1970, but in most of the world, low labor costs undermined the logic of tractors and mechanization.[57] In the rich countries, industrial methods, efficiency, and inputs dominated nearly every aspect of agriculture—even the Dutch tulip harvest. By 1980 each American farmer fed about 80 people; each Australian farmer somewhat more.[58]

Mechanization revolutionized agriculture and agricultural ecology. Tractors and harvesting combines (introduced in the 1920s) made better sense on big fields, so farmers began to eliminate hedgerows and replace

[56]On Denmark, Nielsen 1988; on Ireland, Walsh 1992; on Hungary, Gunst 1990. In Japan in the 1960s, power tillers ("walk-behind tractors") helped ease the farmer's lot in rice fields. Power tillers spread to South Korea and elsewhere from 1970.

[57]Mechanization failed in several cases in Africa (e.g., Jedrej 1983). It did not work well in the Transbaikal region of Siberia, where it was introduced in the 1930s, in the context of collectivization (Manzanova and Tulokhonov 1994). On problems in Chinese mechanization, see Tam 1985. On obstacles to mechanization of sugarcane harvesting in Queensland, Cuba, Peru, and the USA, see Burrows and Shlomowitz 1992.

[58]On tractor history, see Grigg 1992:49–51. On mechanization, see Rasmussen 1982 and Mannion 1995:95–104.

complex field mosaics by larger fields. Farmers turned increasingly toward those crops that machines could harvest.[59] They specialized more and more in single crops, because each crop required its own set of machinery, thus replacing patchwork patterns with monocultures.[60] This in turn meant that insect and other pests had to be controlled by new means, mainly chemical pesticides.[61] Monocultures also depleted specific soil nutrients faster, requiring more chemical fertilizers. The efficiency of farm machinery meant that farmers could reliably prepare their fields thoroughly and in a timely fashion, which in itself improved yields significantly. Lastly, mechanization meant no land need be devoted to feed for working farm animals, which in the United States in 1920 had taken about a quarter of all cropland.

The social consequences of mechanization were at least as great. Machines expelled farm labor, providing the workforce for industrialization. In the United States the farm labor force, nearly half the population in 1920, plummeted to 2 to 3 percent by 1990. The accelerated migration of rural southern blacks to northern cities—a landmark in U.S. history—derived in large part from mechanization of the cotton harvest, begun in the late 1940s. In the Soviet Union, where mechanization lagged, especially in potato and fruit harvesting, the agricultural population remained as high as 30 percent into the 1980s. Farm mechanization added a decisive boost to modern urbanization.

It helped farms grow too, through economies of scale. American farms, constant in average size from 1890 to 1930, tripled in size from 1935 to 1985. In the USSR, collectivization brought gigantic farms. Collectives averaged 1,600 hectares in 1940 and 11,000 in 1968. In 1977, state farms in the Soviet Union averaged 40,000 hectares, three times the size of Washington, D.C. These gargantuan dimensions reflected ideological commitments, made feasible by the enhanced labor productivity of mechanization.[62]

---

[59]Harvesting absorbed about half of farm labor before mechanization, and is the key operation to mechanize. If one cannot mechanize the harvest, then saving labor on planting or other operations makes less sense.

[60]In the USA, horse- and steam-powered machinery had this effect in some places as early as the 1870s: in Minnesota a prosperous farmer named Oliver Dalrymple plowed furrows 10 km long (Rasmussen 1982).

[61]Pests' predators by and large did not find monocultures appealing habitat, and did not follow pests into unbroken fields in adequate numbers (Andreas Kruess and Teja Tscharntke, as reported in *Science News,* June 11, 1994:375).

[62]Bairoch 1989:327 shows the evolution of labor productivity in Western agriculture: it tripled in the 19th century, then increased 13-fold in the 20th century (to 1985), with almost all the increase coming after 1950.

The international consequences of farm mechanization, modest in the overall scheme of geopolitics, favored the big grain-producing countries. Machines shattered production bottlenecks in lands where farm labor was short, such as North America and Australia, boosting production and prosperity. Mechanization also helped countries already positioned to take advantage of it, those with big fields, flat farmland, and climates suitable for grain. Mechanization did nothing for lands necessarily carved into tiny fields, for farms on steep slopes, or for the banana harvest. It also did nothing for societies that lacked, and could not build, the infrastructure of motorization—repair shops, spare parts, oil supply systems, and so forth. On balance, farm mechanization helped the United States, Australia, and Canada most; Argentina and the USSR considerably. Among great powers, China and Japan benefited least. Both in agricultural ecology and in international affairs, farm mechanization helped select the winners—among crops, pests, and nations—in the twentieth century.

THE GREEN REVOLUTION. Mechanization dovetailed with the Green Revolution, a crucial departure in agriculture that depended centrally on plant breeding. The Green Revolution was a technical and managerial package exported from the First World to the Third beginning in the 1940s but making its major impact in the 1960s and 1970s. It featured new high-yielding strains of staple crops, mainly wheat, maize, and rice. Plant geneticists selected these strains for their responsiveness to chemical fertilizer and irrigation water, for their resistance to pests, and eventually for their compatibility with mechanized harvesting. Success required new inputs, new management regimes, and often new machines. The great triumph came with dwarf wheat and rice, which could hold up a heavy, grain-packed head without bending or breaking the plant stalk. Like the great political revolutions of the twentieth century, the Green Revolution drew intellectually mainly from the Western world, changed its forms when it spread elsewhere, and led to unexpected consequences.

Dwarf wheat, although finally created in Mexico with American money and expertise, had deep roots anchored in the work of an Austrian monk and Japanese agronomists. The mathematical genetics of Gregor Mendel (1822–1884), lost for decades after he published his work, was rediscovered in 1900. Scientists around the world took note, including those at work on rice and wheat breeding at the agricultural research stations sponsored by Meiji Japan. There the Ministry of Agriculture and Forestry had organized crop breeding research in the 1880s. At a time when land hunger turned many peasants into emigrants and politicians into imperialists, the ministry sought rice and wheat breeds suitable for Japanese circumstances: scarce land, plenty of night soil. In 1925 its research succeeded, producing after many crossbreedings of Japanese and American wheats, a semidwarf wheat known as Norin 10. (Norin was the ministry's acronym.) The ministry distributed it to farmers in 1935, but World War II came before it had much impact on Japan's food supply. In 1946 an agronomist with the U.S. Army noticed Norin 10 and imported it into the United States, where it was further crossbred with wheats in the state of Washington. Norin 10 did not solve prewar Japan's food problems, but it eventually changed the world.[63]

American farmers and plant breeders were hard at work on hybrid maize, hoping to concoct higher-yielding and disease-resistant strains. The great Charles Darwin had been among the first to dabble in crossbreeding maize, publishing his results in 1876. American disciples furthered the work and by 1918 had developed the double-cross, the basis for all subsequent hybrid maize. In 1930 only 1 percent of U.S. maize acreage was sown to hybrids, but the USDA converted to the new gospel in the 1930s. By 1939 a sixth of U.S. corn was hybrid, by 1950 three-quarters, and by 1970 over 99 percent. U.S. corn yields rose to three or four times the levels of the 1920s.[64]

The first farmer to make hybrid corn a commercial success was Henry Wallace in the 1920s.[65] Wallace, later Franklin Roosevelt's Secretary of Agriculture and Vice President, took a special interest in Latin America. As a successful farmer who liked to be called the "father of industrialized agriculture," he saw great opportunity for applying the fruits of modern genetics to the venerable techniques of farming in Mexico and points south. He helped persuade the Rockefeller Foundation to finance a wheat and maize research center in Mexico (1941–1943), which soon

[63]Hayami and Yamada 1991; Hayami 1975. Italian crop breeders also enjoyed early success, using Japanese varieties from 1912; by 1932 a quarter of Italian wheat—mostly in the north—was early-ripening breeds derived from crosses with Japanese wheat. These Italian wheats by the 1970s were in wide use in the Mediterranean world (Dalrymple 1974:10–11).

[64]Mangelsdorf 1974:211–14. In a double-cross, four varieties are combined in two "generations," selecting for the most desirable characteristics. See also Fitzgerald 1990.

[65]Wallace (1888–1965) was a farmer, publisher of an agricultural journal, and son of Warren Harding's Secretary of Agriculture. Wallace became FDR's Vice President in 1941, but was replaced by Truman four years later. He fell out with Truman, partly over foreign policy, and subsequently ran for the Presidency several times as leader of the Progressive Party.

hired another son of the Iowa soil, Norman Borlaug (born 1914).[66] Borlaug had a fresh Ph.D. in plant pathology when he arrived in Mexico in 1944. By 1953 he had set to work combining Norin 10 with Mexican and American varieties. Over the years he and his associates created new wheat strains that proved enormously responsive to heavy dosages of nitrogen and timely ones of water, and which in some cases were highly resistant, at least initially, to pests and plant diseases.

Borlaug was the father of Mexico's Green Revolution. He won a Nobel Peace Prize in 1970 for his plant breeding, and has a street bearing his name in Hermosillo in northwestern Mexico, the heartland of high-yield farming in Mexico. With the help of Ford and Rockefeller money, the United Nations Food and Agriculture Organization (FAO), U.S. Agency for International Development (AID), and other organizations, the revolution spread from Mexico. Its greatest progress came in southwestern Asia's wheat belts, from the Indian Punjab to Turkey, beginning in 1963 when Borlaug sent dwarf wheats to Indian crop-breeding stations. By 1968, 18 countries sowed dwarf wheat.[67]

Similar developments took place in the Philippines after the 1960 birth of the International Rice Research Institute (IRRI), also sponsored by the Rockefeller Foundation.[68] Using dwarf rice strains first selected by Japanese breeders in the 1920s in Taiwan (then a Japanese colony), rice geneticists created high-yield rice varieties that combined the best features of tropical (indica) and temperate (japonica) rice. By the late 1960s, new breeds from IRRI carried the Green Revolution to the world's great rice basket, the broad arc from Bengal to Java to Korea. China developed its own high-yield rice, beginning in 1959. It came too late to mitigate the great famine of 1959 to 1961, which killed some 25 million to 30 million people—a political more than an ecological event in any case—but it did arrive in time to help Chinese agriculture weather the storm of the Cultural Revolution (1966–1976).[69]

Borlaug saw the Green Revolution as humankind's best hope to feed rapidly growing populations, as perhaps it was. But its geographic spread

[66]Jennings 1988 deals with the politics of the Rockefeller initiative. Apparently Henry Wallace and the top managers of the foundation hoped to support the president elected in 1940, Avila Camacho, from whom they hoped for some compromise with respect to American property (including that of Standard Oil) nationalized by the previous president, Lázaro Cárdenas. Wheat was then a minor crop in Mexico. See also Fitzgerald 1986.

[67]Lupton 1987:68–9.

[68]Anderson 1991.

[69]Dalrymple 1974:10–15, 73–75.

suggests it had other attractions as well. The Green Revolution had little impact on sub-Saharan Africa, except for higher maize yields, notably in Zimbabwe.[70] It received its greatest support outside of Mexico on the frontiers of the communist world from Turkey to Korea, and recommended itself as a means to blunt the appeal of socialist revolution, at its height in the 1960s. The rice program in particular originated from American anxieties about the possible spread of Chinese communism after 1949. Meanwhile, socialist societies—China, Vietnam, and Cuba at least—embraced the idea of scientifically improved crops with equal vigor. In several of its manifestations, then, the Green Revolution was a child of the Cold War.

The Green Revolution also appealed keenly to most of the influential segments of society within Asian and Latin American countries. It promised to augment the incomes of landed elites and, where this was an issue, make land reform less urgent. To state bureaucracies it seemed to show a way to urban industrial society, and hence to wealth and power, without the risks of alternative paths. A more efficient agriculture, particularly an export-oriented one, could build up capital needed for industrialization and at the same time get labor off the land and into factories. Achieving this without the brutal costs of Soviet methods, or the noose of huge foreign debt, made excellent sense to the influential ministries of Mexico or Indonesia. Furthermore, the Green Revolution promised independence from American food aid, which recipients normally suspected as a political tool.[71] For all these reasons, both in the United States and in Latin America and Asia, the Green Revolution in the 1960s and 1970s was a technology package whose time had come.

Its impact was sudden, substantial, and like most revolutions, not quite what its instigators had in mind. Dozens of countries managed to keep food production ahead of population growth, mainly thanks to high-yield wheat and rice. By 1970 about 10 to 15 percent of the Third World's wheat and rice area was under the new varieties. By 1983 the proportion was over half, and by 1991 three-quarters.[72] In China, high-yield strains accounted for 95 percent of rice and maize by 1990.

---

[70]Hybrid maize research began in Southern Rhodesia (now Zimbabwe) in 1930. A variety (SR-52) released in 1947 proved successful on the commercial farms (mainly white-owned) after 1950. SR-52 raised yields in southern Africa by about 50%, but the expensive inputs limited its use among poorer peasants. No other Green Revolution crop had significant impact in sub-Saharan Africa (Jahnke et al. 1987; see also Low 1985).

[71]This occurred especially after the passage in 1964 of U.S. Public Law 480, which connected food aid to improving attitudes toward the United States.

[72]Tolba and El-Kholy 1992:296; WRI 1996:226.

This dissemination of new breeds amounted to the largest and fastest set of crop transfers in world history. Bountiful yields became routine between 1960 and 1990, with effects more sudden and sizable than those of history's previous agricultural turning points. From the dawn of agriculture until the seventeenth century, yields (in Europe at least) increased only about 60 to 90 percent. The first "agricultural revolution" beginning in England around 1680, doubled English yields within 70 to 90 years. Most other European societies followed suit. Then between 1860 and 1910, yields doubled again—much less in the United States, more in Europe. Meanwhile, outside of Europe and the USA, yields and labor productivity stagnated or declined between 1800 and 1950, contributing to the prevailing inequalities in wealth and power.[73] With its basis in crop breeding and the transfer of successful strains, the Green Revolution merits comparison with the great historical crop introductions, such as the arrival of American food crops (maize, potato, cassava) in Eurasia and Africa after 1492, the importation of Southeast Asian plantains into tropical Africa, and the Arab introduction of citrus and sugarcane to the Mediterranean world after A.D. 900. Table 7.3 gives an impression of the phenomenal changes in crop yields since 1960 in the Third World.

As these yield figures show, the Green Revolution created the promised fields of plenty. It did much else besides. Ecologically it combined with mechanization to promote monoculture. Since farmers now had to purchase seed rather than use their own, and because they needed fertilizers and pesticides specific to a single crop, they saved money on inputs

TABLE 7.3 YIELD HISTORY IN 93 DEVELOPING COUNTRIES, 1961–1992

| Crop | Yield (kg/ha) | | | | Increase Factor |
	1961–1963	1969–1971	1979–1981	1990–1992	1961–1992
Wheat	868	1,153	1,637	2,364	2.7
Rice	1,818	2,218	2,653	3,459	1.9
Maize	1,157	1,456	1,958	2,531	2.2

Source: Adapted from WRI 1996:226.

[73]Bairoch 1989.

by buying in bulk for one crop. Monoculture, as explained earlier, invites pest problems. Often even the initially pest-resistant crops eventually proved vulnerable to one or another infestation. Hence farmers turned to heavier and heavier doses of pesticides. This efficiently selected for resistant pests—as antibiotics did for bacteria.[74] Meanwhile, most of the pesticides missed their targets and ended up elsewhere, sometimes in water supplies, human tissues, and other awkward places. The WHO estimated in 1990 that pesticide poisoning killed about 20,000 people per year, mostly in cotton fields. Roughly a million people (as of 1985) suffered acute poisoning, two-thirds of them agricultural workers.[75] The vast fertilizer requirements of the Green Revolution led to eutrophication of lakes and rivers. The necessary irrigation helped drive the huge dam-building programs of China, India, Mexico, and elsewhere (as described in Chapter 6). The Green Revolution also altered the species and genetic diversity of agriculture: it extended the sway of rice, wheat, and maize, reducing the use of lesser crops not so responsive to nitrogen- and water-rich diets; and it vastly reduced the varieties of rice, wheat, and maize in wide use. Before the Green Revolution, farmers raised thousands of strains of wheat around the world. After it, they increasingly used only a few. In this respect, the Green Revolution was a gamble that scientific agriculture could protect a few high-yield strains from pests and diseases. By and large it has, by staying one step ahead of the evolution of pests, as antibiotics did with pathogens.[76] It was also a gamble that oil and water would remain cheap enough to satisfy the energy gluttony and bottomless thirst of the new agriculture. So far, this too has worked, within limits.[77]

The social consequences of the Green Revolution brought more surprises. In many settings it did not defuse agrarian tensions. In Mexico and the Indian Punjab, for example, the Green Revolution strongly favored farmers with reliable access to credit and water. Some of the less fortu-

[74]Resistant pest species increased about four-fold between 1955 and 1988 (estimated from Tolba and El-Kholy 1992:295).

[75]Ibid. p. 296; see also Pimentel and Lehman 1993.

[76]A partial exception came in 1970 when a fungus especially well adapted to U.S. corn destroyed 15% of the harvest, prompting farmers to vary the genetic profile of their seed in 1971 (Mangelsdorf 1974:213).

[77]See Freeman 1993 for the limits as seen by a booster (a former U.S. Secretary of Agriculture). Pimentel and Heichel 1991 show that in energy terms high-yield crops are about 25% as efficient as hand-and-hoe agriculture, and half as efficient as ox-and-plow farming. Indeed, modern U.S. farming burns far more calories than it produces when one factors in the energy requirements of making fertilizer.

nate drifted to the cities, some went to work for the more successful, others went to the United States or the Persian Gulf to work for wages—in cases to accumulate enough capital to become prosperous peasants—Green Revolution *kulaks*—themselves. As a rule, though not without exceptions, the Green Revolution promoted income inequality among farmers. Where alternative employment for the losers was hard to find, as in Punjab or highland Ethiopia, social frictions intensified into overt class and ethnic or religious conflict. A sampling of the literature suggests that the social effects proved more favorable in lands raising rice than those with wheat.[78]

The Green Revolution, like farm mechanization, selected winners internationally as well as ecologically and socially. South Korea, China, India, and to a lesser extent Mexico improved their agricultural balance of payments, reduced or eliminated food dependence, and, whatever the ecological or social costs, improved their international economic and political position. Countries that could not create favorable conditions for the Green Revolution—those with too little water or underdeveloped credit markets—suffered in comparison. Broadly speaking, this meant sub-Saharan Africa sank in the scales against Asia and Latin America. As a Cold War tool of the West—which in part it was—the Green Revolution served its purpose, although high-yield rice strengthened communist China as much as it did Asia's island fringe, which America relied upon to contain China.

The Green Revolution did something—but not much—to empower Latin America and tropical Asia vis-à-vis the West and Japan. It helped in the industrialization drives of Taiwan, South Korea, Indonesia, and the other "Asian tigers." It made India a food exporter. But while the Green Revolution made Third-World agriculture more land- and labor-efficient, it could not match the productivity increases ongoing in the West and Japan. In 1950, agriculture in the West was seven times more labor-efficient than in the Third World; in 1985, 36 times more—and about 36 times more prosperous. The Green Revolution did not engineer an income redistribution toward Third World farmers. Nor did it achieve food independence except for a few countries. Until 1981 the Third

[78]Critiques of the social implications of the Green Revolution are numerous. See Shiva 1991b; Thandi 1994 (on Punjab); Hazell and Ramasamy 1991 (on Tamil Nadu, where things went well); Alauddin and Tisdell 1991 (on Bangladesh, where they did not); Simonian 1995:170–72 and Sonnenfeld 1992 (on Mexico). On income distribution specifically, see Sharma and Poleman 1993 and David and Otsuka 1994. Bezuneh and Mabbs-Zeno 1984 write on Ethopia's Green Revolution and its impetus to unrest and political revolution in the 1970s.

World had long been a net exporter of food; after 1981 it was a net importer.[79]

These facts derive from the continuing agricultural revolution in the West and Japan from which the Green Revolution was an offshoot; and from agricultural and trade policy. Take Britain, for example, a major food importer since the repeal of the Corn Laws in 1846. British agricultural yields stagnated between 1890 and 1940. Subsidies and protection began in the 1930s. Meanwhile, the biological, chemical, and mechanical transformations of modern agriculture took hold, and yields rose beginning in 1942. The difficult food situation in Britain during World War II and the immediate postwar years predisposed all governments, Labour and Conservative, to favor agricultural subsidies, firmly buttressed in 1947, and to pursue higher yields through scientific agriculture. Yields doubled or tripled by the 1980s. To the dismay of devotees of the principle of comparative advantage, Britain, which grew 30 percent of its grain in 1936, managed self-sufficiency in 1986.[80] Similar miracles occurred in the fields of much of Europe, Japan, Australia, New Zealand, and North America after 1945.[81] The Soviet Union partly missed out because modern genetics rankled the socialist sensibilities of Stalin and Khrushchev, delaying progress in plant breeding until the mid-1960s. Before 1960 the USSR partook of mechanization and irrigation, but only after 1965 did it undertake conversion to the doctrine of genetic manipulation and heavy use of nitrogen. Thus the total geopolitical effect of modern changes in agriculture improved the relative position of the West and Japan slightly, that of China, the Asian tigers (South Korea, Taiwan, Malaysia), and Latin America even more slightly, while contributing to the relative decline of Soviet status, and to the weakness of Africa.[82]

[79]Bairoch 1989:346.

[80]Blaxter and Robertson 1995. The hallowed principle of comparative advantage states that countries maximize their income by producing only what they produce most efficiently and trading for all other goods. This Britain resolutely did not do, preferring food security to income maximization.

[81]See Cochrane 1993; Hayami and Yamada 1991. In the century 1880–1985, Japanese agriculture increased labor productivity 16-fold despite the requirements of rice. Land productivity (yields) improved by a factor of 4.4 (Hayami and Yamada 1991:253–4).

[82]In addition to the works cited above, I consulted Burmeister 1990, and Freeman 1993.

Conclusion

The general transformation of farming after 1940, of which mechanization and the Green Revolution were parts, both shaped the twentieth century and reflected its dominant trends. It was energy- and knowledge-intensive. It replaced simpler systems with more complex ones, involving distant inputs and multiple social and economic linkages. It reduced family and regional autonomy, enmeshing farmers in a world of banks, seed banks, plant genetics, fertilizer manufacturers, extension agents, and water bureaucrats. It transplanted what worked in the West and Japan to other societies. It sought to harness nature tightly, to make it perform to the utmost, to make it maximally subservient to humankind or at least some subset thereof. And it sharply increased output, making us dependent upon its perpetuation. As of 1996, to feed ourselves without these changes, we would have needed to find additional prime farmland equal in area to North America.[83]

Lacking such a spare continent, the human race ended the twentieth century in a rigid and uneasy bond with modern agriculture. Our recast agroecosystems depended on social and international stability to safeguard required flows of inputs. Our social and political systems required the perpetuation of these agroecosystems.

The modern agricultural revolution was nearly as important as the new regime in human-microbial relations in shaping the twentieth century. Both fundamentally affected the well-being, health, and security of life for billions of people. Both helped govern the ongoing redistributions of power and wealth among classes and nations. Both represented a drift toward ever greater complexity—and potential vulnerability to disruption—in the systems that underpin modern life.

The fact that we are not more often food for microbes depends on the precarious balances of modern public health; that we in turn have as much to eat as we do (questions of distribution aside), depends on the no less precarious balances of modern agriculture. "Though you drive out nature with a pitchfork, yet it will always return." So thought the Roman poet Horace.[84] Is his wisdom now out of date?

---

[83]This figure comes from Dennis Avery, cited in the *Economist*, 16 July 1996:23. Other solutions include, of course, more equitable distribution of food and an alternative technical means of making food production more land-efficient.

[84]From Epistles 10:24: "Naturam expellas furca tamen usque recurret."

8

The Biosphere: Forests, Fish, and Invasions

In the realm of Nature there is nothing purposeless, trivial, or unnecessary.

—Maimonides, *The Guide for the Perplexed*

The heady changes in human health and agriculture meant more people lived longer in the twentieth century. That, together with more powerful technologies and more tightly integrated markets, encouraged faster harvesting of forests and fish. The frenetic pace of twentieth century economic activity—long-distance trade especially, but the feverish five-year plans too—linked ecosystems more systematically than ever before, and with biological consequences that we can in many cases as yet only dimly perceive. In some cases, however, the consequences are as plain as day.

Forests

Forests experienced another bad century, declining in area and quality.[1] Cropland expansion played a central role in forest disappearance. By and large the sharp changes took place in the tropics and in northern, or boreal, forests. The long history of deforestation, from the dawn of agriculture to the present, accounts for a reduction in global forest area of somewhere between 15 and 45 percent; estimates vary maddeningly.[2] In Africa and monsoon Asia, only about one-third of the forests of 10 millennia ago still remain. In the Americas, on the other hand, roughly three-quarters still stand; in Russia two-thirds. At the end of the twentieth century, big blocks of forest stood in only three places in the world: the Amazon and Orinoco basins of South America; across northern North America, from Labrador to Alaska; and across northern Eurasia, from Sweden to Sakhalin. Great swatches once existed but disappeared into patchwork remnants in four places: from central India to northern China; Madagascar; Europe and Anatolia; and the Atlantic coast of Brazil. Giant blocks of forest now much shrunken and degraded stand in tropical Africa and eastern North America. Of this monumental forest clearance, perhaps half took place in the twentieth century. Nearly half of this was cleared in the tropics between 1960 and 1999.[3]

The two main reasons people remove forests are to use the timber or to use the land. Before 1860 in North America, 90 percent of deforestation was driven by conversion to farmland or pasture.[4] Similar proportions probably obtained elsewhere. Since then, however, logging has emerged

[1]Quality in this case is measured as total biomass, not that generated per year. I shall use the term "degradation" to describe decline in forest biomass. Such terms require definition. A regenerating cutover forest may grow additional wood faster than a mature forest, but it does not provide the same wildlife habitat or hydrological stability.

[2]Meyer 1996:61 offers 15%; Williams 1994:106 suggests 17.6%; WRI 1997 gives 46%. My Table 7.1 implies 26%. See Mather 1990 and Williams 1994 for discussions of estimates. Williams 1990c and Richards 1990b differ markedly within the same volume on the scale and scope of historical deforestation.

[3]WRI 1997 says 450 million ha of forest were lost in the tropics from 1960 to 1990 (an FAO estimate). My Table 7.1 implies nearly two-thirds of total deforestation took place in the twentieth century (1 billion of 1.7 billion ha). Williams 1990c, 1994, estimates total woodland and forest clearance at about 9 billion ha, following E. Matthews. Richards 1990b:164 estimates a net forest loss during 1700–1980 at 1.2 billion ha. My estimates lie in between, but are somewhat closer to Richards'.

[4]Mather 1990:45.

after Dean 1995; Monteiro and Kaz 1992

ARCTIC OCEAN

PACIFIC
OCEAN

ATLANTIC
OCEAN

5°S

10°S

15°S

Bahía

20°S

25°S

Rio de Janeiro

São Paulo

ATLANTIC OCEAN

30°S

Atlantic Forest of Brazil

In 1500   In 1990

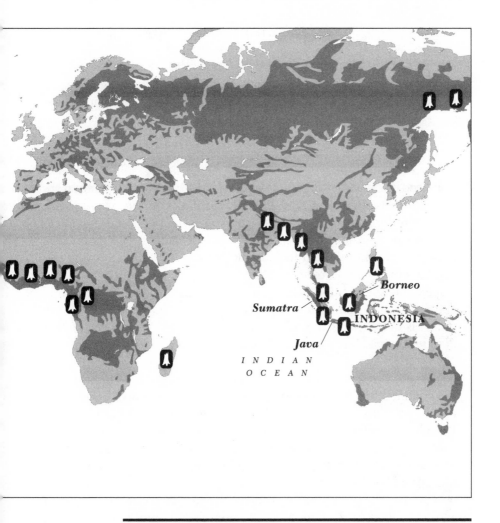

7. Changes in World Forests, c. 1920 to c. 1990

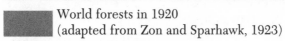
World forests in 1920
(adapted from Zon and Sparhawk, 1923)

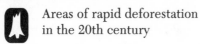
Areas of rapid deforestation
in the 20th century

as a major force in deforestation. At the end of the twentieth century, logging caused almost all deforestation in temperate and boreal latitudes but less than half of it in the tropics, where forest conversion proceeded fastest. Over the course of the century, logging motivated perhaps a sixth of the deforestation and land conversion the rest.[5]

While the general trend of the twentieth century was one of forest loss, sharp distinctions existed within that. By and large, temperate forests stabilized or expanded after 1910 or 1945 at the latest, while tropical and boreal forests shrank, most rapidly after 1960. That was true because temperate forests had declined sharply in previous centuries and could rebound because of the confluence of three factors: slow population growth, less land required for farming because of yield improvements, and the emergence of overseas sources of supply for timber. In this last respect, the stabilization of forest area in temperate lands promoted the deforestation of the tropics. To illuminate these matters, consider two transects of the globe: the Atlantic forests of continental-sized countries in the Americas, and a pair of Asian archipelagoes.

NORTH AMERICA AND BRAZIL. The Atlantic coastlands of the American continents in 1500 supported great forests. Amerindian burning had reduced them slightly from their maximum extents. But after 1500 in Brazil's coastal forest, and after 1607 in North America's eastern woodlands, a long siege by colonial settlers began. In North America, farmers, with some help from loggers, cleared huge areas in the Canadian Maritimes, Quebec, Ontario, and the eastern 30 U.S. states. By 1910 they had occupied all the promising farmland of the eastern part of the continent and cut over all the good timberlands, from the cedars of Louisiana to the white pine of New Brunswick. A quarter of the cut in 1900 went for railroad ties, which before 1920 had to be replaced every few years. In all, more than half of these eastern forests disappeared between 1607 and 1920.

After about 1920, however, the eastern woodlands returned. The process began as early as 1840, when midwestern and Ontario farms drove older ones in New England and the Maritimes out of business. Farm abandonment in the East continued rapidly as the railroads girded

---

[5]In the 1980s, the FAO estimated that logging accounted for two-thirds of moist tropical forest destruction (Westoby 1989:153). Fuelwood and charcoal extraction (roughly 1 billion m³ in 1960 and 1.8 billion m³ in 1992), can (rarely) eliminate forest. Data from FAOSTAT, on the FAO web site, 22 September 1997. Even lands of ancient settlement, such as Iran, lost much of their forest cover only recently. See Planhol 1969.

the continent. Then, from the 1930s, more marginal cropland was abandoned to forest regrowth as farming yields rose. The toll of the lumber business also declined: per capita timber use in the United States peaked in 1907, and total use of forest products declined modestly from 1910 or so, because fuelwood use plummeted after 1935, and because iron, steel, and plastic replaced wood in many uses. Finally, fire suppression lowered the loss to forest fires by about 90 percent between 1930 and 1960. For all these reasons, the eastern woodlands regenerated—and total forest area in North America stabilized after about 1920, as eastern growth roughly equaled clearances in the West.[6]

In Brazil things worked out differently. Agriculture, especially sugar, slowly chipped away at the million square kilometers of coastal forest until 1850. Then, in the hinterlands of Rio de Janeiro and São Paulo, coffee joined the assault. Brazilians firmly believed coffee needed forest soils. "The Green Wave of Coffee," wrote Monteiro Lobato, a prominent writer of the early twentieth century, "produces only at the cost of the earth's blood. It is . . . insatiable of humus."[7] Railroads, mines, fuelwood removals, and by the late 1920s a timber trade—all took a share as well. The coastal forest, from Recife to the southern state of Rio Grande do Sul (still nearly 400,000 km^2 in 1900), shrank faster and faster. Agriculture never slackened its demand for new land, as population growth remained strong in Atlantic Brazil, and the land tenure system assured a steady supply of landless peasants. Fuelwood and charcoal consumption remained high—Brazil had almost no fossil fuels. After 1950 dam building, Brazil's solution to its energy constraints, flooded out additional forest. By 1990, only 8 percent of Brazil's Atlantic coastal forest remained. The technological and social changes that halted forest destruction in North America did not happen, or happened too faintly, to arrest deforestation in coastal Brazil. In consequence of continued land hunger and other concerns, Brazil's government systematically opened up Amazonia to colonization in the 1960s. Between 1960 and 1997 roughly 10 percent of that vast forest, the world's largest rainforest and its most botanically diverse province, became pasture, farmland, or scrubland.[8]

[6]The total stock of wood in U.S. forests grew faster than the total harvest from c.1950 (MacCleery 1994) or 1963 (Mather 1990) onward. The North American material is from Lower 1973; MacCleery 1994, Marchak 1995; Williams 1988, 1990c. U.S. forests shrank about 1% in the 1980s.

[7]Quoted in Dean 1995:245.

[8]Dean 1995; McNeill 1988. On Latin America generally, see Houghton et al. 1991.

JAPAN AND INDONESIA. The forest histories of Atlantic North America and Atlantic Brazil formed a contrast but had no significant links. In the cases of Japan and Indonesia, sylvan histories became meshed: Japan's retention of forest cover depended on timber imports from Indonesia, among other places.

Japan has a singular forest history.[9] Its green mantle shrank and grew back over the centuries, depending on economic cycles and national policies. From about 1780 to 1860 careful forestry had checked deforestation. But by the late nineteenth century, with the intense drive on the part of the Meiji government to industrialize the country, Japan's forests came under great pressure once more. By 1900 virtually no old-growth forest existed except on the sparsely populated northern island of Hokkaido. Efforts to safeguard remaining forests fell prey to the demands of militarization in the 1930s and the extreme fuel shortages of the war years. By 1945, Japan's forests were heavily overexploited, and postwar reconstruction of bombed-out cities intensified the search for timber.

After 1950 a great reversal took place, and Japan once more became a "green archipelago." This reversal derived from two causes. First, in the 1950s, Japan shifted away from charcoal and fuelwood to fossil fuels in its energy system; charcoal use was banned in the 1960s. Second, Japan began to import timber in a large way, permitting a more or less free market. Softwoods came from Oregon, Washington, and British Columbia, and at times from Siberia. When the government abandoned its tariff protection of domestic lumber, Douglas fir undercut domestic log prices. Hardwoods came from Southeast Asia—first from the Philippines and Malaysia, then after 1965 from Indonesia. Japan became the world's largest importer of timber and pulpwood, and its companies created a unified Pacific Rim market in forest products. In this way Japan alleviated pressure on its own forests, which by the 1980s covered a larger share of the national territory (67 percent) than in any other temperate country except Finland.[10]

Indonesia is a sprawling equatorial archipelago, from end to end as long as the continental United States. Parts—eventually all—of it came under Dutch control from 1610 to 1949. It once featured widespread tropical forest, part of an enormous block that long ago stretched with gaps from the Ganges to the Yangtze, and to northern Australia. The Indian and Chinese chunks of this block shrank back before the axe and

[9]Totman 1989.
[10]Cox 1988; Marchak 1995:117–42.

plow centuries ago. The portions in Thailand and peninsular Malaysia did so in the twentieth century, mainly after 1955.[11] The biggest bite came out of Indonesia's forests after 1965.

Systematic depletion of Indonesian forests dates at least to 1677, when the Dutch East India Company began building ships of Java teak, one of the world's most durable timbers. The rich volcanic soils of Java and Bali encouraged population and agricultural growth. By 1930 Java had 42 million people, compared to 3 or 4 million in 1600, and wet-rice cultivation and teak cutting had eliminated much of the natural forest. Japanese occupation (1942–1945) led to record-high teak and fuelwood extractions. War, occupation, revolution, insurgencies, and civil war (1941–1967) trimmed back Java's forests, and continued demographic growth (to 105 million by 1985) did too. By the mid-1960s, Java's forests consisted of second-growth swiddens[12] and replanted teak from the Dutch era of scientific forestry (c.1860–1941). But the larger islands held plenty of tall timber.[13]

Sumatra and Borneo, large islands with poor soils and historically scant populations, still supported vast forests. One Dutch official in the 1920s noted the fine timber in eastern Borneo, but expected there would never be enough people to cut it. That changed. After a bloody coup and civil war in 1965–1966, General Suharto (1921–) replaced General Sukarno (1901–1970) as president of Indonesia. Principles of economic nationalism and hostility to foreign investment waned, paving the way for capital-intensive forestry. Simultaneously, logging technology advanced so as to make previously remote stands accessible. These new conditions—and the near exhaustion of marketable timber in the Philippines—provoked a feverish assault on the forests of Indonesia's outer islands.

The arrangements under which this happened played a major role in shaping and maintaining the Indonesian state. Emperors from Rome to China had long ago rewarded faithful soldiers with land. Suharto used logging concessions in the same way. His officers, and a few other chosen friends, acquired concessions, joined forces with foreign firms for capital and expertise, and made their fortunes by exporting logs to the sawmills and pulp factories of Japan, Singapore, Korea, and Taiwan. Indonesia soon became the world's largest exporter of tropical timber. Suharto and his generals saw that making the plywood and pulpwood in

[11]Brookfield et al. 1990; Feeny 1988.
[12]That is, low forest regrowing in land once cleared and farmed for a few years.
[13]Java's forest history is from Durand 1993, Peluso 1992, and Potter 1996.

Indonesia would make them richer still, so in the 1980s new laws reduced and then banned log exports. But the felling continued apace; Indonesia by 1982 led the world in plywood exports. The 20-year duration of concessions discouraged replanting, despite laws after 1980 requiring it, so the logging frenzy left behind a patchwork of scrub and grass, with some acacia and eucalypt plantations.[14] By 1990 about a third of Indonesia's forests had vanished. In 1992 when the U.N. Conference on Environment and Development issued a nonbinding resolution suggesting a ban on tropical logging by the year 2000, Indonesian concessionaires redoubled their efforts so as to make sure they got their assets to market in time. The lucky ones converted their allotted trees into money before the huge fires of 1997–1998, which cast a pall over Indonesia and Malaysia for months and turned their trees into ash and smoke.[15]

The stories of Indonesia and Brazil form but a chapter in one of the central events of our time: the great clearance of tropical forests. Tropical forests—a varied category—once covered about 1.5 billion hectares spread mainly among Latin America, West and Central Africa, and southern and Southeast Asia. By the mid-1990s about one-third of that area had been converted to other land cover. Most of this forest removal—over an area about the size of India—occurred between 1960 and 1990.[16] Africa lost about half its tropical forests, Latin America not quite a third. The late twentieth century was a great age of deforestation, like that of the Roman Mediterranean, Song China, or North America in the railroad age. The scale in the late twentieth century was larger, the ecological effects quite different, and the technologies employed radically different, but the motives much the same: arable or grazing land and marketable timber. Alarm about the obliteration of forest peoples, the lost ecosystem services, and the contribution of deforestation to greenhouse gas accumulation, while considerable after about 1980, weighed little in the balance against these motives. The political power of the beneficiaries of deforestation was far too great.[17]

---

[14]Potter 1996 says 11% of timber concessions complied with the requirement to reforest.

[15]Marchak 1995:237–68; Brookfield et al. 1990; Westoby 1989:133–8; Potter 1996; Dauvergne 1997.

[16]WRI 1996:201; NRC 1993b:34–5. About 8% were cleared in the 1980s alone.

[17]In Indonesia, for example, the trade associations for the sawmillers, loggers, plywood manufacturers and furniture makers were all headed by Suharto's golfing partner, Bob Hasan. In consequence, these groups were rarely denied the legislation they wanted (Bresnan 1993:267). On Philippines deforestation and politics, see Kummer 1991.

In West Africa, baobab trees carried cultural and social significance as meeting places and sites where local rulers held court. To fell this baobab, these men injected poison into its roots, waited for the tree to die, then knocked it over with a bulldozer. They were preparing the land for cotton planting in Mali, around 1950. One of the largest, and least successful, development projects of the late colonial era was the French effort to promote cotton in Mali. Around the tropical world after 1950 the quest for new farmland felled millions of trees.

Whaling and Fishing

The driving forces behind tropical deforestation—money to be made, mouths to feed—also recast the hydrosphere's biota. Governments and international organizations recognized some of the ill effects of the unchecked harvesting of fish and whales, as of timber, but they did little about it. Fisheries are notoriously hard to manage.

As an open-access resource, fisheries were routinely overfished. Fishermen had no incentive to let fish remain in the sea where others might catch them. While all fishermen would be better off if they could refrain from catching young fish, which jeopardizes future fish stocks, any fisherman who did so lost fish to others who did not. Since many fish and whales are migratory and none respect territorial waters, efforts to keep fish for oneself or one's compatriots brought only modest results. For most of the twentieth century, these difficult facts kept the fisherman's life hard and poor. For commercially viable fish and whales it made life short.

WHALING IN THE SOUTHERN OCEAN AFTER 1904. Few creatures felt the rough hand of the rogue primate quite like whales.[18] Whales enjoyed unusual peace—they had few predators—for their first 50 million years. Whaling began in prehistoric times, and until the end of the nineteenth century was pursued by souped-up Paleolithic means: chase the prey and throw a spear—a harpoon—into it. On one whaling ground after another, intrepid whalers hunted their quarry to the point where too few were left to reward the chase. Vikings and Basques, the pioneers of deep-sea whaling, began the process in the North Atlantic. Dutch and English whalers brought Spitzbergen and Greenland bowhead whale populations near to extinction between 1610 and 1840.[19] The Industrial Revolution quickened the hunt for whales. Sperm oil became especially useful in lubricating machinery. Manufacturers found many uses for baleen—the plastic of the nineteenth century—in corsets, umbrellas, stays, and the like. Also called whalebone, baleen is the hornlike substance forming the plates through which toothless (baleen) whales strain their meals. Americans carried the hunt to the wide Pacific and dominated it between 1820 and 1860. By 1860 most of the easily hunted whales—sperm and right whales—were gone. The last large grounds of bowhead whales, in the Bering Sea, were hunted out by Americans by 1890, bringing famine to Aleuts and Chukchis, native peoples of the Bering Sea coasts, for whom the whales were essential food. By 1900, whaling was "a dead or dying industry around the world."[20] Plenty of whales remained, but they were harder to catch.

The survivors were mainly those species of rorquals (large baleen whales) that swam too fast to be caught by rowed catcher boats and, inconveniently for the whalers, sank when killed. These included the blue whales, the largest creatures in the history of life on earth, as well as the

[18]The best histories of whaling are Ellis 1991; Kock 1995; and Tønnessen and Johnsen 1982. On the Southern Ocean specifically, Knox 1994 is excellent.

[19]Hilborn 1990:375–6. Some evidence suggests the Japanese engaged in deep-sea whaling a millennium before the Basques.

[20]Hilborn 1990:378; on the Bering Sea bowheads: National Research Council 1996:187–88.

smaller fin, sei, Bryde's and minke whales. In 1900, over a million rorquals feasted on vast shoals of krill in the bone-chilling waters of the Southern Ocean, the richest whaling grounds in the world.[21] But new technology soon opened the climactic chapter in the thousand-year whale hunt.

Between 1864 and 1868, a Norwegian whaling captain, Svend Føyn, developed the harpoon cannon, which fired explosive grenades into whales. Føyn (1809–1894) was an unsentimental creature who pursued whales from Spitzbergen to Antarctica with an Ahab-like determination. In a case of life nearly imitating art, Føyn almost met Ahab's fate—Moby Dick's pursuer was killed by a harpoon line that caught round his neck and whisked him into the Pacific—when a harpoon line caught round his ankle and yanked him into an icy sea. When rescued he said only, "I lost my cap." Mounted on the bow of steam-powered catcher boats, Føyn's cannon permitted whalers for the first time to kill rorquals consistently. Norwegians refined this technology in the North Atlantic, depleted rorqual stocks, and amassed the skills and capital that allowed them to dominate modern whaling until 1950. They opened the Antarctic whale hunt in 1904, and by 1910 it outstripped the established northern ones. Humpback whales, the easiest to catch, provided the initial bounty but declined after 1911. More whalers flocked south when Iceland, then a Danish territory, banned all whaling using Icelandic bases in 1915. The whalers used subantarctic islands such as South Georgia and the South Shetlands as their land bases. But soon the British governor of these islands, Sir William Allardyce, foreseeing that the whale stocks of the Southern Ocean might go the way all others had gone before, placed restrictions on whaling.

This constraint sparked the imagination of Petter Sørlle, a Norwegian whale gunner, who designed the stern slipway, making possible the factory ship, which could take a 100-ton blue whale on board and render it into oil and bonemeal within an hour. The first seagoing slaughterhouse sailed in 1925, eliminating the need for licenses from and duties to British authorities. Norwegians also developed the technique of injecting compressed air into dead rorquals so they would float until towed into the maw of a factory ship. With these advances, a new era of profitable slaughter began. Britain rivaled Norway in the 1920s, and Argentina, the United States, Denmark, and Germany also joined the hunt by the 1930s, the peak

[21]The Southern Ocean refers to the waters south of about latitude 60°S, which are physically and biotically distinct from those of the adjoining Atlantic, Indian, and Pacific Oceans to the north.

In 1903, when this photo was taken, whaling was still an artisanal craft, although persistent whalers nonetheless had managed to bring the populations of many whale species near to extinction. Sperm whales were rare by 1860, but after 1904 whalers armed with twentieth-century technology tapped the mother lode of whales, the Southern Ocean encircling Antarctica. This revived the world's whaling fleets and provoked a sustained assault that brought several species to the brink of extinction by 1960. Here an 8-ton sperm whale meets its fate.

decade for catching blue whales. Japan in 1934 and the USSR in 1946 entered the field and eventually displaced Norway in the Antarctic hunt.

The profit derived mainly from whale oil.[22] Rorquals' oil, actually a fatty acid, had three main applications. With the hydrogenation process invented in 1903, whale oil could be made into margarine, a popular substitute for butter since the 1920s, and into soap. A by-product of boiling whale blub-

---

[22]Whale meat also found a market in Japan and Korea. It was a regular feature of school lunches in Japan after 1947. Arguments about its centrality to Japanese culture and cuisine featured prominently in Japanese attempts to enlarge their whaling quota in international negotiations and to avoid the restrictions that eventually came. When quota exemptions were awarded in the 1980s to "aboriginal peoples" for whom whaling was thought to be integral to cultural survival, the Japanese sought to be redefined as aboriginals.

Whaling by 1939 had become an industrial enterprise, as suggested in this scene from waters between Australia and Antarctica. Factory ships with stern slipways could bring entire whales aboard and render them into whale oil and bonemeal in short order. Whales could not breed fast enough to keep pace with industrial methods of whaling, and their numbers plummeted in the first 75 years of the twentieth century.

ber into oil is glycerin, needed for nitroglycerin which is used in dynamite. The armaments races preceding World War I quickened the market for glycerin and helped steel the determination of whalers. Millions of whales became, through ingenious chemistry, margarine, soap, and explosives.[23]

The most valuable whales were the biggest. Between 1913 and 1938, blue whales became scarce, so whalers moved on to fin whales, which by the late 1950s became hard to find, prompting the use of sonar (invented for antisubmarine warfare in World War II) and aircraft for whale spotting. Whalers progressed, one by one, to smaller and less valuable whale species, until only minkes survived in anything like pre-1904 numbers. In 1900 there were about 150,000 to 250,000 blue whales in the Southern

[23]Bonemeal from whales was also used as fertilizer and chicken feed, but oil was the moneymaker. Sperm oil (actually a wax) was used throughout the twentieth century as a lubricant, notably in the automatic transmission of luxury cars. The USA maintained a strategic reserve of whale oil into the 1980s.

Ocean; in 1989, about 500. Of fin whales, a population of perhaps 750,000 in 1900 stood at 70,000 by the time of the 1982 moratorium on whaling, and 20,000 by 1989. In previous centuries, whalers had depleted whaling grounds one after the other. In the twentieth century whalers found the mother lode of whales, and depleted it species by species.[24]

Concern over dwindling whale stocks existed from early in the history of modern whaling, as Allardyce's and Iceland's actions show. By 1935 the clear decline in blue whales led to regulations overseen by the League of Nations. These had scant effect. The International Whaling Commission (IWC) followed in 1946, an association of most whaling nations that protected the price of whale oil—not whales—by allotting quotas among themselves. This amounted to foxes guarding the henhouse. In 1964 the IWC shifted towards preservation of whale stocks, and decided humpbacks should be protected. In 1965 it accorded blue whales the same status. Renegade nations and pirate whalers prevented any effective limit to the hunt until the 1970s. The Soviets apparently regarded IWC quotas as mere bourgeois morality, and built a fleet designed to hide an illicit catch that amounted to more than 90,000 whales (1949–1980), including protected humpbacks and blues.[25] Among the great pirate whalers in the 1950s was Aristotle Onassis, whose ships used helicopters to find whales and who hired blacklisted Norwegian Nazi collaborators to kill them.[26] After 1963, whale scarcity and conservationists' agitation drove most of the whaling fleets out of business, leaving the Japanese and the USSR to dominate. In the 1960s, Japanese whalers operated from bases in Vancouver Island, Peru, Brazil, and Newfoundland, as well as home waters. They, together with the Norwegians and Icelanders, proved ingenious in avoiding the moratorium agreed upon in 1982 and in force from 1986: a few thousand whales killed "for scientific purposes," and thus exempt from the moratorium, ended up in sushi bars. Nonetheless, after 1986, whaling amounted to a small fraction of its earlier bustle. Populations of

[24]Kock 1995:28; Knox 1994:343–6; Payne 1995:269; Tonnessen and Johnsen 1982:751. All figures ultimately derive from the International Whaling Commission.

[25]Payne 1995:298–99. The USSR in 1959–1961 built two factory ships the size of World War II aircraft carriers, complete with devices to prevent foreign aircraft from detecting clandestine catches. Kock 1995:29 presents the figures, 1949–1980, including 46,000 humpbacks and 8,000 blues. Officials in the Fisheries Ministry, the KGB, and perhaps elsewhere sold this poached whale meat to Japan.

[26]In the early 1950s, Onassis financed and organized whaling in the Pacific in violation of the IWC system, using flags of convenience from nations that were not IWC members (Ellis 1991:431–3; Tønnessen and Johnsen 1982:534–8.

most whale species appeared to be growing after 1990. Given favorable conditions—a tall order in light of the history of whaling—whale numbers, except the slowbreeding blues, might recover in 60 to 100 years and escape their brush with extinction.[27]

In the Southern Ocean, Svend Føyn's legacy killed 1.5 million whales and lowered the whale biomass from about 43 million tons to about 6 million between 1904 and 1985. This left a lot of krill up for grabs. As humpbacks and blues were brought to the edge of extinction by the late 1930s, fins, seis, and minkes had more krill to eat and presumably reproduced faster. As fin whales became the next target (1935–1960), more krill were left for seis and minkes. And so it went, each species briefly the beneficiary of the decline of the whalers' previous target. Throughout, the whale hunt was a boon for penguins and seals, whose numbers rose, at least after 1950, in consequence of the near obliteration of whales. Unless human krill fishing closes it, their expanded niche should last for decades to come. Given such a headstart, they could very well prevent whales from recolonizing the Southern Ocean, even if humans permit it.[28]

Whalers in the twentieth century, as in centuries before, killed the goose that laid their golden eggs because it made economic sense to do so. Whales reproduce slowly, so it was uneconomic to milk the resource and preserve it. Economic rationality required killing all the whales as fast as possible and investing the proceeds in something that grew faster: stocks, bonds, even savings accounts. Even if problems of an open-access resource are solved, whales will never be far from extinction whenever pure economic logic takes precedence.[29]

MARINE FISHERIES SINCE 1900. Whales avoided total extinction in large part because of their unusual appeal to conservationists. No fish enjoys such status. In the twentieth century the seas sustained humankind as never before. We ate more fish than our ancestors did, in total and per capita. We converted yet more into fertilizer and animal feeds. New tech-

---

[27]Kock 1995:19–23, Gambell 1993, and Stoett 1997 summarize the history of whaling regulation. See Kalland and Moeran 1992:93 on Japanese global operations.

[28]On ecosystem effects, Knox 1994:349–55. Neither Knox nor any other author I consulted mentions the impact of one whale species' decline upon the next; that is my supposition, based on Knox's comments about krill availability. Russian trawlers in 1972 began harvesting krill, and other fishing fleets followed suit. It is also possible that the hole in the ozone shield over Antarctica will undermine the primary production of the Southern Ocean, leaving less krill for all creatures. On the seal population history on South Georgia, see Hodgson and Johnston 1997.

[29]Knox 1994:345–6.

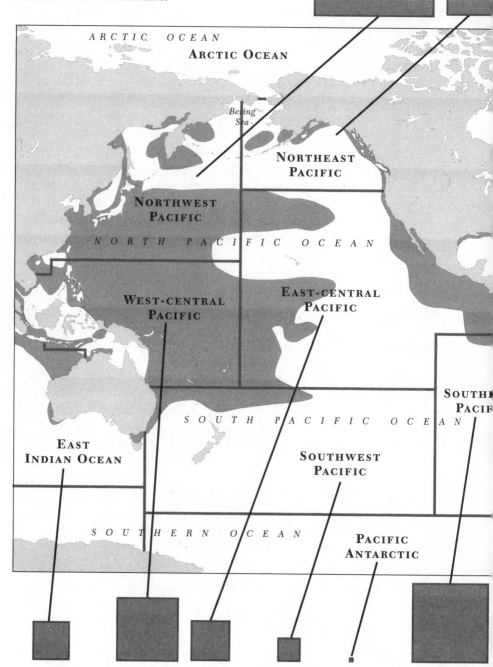

8. Active Fisheries in the Late Twentieth Century

- ▮ Oceanic fisheries
- ▬ FAO divisions of world's oceans
- **ARCTIC** Ocean division names

ARCTIC   OCEAN
ARCTIC OCEAN

Bering Sea

NORTHEAST PACIFIC

NORTHWEST PACIFIC

NORTH   PACIFIC   OCEAN

WEST-CENTRAL PACIFIC

EAST-CENTRAL PACIFIC

SOUTH PACIF

SOUTH   PACIFIC   OCEAN

EAST INDIAN OCEAN

SOUTHWEST PACIFIC

SOUTHERN   OCEAN

PACIFIC ANTARCTIC

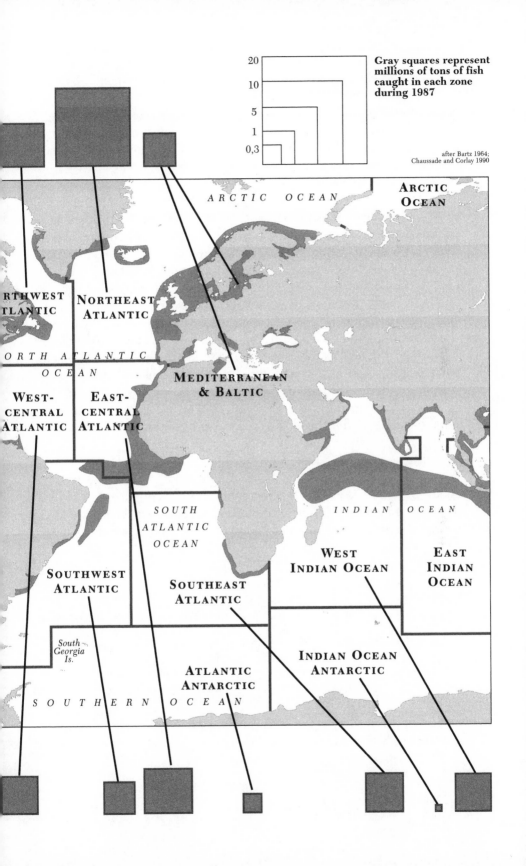

Gray squares represent
millions of tons of fish
caught in each zone
during 1987

20
10
5
1
0,3

after Bartz 1964;
Chaussade and Corlay 1990

ARCTIC OCEAN

ARCTIC
OCEAN

RTHWEST
TLANTIC

NORTHEAST
ATLANTIC

ORTH ATLANTIC
OCEAN

MEDITERRANEAN
& BALTIC

WEST-
CENTRAL
ATLANTIC

EAST-
CENTRAL
ATLANTIC

SOUTH
ATLANTIC
OCEAN

INDIAN OCEAN

WEST
INDIAN OCEAN

EAST
INDIAN
OCEAN

SOUTHWEST
ATLANTIC

SOUTHEAST
ATLANTIC

South
Georgia
Is.

ATLANTIC
ANTARCTIC

INDIAN OCEAN
ANTARCTIC

SOUTHERN OCEAN

nologies and cheap energy made this bumper catch possible. By the 1990s roughly 8 percent of the net primary productivity of the oceans went into the fish we caught.[30] The catch involved routine overfishing, which no management regime could stop. The ecological effect in many cases was to lop off the top link in the marine food chain—large predator fish—and turn the seas over to their prey.[31]

Fish are hard to count, so no one ever knows marine fish populations accurately. Landings can be counted, but whether their fluctuations correspond to changes in fish stocks, and whether changes in fish stocks are the consequence of fishing, are often hard to say. This information conundrum exacerbates the management conundrum of an open-access resource.[32] One thing is certain: fishing is important. About a billion people rely on fish for their main source of animal protein. Official figures (Table 8.1) show that about three-quarters of the world's fish catch comes from marine fisheries. These figures neglect the discarded by-catch, which in the 1990s amounted to about a third of the landed (recorded) catch. They also neglect the fish never reported, the "black" catch, which might add another third or half to the totals.[33] The figures show a burgeoning catch from the late 1940s to 1973, then slower growth. Part of the early bonanza derived from the effects of World War II, which swept fishing fleets from the sea and allowed stocks to recover where they had formerly been reduced, notably in the North Atlantic. The heady expansion of the 1950s and 1960s came mostly from the migration of fishing fleets into the rich waters of the South Atlantic and South Pacific. By the 1980s, fishermen landed as much in two years as their ancestors had in the entire nineteenth century. The figures suggest that the sea surrendered more fish in the twentieth century (around 3 billion tons) than in all previous centuries combined.[34]

[30]Pauly and Christensen 1995.

[31]Caddy 1993 addresses this theme for enclosed seas.

[32]Rothschild 1996. Different species of fish have very different population dynamics. Caddy and Gulland 1983.

[33]The *Economist*, 19 March 1994:24, cites 30–50% for the black catch. See Speer et al. 1997:5–7 on the by-catch in American fisheries.

[34]The assumptions behind this guess are legion. I arrived at the number 3 billion tons for the 20th century by using the data in Table 8.4 and interpolating unabashedly. Using Hilborn's estimate of 1.0 million tons per year in 1800, I calculate that it would take 28 centuries to achieve the cumulative catch of the 20th century. But the catch in 1800 included many fishing grounds (e.g., the Grand Banks) that before 1450 were untouched. Prior to shipbuilding and navigation advances of the 15th–18th centuries, only a fraction of the catch of 1800 could have been achieved.

TABLE 8.1 GLOBAL FISH CATCH, 1800–1996

| Date | Fish Catch (million metric tons) | | | Total |
	Marine Catch	Inland Catch	Aquaculture	
1800	≈1			
. . .	. . .			
1850	≈1.5			
. . .	. . .			
1900	≈2.0			
. . .	. . .			
1938	≈22			
1945	≈13	≈5		≈18
1950	≈15			
1958	≈29			
1961–1963	33			
1964–1966	40			
1967–1969	47			
1970–1972	51			
1973–1975	51			
1976–1978	54			
1979–1981	56			
1982–1984	60	6	7	73
1985–1987	68	6	9	83
1988–1990	71	6	12	89
1991–1993	68	6	15	89
1994–1996	74	7	21	101

Sources: Hilborn 1990; FAOSTAT (online database maintained by U.N. Food and Agriculture Organization).

Note: slightly higher figures, 1950–1994, appear in UNFAO 1997:4 (fig. 1). The trends are consistent even if the precise figures differ. FAO has published several mildly contradictory data sets on landings during 1950–1994.

The aggregate figures conceal a persistent pattern. Numerous important fisheries collapsed in the twentieth century, generally the most valuable ones. The catch of the 1980s and 1990s included great shares of previously uneconomic fish ("trash fish") sought out because cod, herring, haddock, and tuna, among others, became harder to find.

Local fishery collapses are nothing new. Regulations designed to prevent overfishing appeared in the thirteenth century and perhaps earlier. France and Britain attempted an international agreement to curtail fishing in 1839, presumably in response to declining yields.[35] But larger collapses awaited the twentieth century, with the power of its markets, or planned production quotas, and technologies. North Sea fisheries and the Maine lobster fishery collapsed in the 1920s. The Japanese pilchard fishery, the world's largest in the 1930s, collapsed in 1946–1949, recovered exuberantly in the mid-1970s,[36] and crashed again in 1994. The California sardine fishery, immortalized in literature by John Steinbeck in the 1930s, declined sharply after 1945 and died out in 1968. In that same year the Atlanto-Scandian herring fishery collapsed. So did the northeastern Atlantic cod. Sometimes populations recovered only to crash again, as with Japanese pilchard. Sometimes there was (as yet) no recovery, as with California sardines or North Atlantic cod.[37] Wherever modern fishing methods were applied, sooner or later high fishing pressure, combined with a natural downturn in fish stocks, led to a crash. The North Atlantic, lately the world's second leading fishery, saw this happen early and often in the twentieth century.[38]

On the world's richest fishing grounds, it happened only once. In the mid-1950s, out-of-work fishermen and boats, veterans of the dying sardine fishery, headed south from California to Peru and set up shop in the Humboldt Current. There cold, oxygen-rich water supports phytoplankton in profusion, which in turn supports shoals of anchoveta and Chilean jack mackerel. Peruvians had fished these waters for centuries, but not with power boats, big nets, and airplanes to find the fish. By 1962, Peru landed more fish tonnage than any country in the world, nearly 7 million tons. The fishery peaked between 1967 and 1971 at 10 to 12 million tons, 20 percent of the world's total. Anchoveta, converted into fish meal and

[35]Graham 1956:497. [36]Yamamoto and Imanishi 1992.

[37]The Atlantic cod catch in the 1990s stood at about a quarter of its 1968 peak; cod stocks in the western North Atlantic stood at about 10% of their historical average (WRI 1997:297).

[38]On these collapses, see the chapters in Carré 1982, Glantz 1992, Hilborn 1990, McEvoy 1986, Speer et al. 1997, and UNFAO 1997.

fish oil, anchored Peru's foreign trade, providing a third of its foreign exchange. In 1972 the bottom fell out: 4.7 million tons, then 2 to 4 million tons per year for 15 years. The anchoveta collapse lowered world fish production by about 15 percent. This disaster hamstrung Peru's economy, contributing to the turbulent politics of the 1970s and 1980s, which featured relentless inflation, mass unemployment, and the emergence of violent revolutionary groups.

The collapse coincided with an El Niño in 1972, and the absolute nadir of the catch (1.5 million tons) came with a strong El Niño in 1982–1983. These periodic short-lived fluctuations in Pacific currents, which bring warmer, nutrient-poor water to the Peruvian coast, clearly had a role in sinking the anchoveta fishery. But overfishing had a hand in it too, and slowed recovery after El Niño conditions abated. As with the California sardine collapse, both natural and social causes combined in Peru's misfortune.[39]

While the anchoveta collapse is the most spectacular in the history of fisheries, after the 1970s a general, if less acute, crisis afflicted almost all the world's fishing grounds. The great growth in fisheries since 1945 had come from two sources: finding new fisheries and intensifying fishing effort in old ones. By and large, the new ones were in the Southern Hemisphere, like the anchoveta. But the pilchard fishery in the southeastern Pacific, which opened up in 1971 and thrived in the 1980s, proved to be the last. Thereafter no significant fisheries remained unfished. This, then, was a turning point in the history of fishing, analogous to the closing of a frontier on land.

After 1971, maintenance of fish yields, let alone growth, required ever more intensive effort. This came from government subsidies (see below) and improved technologies. New technologies included sonar and satellite imagery for finding fish, and bigger and better trawlers and drift nets for catching them.[40] Fishermen stalked reef fish with cyanide and dynamite. All this was an intensive effort indeed, and very efficient at catching those species of fish that school. A behavior that evolved among about a quarter of all fish species as a survival mechanism became nearly suicidal under relentless human predation.

The subsidies and new technology after 1970 brought one fishery after

[39]Caviedes and Fik 1992. El Niño returned in 1998, lowering the anchoveta and Chilean jack mackerel catches by 50–75%.

[40]Drift nets, used primarily by Japan, Taiwan, and Korea in the Pacific, were banned in 1992, although occasionally they were still used after that, notably by Italian fleets. They could be 50–70 km long, be up to 40 m in depth, and catch everything in their path.

another into decline if not collapse. By the 1990s more than two-thirds were fully exploited or overfished.[41] The best hope for catching more, ironically, lay in fishing less: netting fewer immature fish and letting more grow to full size and to reproductive age. In the 1990s, fishermen lost about 9 million tons a year because of overfishing. Their ingenuity and determination resulted in higher catch totals but masked a deterioration in most fisheries.[42]

The general pattern of boom and bust, more pronounced in marine fisheries than in most extractive industries, derived from the interplay of natural fluctuations and social conditions. Initially, as one or another fishery collapsed, a new one (or two) was found. This underpinned the great growth, especially from 1950 to 1971. These conditions encouraged fishermen and fishing nations to abjure almost all regulation. The absence of regulation assured that busts would follow booms, because any success attracted new fleets. As these fleets became more technologically sophisticated, they became more expensive. The fixed capital sunk into a fishing fleet could not be left idle: it had to fish day and night in all seasons, depleting fish stocks to the point where the sale of its catch could no longer cover its operating costs. These economic realities brought frequent collapses in the North Atlantic and North Pacific.

The collapses brought subsidies, which deepened the problem, and they brought regulation, which did not solve it.[43] In the mid-1970s, following Peru, almost all fishing nations proclaimed 200-mile territorial waters, called exclusive economic zones, or EEZs. This nationalized all the continental shelves, although some countries, like Namibia, could not enforce their EEZs and saw "their" fish scooped up by long-distance fleets from eastern Europe and East Asia.[44] Within these zones over-

[41]UNFAO 1997:36, 65. Some inland seas bucked the trend. Pollution, in the form of eutrophication, enhanced fisheries in the Mediterranean, where they reached their record high in 1991, and in the Baltic. In the Baltic, higher catches chiefly resulted from harvesting fish that once went to seals; only 20–40% was due to eutrophication (Elmgren 1989). Other onceflourishing inland sea fisheries declined. The Caspian fishery was a victim of pollution; the Aral Sea fishery disappeared in 1982 as the sea shrank (see Chapter 6).

[42]Paul 1987; UNFAO 1997; Troadec 1989.

[43]Ludwig et al. 1993 argue that booms inevitably produce overcapitalization and excess capacity in fisheries because the size of the resource is always uncertain. Governments cannot bear the political price of shedding capital and labor from fishing fleets when booms collapse, so subsidies emerge. The authors this pattern regard as inevitable.

[44]Between 1965 and 1980, Namibia's pilchard catch went from half a million tons to zero (Sparks 1984).

fishing continued: domestic fleets simply replaced foreign ones, and many countries invested heavily in fishing fleets to exploit their EEZs. When Canada declared its EEZ in 1977, for example, it acquired firm dominion over the Grand Banks, which had been an international fishing ground since the sixteenth century. To exploit it and to boost the economy of Newfoundland, Ottawa in the 1980s subsidized a great expansion of the fishing fleet and a technological upgrade as well. Newfoundland fishermen became too efficient for their own good, and Ottawa soon had to pay them not to fish. The catch, nearly a million tons a year in the 1960s, plummeted after 1990, obliging the federal government to declare a moratorium in 1992, which put 25,000 Newfoundlanders out of work and on the dole. They waited, to date in vain, for the cod to return. Subsidies worldwide—about $50 billion a year by 1995—helped double global fishing fleet tonnage from 1970 to 1995. Further management techniques emerged, the most successful pioneered by New Zealand.[45] None of these, however, could prevent the effects of overcapitalization, overcapacity, and overfishing: conversion of the high seas into a domain of uncommercial species.

The human appetite for seafood was hard on fish that schooled but a boon for those populations that could be cultured. Aquaculture had existed for centuries, but only after 1980 did it become big business.[46] World fish farming yielded 5 million tons in 1980 and 25 million in 1996. China, which refined carp farming exquisitely under the Song dynasty (A.D. 960–1279), accounted for more than half the world's aquaculture total as of 1995, and by 1988 had become the world's largest producer of fish. Roughly 80 percent of aquaculture took place in Asia, about two-thirds of it in inland freshwater ponds stocked mainly with carp. The remainder featured shrimp, prawns, and salmon in coastal cages.[47] Only a few species can be farmed, so the development of aquaculture, like Green Revolution farming, involved maximizing production of a few species. Moreover, fish farming did not much reduce the pressure on marine capture fisheries because aquaculture used fish meal as feedstock. So, in effect, in order to eat

[45]This was individual tradable quotas (ITQs), which were intended as a way to privatize the fish within the nationalized waters. Many countries followed suit after New Zealand experimented with this in the early 1980s. On recent fisheries management efforts, see Hannesson 1994.

[46]Carp ponds existed in China from at least 2,500 years ago (Landau 1992:4).

[47]The shrimp farms were the main factor in destroying coastal mangroves in southern and Southeast Asia after 1982 (Gujja and Finger-Stich 1996). On Asian aquaculture, see Nakahara 1992, and Liao and Shyu 1992.

shrimp and salmon, people caught large amounts of anchovy and men-haden. If fisheries experts are right that marine capture is at or near its theoretical limit, and if aquaculture continues its trend of recent decades, in 10 to 15 years, more fish will be cultured than caught. A transition equivalent to the Neolithic agricultural revolution on land is happening in the seas in the late twentieth and early twenty-first centuries.[48]

Biological Invasions

The swirling currents of trade and migration made the twentieth century one of the great epochs of biological invasion, or bioinvasion. When one creature somehow establishes itself in a new place, thrives, and upsets previous balances, it has pulled off a bioinvasion.[49] Most of the emergent diseases of the twentieth century, as noted earlier, are in fact bioinvasions. They can happen without human intervention. In recent centuries they happened more often because of rising transport and trade. The burgeoning scale of the grain, cotton, and wool trades vastly enhanced plant migration, as seeds hitched rides in sacks and bales. Hu-mankind has long been far and away the most efficient agent of plant dis-persal. No large animals had spread throughout the world before humankind blazed the trail; all animals that are now widespread owe their success to their compatibility with human enterprise.

The history of biological invasion accelerated sharply with European in-tercontinental voyaging (1492–1788) and with the labor migrations that followed. But in the twentieth century the sheer quantity of traffic among regions, and especially among continents, afforded far more invasion op-portunities. And the frequency and intensity of perturbations to ecosys-tems sharply improved the odds of success, for disturbed ecosystems are especially liable to successful invasion. From the detached viewpoint of geological time, one of the main distinctions of the twentieth century will prove to be its promotion of bioinvasions.[50]

[48]On aquaculture, see Landau 1992, and UNFAO 1997:11–13.

[49]On the theory of bioinvasions, see Elton 1958, Groves and Burdon 1986, and Williamson 1996.

[50]Burney 1996. About 10% of introduced species become established permanently, and about 10% of those become economically significant pests (Williamson 1996).

Many invasions began as intentional introductions, especially in the period 1850 to 1920, when European, Chinese, and Indian overseas migrants tried to bring familiar species with them. Australia and New Zealand featured "acclimatization societies," citizen groups unsatisfied by the antipodean flora and fauna. They made it their business to import chosen species, usually familiar ones from Britain. An American who allegedly felt the United States ought to have every bird mentioned in Shakespeare released 80 pairs of starlings into New York's Central Park during 1890–91, to which event America owes its many millions of starlings today.[51] Fur-bearing creatures made popular introductions, especially into Europe. Hermann Göring, Hitler's fashion-conscious Air Marshall, had American raccoons brought to Germany in hopes of making raccoon coats. His raccoons eventually escaped captivity and colonized the Mosel valley and parts of Luxembourg and Holland.[52] Intentional introductions usually failed, but some, like starlings, succeeded too well.

The most disruptive introductions, however, were accidental, stowaways on cargo ships, airplanes, or military transports. The most costly in economic terms were insect pests. The boll weevil, a well-documented alien from Mexico, ravaged the U.S. cotton crop after 1890. The fire ant, a fierce Brazilian which arrived in Mobile, Alabama, in the 1930s, became the target of a costly and counterproductive extermination campaign in the southern United States. The gypsy moth, an unruly guest from France, was introduced to suburban Boston in 1869 by an artist and entrepreneur, E. L. Trouvelot, who hoped to cross it with silkworms and establish a textile empire. Trouvelot's moths escaped and have plagued the deciduous trees of eastern North America ever since. By one calculation, introduced insects cost the United States about $100 billion between 1906 and 1991; by the mid-1990s they cost well over $100 million

[51]Tenner 1996:119–21 doubts the Shakespeare fixation of Eugene Schieffelin; Williamson 1996:46–7 does not. Starlings took 60 years to reach the Pacific coast and Alaska. Elton 1958:22–4; see also Long 1981. Starlings spread a lung fungus among humans, cause crop damage, and were the object of a major control program in California in the 1960s. They are a nuisance in Australia too.

[52]Druett 1983:95. The North American muskrat and mink did well in Europe too. The muskrat, first brought from Canada to Bohemia in 1905, became a pest, undermining earthen dams and banks and creating marshes. One British writer thought releasing muskrats a more vile act than releasing man-eating tigers. Muskrats by 1952 were feral from France to Japan— all within a half-century! (Williamson 1996:89–92; De Vos et al 1956).

each year. Invasive species of all sorts cost Americans about $123 billion annually by 1999, or $500 per person.[53]

Invaders that succeeded too well and became pests achieved "ecological release," by leaving behind the constraints on their own population growth—diseases, predators, competitors—in their homeland. More exotics achieved this nirvana in the twentieth century than ever before. The most dramatic case was that of the humble rabbit.[54]

THE EUROPEAN RABBIT. *Oryctolagus cuniculus,* which apparently originally derived from Spanish stock, has proved to be a global conquistador.[55] With human help, it colonized Europe in ancient and medieval times at a snail's pace, taking 700 years to occupy the British Isles. Its food and fur value led to its intentional introduction around the world in modern times.

The greatest rabbit invasion took place in Australia. After several had failed before him, a squatter named Thomas Austin successfully introduced rabbits in Victoria in 1859. In six years he killed 20,000 rabbits on his land. By 1870, rabbits were a pest in much of the country. Graziers were busily and unwittingly building a rabbit paradise, sowing grass, making ponds (for livestock), and killing off all the kangaroo, wallabies, possums, dingo, and birds of prey they could, thus eliminating rabbit rivals and predators. They drove to extinction the rat kangaroo, which left behind a network of tunnels perfect for rabbit burrows. Rabbits aroused the ire of graziers because they ate the grass that otherwise fed sheep. It took about 60 years (1859–1920) for rabbits to colonize all of Australia that suited them. Western Australia built a 2,100-kilometer fence from 1902 to 1907 to safeguard the wool clip, then Australia's most valuable export. But rabbits easily breached this Maginot line, latterly called the "bunny fence," and munched their way across the country. By 1950, Aus-

[53]These costs were calculated by USOTA 1993, and by the U.S. Commerce Department as reported in the *Washington Post,* 3 February 1999:A15. On fire ants, see Tenner 1996:110–3; and Conniff 1990. American fire ants are vicious creatures whose stings can kill livestock and children. They attack almost any animal in the States, but are much more peaceful at home in southern Brazil. The USDA tried various insecticides from the late 1940s but quit in 1978—after killing countless birds, reptiles, and innocent insects—leaving far more fire ants than before. By obliterating many of its competitors, the USDA paved the way for faster fire ant colonization.

[54]An attempt at a general history of biological invasions is Di Castri 1989; see also Sykora 1990 and Thellung 1915.

[55]Thompson and King 1994; De Vos et al. 1956; Jaksic and Fuentes 1991 (on Chile).

tralian grass fed about half a billion rabbits, which to graziers meant 50 million fewer sheep. Then officials introduced a Brazilian rabbit disease, myxomatosis, which killed 99.8 percent of Australian rabbits. But it did not kill them all, and those few that survived were immune to the disease. So were their descendants, which numbered about 100 million in the 1990s.[56]

Rabbits almost matched their Australian success in New Zealand and may yet in Argentina. In New Zealand, where they first appeared around 1864, they enjoyed almost a century of nearly unmolested bliss—plenty of grass and few predators—until the late 1940s. Then New Zealand mounted an airborne poisoning campaign that killed millions, to the delight of stockmen. Rabbits established a toehold in South America through intentional releases in Chile in the late nineteenth century, and eventually colonized some of the islands in the far south. Between 1945 and 1950, in an unrecorded epic recalling the story of San Martín's army, rabbits marched over the Andes's high passes to Argentina, to behold a welcoming sea of grass. They pushed out their Argentina frontier at about 15 to 20 kilometers per year, and by the mid-1980s, despite a setback in the form of introduced myxomatosis, they colonized about 50,000 square kilometers, roughly equivalent to the area of Slovakia or West Virginia.

The rabbit did poorly in North America, establishing only a handful of enclaves in California and Washington. In Africa it long ago formed a beachhead in Morocco and Algeria, where it attracted enthusiastic hunters, but elsewhere it did not flourish. In South Africa (Natal) a small community was wiped out by ants that attacked infant rabbits. In 1656 a smaller group colonized Robben Island (where apartheid's most famous prisoners were kept) and nibbled vegetation to near oblivion. The European rabbit's greatest successes, outside of Australia, were on islands. Rabbits now occupy 800 islands around the world and dominate some small ones. Of humanity's many fellow travelers, only the rat has proved a more nimble colonizer than the European rabbit.[57]

THE CHESTNUT BLIGHT. In 1897 the U.S. Department of Agriculture created the Section of Foreign Seed and Plant Introduction, which,

[56]Myers 1986.

[57]Of some 63 species of rat, only 2 (*Rattus rattus* and *R. norvegicus*) have spread throughout the world with humankind. Both are Southeast Asian in origin: notwithstanding its name, the Norway rat comes from southern China; both are also large, fast-breeding, and omnivorous, but cautious about new foods. They (and the house mouse, *Mus musculus*) are among the greatest beneficiaries of human colonization of the planet. See Michaux et al. 1990.

in keeping with the spirit of the Progressive Era, sought to improve the American biota. One of its scientists, Frank Meyer, brought 2,500 plants from Asia to the United States between 1905 and 1918. He among others shipped nursery stock of the Asian chestnut, some of which carried a fungus to which the Asian tree is resistant, but the American cousin vulnerable. Spreading from a beachhead in New York, the fungus within 50 years covered the hardwood forests in the eastern United States. The chestnut had previously been dominant in large areas of forest, making up more than 25 percent in many stands. Chestnut blight cost the United States a useful rot-resistant timber, preferred for posts and fences; it eliminated pig farming in the Appalachians, which had depended on fallen chestnuts; and it ended the custom of gathering autumn chestnuts for winter roasting. The chestnut's virtual eradication in North America was perhaps "the largest single change in any natural plant population that has ever been recorded by man."[58] Lesser biological invasions that ravaged forests were Dutch elm disease, imported into Britain from mainland Europe in 1927 and into the United States in 1930; and the pathogen that killed about a quarter of Japan's pines between 1900 and 1975, apparently an American import. By and large, bioinvasions of forests were rare and confined to temperate forests, because tropical forests were too diverse for single invaders to have much impact.

GRASSLAND INVASIONS. In the twentieth century, most of the temperate grasslands outside of Eurasia came under more intense occupation and grazing pressure than ever before. As a result, many of them lost their previous vegetation, which was not adapted to life under the hoof. In its place came Eurasian and African grasses, selected over millennia for their compatibility with humankind's favorite ruminants. The extreme examples are the grasslands of Australia, southern South America, and western North America, where the process began by 1800 or so but climaxed in the twentieth century. Between 1889 and 1930, cheatgrass—unpalatable to livestock and a weed in wheat and alfalfa fields—colonized much of America's intermontane West. Tumbleweed, introduced to the Dakotas in the late 1870s from Ukraine, rolled far and wide in North American prairies, becoming a nuisance to farmers and ranchers alike. By the 1940s, herbicides checked the tumbleweed's progress, although later

[58]Anagnostakis and Hillman 1992; Elton 1958:21–4. The quote is by J. L. Harper in von Broembsen 1989. On Dutch Elm disease, see Karnosky 1979.

it flourished on Nevada's nuclear test sites. One alien (but less trouble-some) species in the American west was called "professor grass" because it escaped from a research station.[59]

AQUATIC INVASIONS. Biological invasions in the world's waterways were also commonplace in the twentieth century, far more so than in any previous time.[60] Some introductions, particularly of edible fish, were in-tentional. Since 1850, people have transplanted about 250 freshwater fish into about 140 countries. Most introductions failed, and only a few altered socioeconomic history. Chinese and Indians, scattered widely around the world in the great migrations of indentured laborers after 1840, often tried to establish carp in the lakes of their new lands, and sometimes suc-ceeded. The thousands of artificial lakes created in the twentieth century offered wide scope for exotic introductions, especially in dry climates where existing species were not adapted to stable water levels. In Lake Kariba (Zambia) a thriving fishery was created in 1974 with the intro-duction of fish from Lake Tanganyika. Throughout the tropics, people have let loose the "aquatic chicken," tilapia (*Tilapia mossambicus* or *niloticus*), a fast-breeding, tasty fish that flourishes amid sewage and brackish water. It was brought experimentally to Tamil Nadu, in south-eastern India, in 1952 and quickly colonized much of the state. It in-vaded West Bengal from Bangladesh (in the 1980s), where it became important as protein for the poor.[61]

In big lakes or even enclosed seas, invasions had big consequences. The Nile perch, a predator the length of a basketball player and the weight of a sumo wrestler, made its debut in Lake Victoria in the 1950s, introduced by someone who fancied it as a sport fish. After a slow start, the Nile perch achieved ecological release. The *T. rex* of freshwater fish, it proceeded to eliminate over half the fish fauna in the lake, some 200 cichlid species—the largest vertebrate mass extinction in recorded his-tory. (The lost cichlids are lamented by evolutionary biologists, as they are the best extant indications of how new species evolve.) The total fish catch, mainly Nile perch, rose sharply after the 1970s, inspiring fish pro-cessing plants and an export trade, mostly to Israel. The ecological changes in Lake Victoria helped the big operators—it takes a big boat to

---

[59]Mack 1986, 1989; Manning 1995:169–90.
[60]Ashton and Mitchell 1989:114.
[61]Sreenivasan 1991; Fernando and Holcik 1991.

catch a Nile perch—more than the small-scale fisherfolk, mainly women, who caught cichlids from canoes.[62]

In the North American Great Lakes, several introductions stirred up the aquatic biota and brought down some fisheries. The lakes are only about 10,000 years old, created by the last glaciation, and so their biotas are not well adapted to the full range of possible perturbations, and hence are especially vulnerable to disruption. Niagara Falls cut the lakes off from the outside world, making them a biogeographical island. The Erie (1825) and Welland (1829) Canals ended that isolation; at least 139 alien species invaded the Great Lakes after 1825. When the St. Lawrence Seaway opened in 1959, the pace of invasion stepped up, because seagoing ships with huge tanks for ballast water now could sail the Great Lakes, and dump their ballast water with its marine menagerie. A few invaders, mostly fish, had important ecological and economic consequences. Two of the most disruptive were the sea lamprey and the zebra mussel.

The sea lamprey is a long, snakelike fish with a suction-cup mouth and rasplike tongue. It feeds by attaching itself to fish and sucking their blood. Its original home was the North Atlantic, but sometime before 1850 it infiltrated Lake Ontario. Niagara Falls kept it out of the other Great Lakes until 1920 or so, when it wiggled into Lake Erie, perhaps via the Welland Canal. Over the next 30 years, the sea lamprey spread throughout the upper Great Lakes, battening on the abundant lake trout and whitefish. The sea lamprey had the same tastes as people and preferred commercially important species. With some probable help from overfishing, the sea lamprey lowered lake trout and whitefish landings by more than 90 percent from 1940 to 1960. Around 1956 in Lake Michigan, lake trout, formerly the main commercial catch, went extinct.[63] A few thousand Canadian and American fishermen lost their jobs to the sea lamprey, and the remainder went after smaller, low-value species (as did the sea lamprey). In 1958 a chemical poison began to control lamprey populations. Combined with pollution control and stocking, this helped some fish pop-

---

[62]Dennis 1996:178–9; Ogutu-Ohwayo 1990; Williamson 1996:124–5. In reducing the cichlid population, the perch has improved life for algae, which (as they decompose) consume the lake's oxygen: The Nile perch may end up a victim of its own success. Cichlids may survive because they evolve so quickly: they may hit upon traits that allow them to live in a Nile perch kingdom, or a low-oxygen lake.

[63]Illinois Department of Energy and Natural Resources 1994, 3:159–60. Authorities stocked Lake Michigan with lake trout from 1965, but the fish still do not reproduce. The lake trout, interestingly, when introduced into Lake Titicaca in South America, achieved ecological release, to the great detriment of native fish.

ulations to recover after about 1970, but commercial and sport fishing on the lakes has yet to recover from the irruption of the sea lamprey.[64]

The zebra mussel is a striped mollusc, native to the Black and Caspian Seas. It hitched a ride to the Great Lakes in ballast water in 1985 or 1986. In 1988 it was discovered in Lake St. Clair, between Lakes Huron and Erie. By 1996 it had colonized all the Great Lakes and the St. Lawrence, Illinois, Ohio, Tennessee, Arkansas, and most of the Mississippi Rivers. The zebra mussel filters water to feed and removes numerous pollutants and algae, leaving cleaner and clearer water wherever it goes. Preferring hard, smooth surfaces, it delighted in the industrial infrastructure of the Great Lakes region, building thick colonies that sank navigational buoys and clogged water intakes on factories, power plants, and municipal water filtration systems. By the early 1990s it had temporarily shut down a Ford Motor plant, a Michigan town's water power supply, and cost the United States about a billion dollars a year. In dollar terms it threatened to become the most costly invader in U.S. history, a distinction previously held by the boll weevil.[65]

The saga of the zebra mussel was but one episode in a costly biotic exchange between America and the Soviet Union. In 1979 or 1980, a comb jellyfish *(Mnemiopsis leidyi)*, native to the Atlantic coast of the Americas, stowed away on a ship bound for a northern Black Sea port. Amid highly polluted and anoxic (oxygen-poor) waters, it found ecological nirvana as few species ever had, devouring zooplankton, larvae, and fish eggs. Nothing in the Black Sea would eat it. By 1988 it dominated life in the Black Sea, accounting by one estimate for 95 percent of its wet biomass. It obliterated fisheries worth $250 million per year, mainly in the Sea of Azov, at a time when Russians and Ukrainians could ill afford it, helping in a modest way to nudge the USSR into the dustbin of history.[66]

So, in the tense Cold War atmosphere of the early 1980s, American ecosystems launched a first strike with the comb jelly and the USSR's biota retaliated with the zebra mussel. The damaging exchange probably

[64]Mills et al. 1993; Regier and Goodier 1992. Whitefish catches after 1980 recovered to pre-lamprey levels; lake trout are no longer caught commercially. Illinois Department of Energy and Natural Resources 1994, 3:165–6.

[65]Mills et al. 1993; USOTA 1993; *Washington Post,* 14 May 1997:H1. The June 1996 issue of *American Zoologist* is devoted to the zebra mussel. Ludyansky et al. 1993 report that the water intake at the Chernobyl nuclear reactor had 1 million to 2 million zebra mussels per square meter in 1981. The boll weevil cost the USA about $50 billion since the 1890s, but lately has been well controlled (Simberloff 1996).

[66]Carlton 1996; Travis 1993.

resulted from the failures of Soviet agriculture, which prompted the grain trade from North America: more trade, more ships, more ballast water.

In any case, the huge expansion of modern shipping, and the wide use (since the 1880s) of ballast water tanks brought about a partial homogenization of the world's coastal, harbor, and estuarine species. It is a distant echo of the "Columbian exchange," the dispersal of organisms that followed 1492. Like the Columbian exchange, its consequences (in human terms) were mixed: the comb jelly did no good, but a stowaway Japanese clam became the basis of a thriving fishery in British Columbia and the state of Washington after 1930. For better and for worse, every seaway since 1880 became a highway for aquatic exchange.[67]

CONTROL OF BIOINVASIONS. "The greatest service which can be rendered to any country is to add a useful plant to its culture." So thought Thomas Jefferson (1743–1826). Keeping the harmful ones out should rank high as well. The economic costs of bioinvasions occasionally mounted so high as to mobilize resistance movements. Before the age of pesticides, this meant biological control: finding a predator or disease that would attack the invader. Normally the search took place in the invader's home territory; and normally—more than 85 percent of the time—it failed. The United States imported over 40 enemies of the gypsy moth after 1890, none of which controlled it. (Nor did lavish applications of DDT.) Some attempts were worse than failures: Late in the nineteenth century, when New Zealand had problems with introduced rats and cats, someone imported weasels, ferrets, and stoats to eat the rats and cats. But they proved disobedient, and joined the rats and cats in a general assault on New Zealand's native birds, most of which went extinct.[68]

Up to 1976 there were 56 more or less successful cases of biological control of plants, of which two were signal successes: the use of a moth against the prickly pear cactus in Australia, and the case of St. Johnswort and the chrysolina beetle. Prickly pear cactus arrived in Australia from North America as an ornamental garden plant but soon escaped, and by the 1920s it covered an area the size of Colorado or Italy. Australian scientists scoured North America for cactus eaters, and in 1925 found *Cac-*

---

[67]Carlton 1985, 1989, 1996; Carlton and Geller 1993. Ballast tanks were invented in the 1840s but came into widespread use only with iron ships in the 1880s. Ballast water is nowadays taken on through steel plates with small holes in them, limiting the variety of organisms that can hitch rides.

[68]King 1984. Jefferson's statement is quoted in Busch et al. 1995:92.

toblastis cactorum, a moth, which rolled back the cactus frontier. St. Johnswort is a European native that in the western United States and Australia became an aggressive weed, called Klamath weed in California. It crowded out useful pasture grasses and made livestock sickly. It appeared in California around 1900 and by 1944 covered 2 million acres of rangeland. Various British and French beetles were enlisted in the struggle against it, with no success. But in 1944, when war inhibited the search for useful European beetles, an import from Australia successfully acclimatized in California, and within 10 years nibbled Klamath weed back to 1 percent of its former domain, to the great satisfaction of California cattlemen. Few biocontrol programs worked as well as these two, because few creatures are choosy enough in their eating habits, although as the science of biocontrol grew more refined, the probability of favorable results perhaps improved.[69]

One recent shining success in biocontrol restored Africa's broad cassava belt to health after 1993. Cassava, a Brazilian root crop imported to Africa in the sixteenth century, by 1990 supported some 200 million Africans. In 1970 a cassava mite, also Brazilian, began to colonize Africa's cassava fields, radiating in all directions from a foothold in Uganda. Unlucky farmers lost half their crops, and no pesticides or other control techniques worked. Then, after 10 years' search in South America, entomologists found the answer: another mite *(Typhlodromalus aripo)* that ferociously pursued and ate the cassava mite. Released first in Benin in 1993, the predator mite quickly checked the havoc caused by the cassava mite, raising cassava yields in West Africa by about a third.[70]

The perfect predator or parasite, one that will attack a targeted pest and nothing else, is always hard to find. The difficulties of biological control made pesticides an appealing option in dealing with runaway exotics. This worked on the sea lamprey, but in many other cases, such as the fire ant and gypsy moth, it failed, and inadvertently poisoned millions of creatures, people included, who happened to be in the way.

In general, the transport revolutions of modern times and the vast scale of human disturbance to ecosystems in the twentieth century have re-

[69]Dahlsten 1989; Groves and Burdon 1986; Williamson 1996:120–4; van den Bosch et al. 1982:21–35. Klamath weed may become a valued crop: in 1996–1997, researchers found its compounds worked as an antidepressant medication.

[70]*Washington Post,* 19 May 1997:A3; *Biocontrol News and Information,* June 1998, at *http://pest.cabweb.org:81/MEMBER/bni/bni19-2/gennews.htm*

fashioned the planet in ways very suitable to invasive species. Fast-breeding, mobile generalists inherited a much larger share of the earth, promoting a grand intercontinental and interoceanic homogenization of flora and fauna.[71] Islands were especially changed—half of New Zealand's flora is exotics—and so were the lands of the European overseas frontiers.[72] The whole process, one of the great biotic revolutions in earth's history, tended to enhance local biodiversity while reducing it globally. In 1958 Charles Elton wrote: "We must make no mistake: we are seeing one of the great historical convulsions of the world's fauna and flora."[73] In the twenty-first century, the pace of invasions is not likely to slacken, and new genetically engineered organisms may also occasionally achieve ecological release and fashion dramas of their own.

Biodiversity and the Sixth Extinction

The twentieth century, a blip of geological time, may be an early stage in a mass extinction, the likes of which have occurred five times since life on earth began. The greatest of these extinction spasms happened about 245 million years ago and swept away perhaps nine-tenths of marine species. The most recent such spasm, about 65 million years ago, ended the career of dinosaurs and paved the way for mammals. Each extinction event spells opportunity for survivors.

Most species that ever lived are now extinct. Somewhere around 14 million (give or take an order of magnitude) now exist.[74] Extinction "background rates," while not constant, suggest that on average over eons one to three species vanish every year. Every 200 years or so, on average, a mammal ought to go extinct (there are currently about 5,000 mammal species). Human exploration, settlement of new lands, forest clearance, and hunting quickened the pace. Since A.D. 1600, at least 484 animals and 654 plants have gone extinct. Most extinctions happened on islands and

---

[71]On what makes a good invader, see Drake et al. 1989.

[72]In central Europe, the proportion of exotics is 10–18% vs. 59% in New Zealand (Sykora 1990; Vitousek et al. 1996).

[73]Elton 1958:31.

[74]This estimate is from Heywood and Watson 1995. Others range from 3 million to 100 million. About 1.75 million have been described sufficiently to be considered "known."

in freshwater lakes and rivers: isolated habitats. The twentieth century extinction rate for mammals was about 40 times the background rate; for birds about 1,000 times background. Roughly 1 percent of the birds and mammals extant in 1900 went extinct by 1995.[75]

Previous extinction spasms derived from unknown causes, but the modern one—if it is that—is different in that its cause is obvious: a rogue mammal's economic activity. Nothing of the sort has ever happened before in earth's history. Most modern extinctions occurred because of habitat loss, although some derived from hunting or predation by introduced species.[76] The world-girdling assault on tropical forests necessarily heightened extinction rates, because roughly half of terrestrial species live in tropical forests. Those forests became increasingly fragmented, creating "islands" of wildlife habitat amid newly created fields, scorched earth, and artificial lakes. Numerous animal populations lost contact with others of their ilk and were thus in the twentieth century consigned to eventual extinction—the living dead—although they may hang on for decades or even centuries. While expert views are numerous, many observers expect 30 to 50 percent of terrestrial species to disappear in the next century or two. If it happens, it will be the sixth great extinction event in earth's history, far faster than any previous one, and unique in its cause. All this belongs to the future: like climate change or ozone depletion, the erosion of biodiversity certainly happened in the twentieth century, but its societal impact as yet remained small.[77]

While reducing the number of species on earth, human agency improved the prospects of a handful of chosen species. These include domesticated plants and animals—our minion biota—and synanthropes such as rats, rabbits, cockroaches, sparrows, raccoons, crabgrass, and the viruses that cause the common cold. Roughly 40 animals and 100 plants

[75]Heywood and Watson 1995:233–4; May et al. 1995:13.

[76]Hunting's impact can be astounding. The passenger pigeon, billions of which flocked over eastern North America in 1850, was extinct in the wild by 1900, and the last representative, Martha, died in a Cincinnati zoo in 1914. In 50 years it went from the most populous bird in North America, and perhaps the world, to extinction, thanks to forest clearance and American marksmen. Even in the tropical forests hunting may play a significant role in extinctions and altering forest dynamics. Redford 1992 estimates that subsistence hunting in Amazonia took about 20 million animals per year in the 1980s and commercial hunting another 4 million, and the total effect of hunting's depredations on Amazonian ecosystems is vast—despite the appearance of intact forest.

[77]Heywood and Watson 1995 summarize much of what is known of biodiversity and extinction. Other handy treatments include Janetos 1997, Kaufman and Mallory 1993, Lawton and May 1995, Leakey and Lewin 1995, and Wilson 1992.

TABLE 8.2 GLOBAL LIVESTOCK POPULATION, 1890-1990						
	Livestock (million head)					
Year	Cattle	Sheep	Goats	Pigs	Horses	Poultry
1890	319	356	52	90	51	706
1910	391	418	83	115	73	828
1930	513	567	153	187	88	1,203
1950	644	631	187	300	69	1,372
1970	1,016	1,001	325	634	81	2,734
1990	1,294	1,216	587	856	61	10,770
INCREASE, 1890–1990						
	406%	342%	1,129%	951%	119%	1,525%

Source: RIVM 1997:95

have greatly expanded their numbers and range through domestication.[78] Human population growth and ceaseless toil remade landscapes and ecosystems so as to maximize the populations of the minion biota. By the 1980s, domesticates accounted for 15 percent of the planet's animal biomass, versus 5 percent for people. Consider for example the population history of the most important domesticated animals, shown in Table 8.2. Except for the horse, which fell from grace when internal combustion engines took its job after 1920, each of these domesticates grew much more numerous in the twentieth century. Sheep, goats, and chickens enjoyed a reproductive and survival riot, multiplying even faster than their human masters. Of course, most of the minion biota was butchered and eaten by people. Its demographic success was contingent upon our tastes and appetites. That success was a direct result of our efforts to turn ever larger portions of the biosphere's energy and materials to our own purposes. The complementary success of synanthropes in the niches we inadvertently created is an indirect result of the same grand process.[79]

[78]Thousands of other plants have been domesticated but on small scales (Heywood and Watson 1995:717–8).

[79]New York City in 1997 was home to an estimated 28 million rats (Washington Post, 4 October 1997).

Conclusion

In his *Theses on Feuerbach,* Marx complained that philosophers only interpreted history, but the point was to change it. In the twentieth century, humankind not only interpreted biological evolution but profoundly changed it. Since at least the first domestication of plants and animals, humankind has influenced biological evolution. But lately that influence has become, in more and more instances, governing. Long ago, cultural evolution replaced biological evolution as the chief agent of change in human affairs. In the twentieth century, human cultural evolution impinged upon, and in cases entirely controlled, the biological evolution of other species as well.

For people, the most powerful changes in the twentieth century were those in disease regimes and in agriculture. These were sharp and fundamental changes in the human condition, affecting billions of people in matters of life and death. The changes in forests, fisheries, bioinvasions, and biodiversity made less difference from the human point of view. Naturally a Dayak tribesman from Borneo, expelled from ancestral land to make room for a clearcut, or a California sardine fisherman thrown out of work might see things differently. So would trees and fish. I have taken the anthropocentric point of view, and within that, the view that billions of people are more important than millions or thousands. Other viewpoints exist.

That said, in the longer run the erosion of biodiversity or the consequences of bioinvasions may yet prove to be more consequential than disease control or the Green Revolution. The range of possible outcomes from these changes are tremendous: they may prove modest, especially if the tendencies of the twentieth century are somehow checked; they may prove large beyond imagination. The history of the twenty-first century will make such matters clearer.

From the very long term perspective, humanity in its high-energy phase (roughly since 1820) may resemble cyanobacteria of 2 billion years ago. Those diminutive creatures, also known as blue-green algae, pioneered new metabolic paths, as we have recently done, and refashioned the world in the process. They excreted oxygen while using hydrogen from water, thus raising the oxygen concentration in the air from 1 part per trillion to the current 1 in 5. This conveniently poisoned most other bacteria, to which oxygen was toxic, and made more room for more cyanobacteria and other oxygen-tolerant creatures. Humankind used more tools than simple oxygen poisoning, but took steps down the same path, toward a biosphere of our making. It is not one of our choosing, however, as we are scarcely more conscious of the process than were cyanobacteria.

PART II

ENGINES OF CHANGE

Toda la historia es solamente la narración del trabajo de ajuste, y los combates, entre la naturaleza extrahumana y la naturaleza humana. . . . [The whole of history is just the narration of the work of adjustment, and the combat, between extra-human nature and human nature.]
—José Martí, *Obras Completas* (1975) 23:44–5

The Cuban intellectual José Martí at the end of the nineteenth century recognized that history unfolded amid environmental changes wrought, in turn, by history. The momentous social, economic, and political changes of the twentieth century took place in tandem with equally great changes in the environment. Why did so much environmental change happen in the twentieth century? Why not before? How did it all happen? Within each of the foregoing chapters, I have offered answers involving several specific cases of environmental change. To answer these questions more systematically, we must turn to the social, economic, and political trends of the century and sketch their labyrinthine links to environmental change. This section of the book attempts just that.

The results will prove unsatisfactory. The relationships between environmental and social changes are dense, reciprocal,

overlapping, and always in flux. Grand theories that ply a sim-
ple answer (entropy, capitalism, overpopulation, patriarchy,
market failure, affluence, poverty) are of little help.

My simplest answer is that in the twentieth century, two
trends—conversion to a fossil fuel–based energy system and
very rapid population growth—spread nearly around the
world, while a third—ideological and political commitment to
economic growth and military power—which was already
widespread, consolidated.

One has to chop up the seamless web of history to write it.
Here I will divide a tangled subject into seven main parts,
grouped into three chapters: population and urbanization
(Chapter 9); energy, technologies, and economics (Chapter
10); and ideas and politics (Chapter 11). The chapter scheme
is a matter of convenience as much as of logic. The connection
between population, which comes first, and politics, which
comes last, was often real and influential for environmental
history, as I hope the final pages demonstrate. All of these
themes, moreover, danced together in a coevolutionary cotil-
lion.

9

More People, Bigger Cities

It was divine nature which gave us the country, and man's skill
that built the cities.

—Marcus Terentius Varro, *De Re Rustica*

Among the greatest distinctions of the twentieth century were its powerful twin surges of population growth and urbanization. These trends reflect billions of human choices, conscious and unconscious, made for
countless reasons. Some of these choices were individual, such as whether
to marry or where to live. Some were political, such as the Australian
choice to build a capital city at Canberra in the 1920s and the decision of
South Africa's National Party to implement apartheid in the 1940s. All
these decisions collectively generated the global trends of population
growth and urbanization. They affected everything in human affairs, to
greater or lesser degrees, from high culture to child nurture to corporate
structure. They affected much that was not human too.

Population Growth

Most discussion of the social forces behind environmental changes is politically charged, and none more so than the issue of population. Debate often boils down to arguments that other people must change their ways to save the earth: Indians and Africans usually argue that population growth matters little; Americans and Europeans that it matters greatly. My view is that it mattered for some varieties of environmental change but not others, and that migration was often more consequential than sheer growth. The issue is anything but straightforward.

The bizarre population history of the twentieth century was the climax (to date) of a long frenzy of reproduction and survival. Chapter 1 considered the very long term; here I will focus on the past 500 years. The globalization brought about by European mariners at the end of the fifteenth century incubated two biological changes significant for subsequent population history. First, their globe-girdling voyages spread diseases among populations that had previously been long isolated from one another. In the short run, this led to catastrophic losses, notably in the Americas and Oceania. Eventually this swirl of infection produced more seasoned immune systems, closer symbiosis between pathogens and hosts, and still later, public health systems, so that the toll from epidemics subsided. Secondly, maritime voyagers distributed food crops far and wide, eventually allowing (within limits imposed by politics and tradition) every region of the world to concentrate on the crops best suited to its ecological and market conditions. Maize spread from its home in the American tropics to East Asia, southern Africa, and the Mediterranean basin. Tropical Africa acquired manioc (cassava) from Brazil, while the Americas got wheat. In sum, the world's food supply improved. By about 1650 these two factors provoked a long-term rise in world population, still in train.

On top of this surge, itself unprecedented in human history, came a tidal wave of population growth. It drew on further improvements in food supply and disease prevention (some described in Chapter 7) and reached a crescendo after 1950. These improvements first caused declines in mortality, to which some societies later responded by restricting their fertility. In between, when mortality had declined but fertility had not, population grew very quickly. Demographers call the whole process "the demographic transition."[1] The rate of growth peaked in the late 1960s (at 2.1% per annum). It so appalled observers that a careful and sensitive one, Kenneth Boulding, seriously suggested establishing tradable permits for having children.[2] Growth slowly slackened after 1970, mainly because women acquired more say in many societies and limited their fertility. By 1996 the total annual increment of population had peaked at about 92 million to 95 million more births than deaths per year.

Population growth quickened and slowed at different times in different places. The demographic transition began first in Europe, where it took more than a century to complete. In East Asia it came only after 1950, but took less than half a century. In Africa, it remained unfinished, for fertility by the late 1990s had only recently and spottily begun to subside. Table 9.1 gives a rough idea of the pace of population growth by world region.

In the period 1850 to 1950 the populations of Africa, Asia, and Europe roughly doubled. Meanwhile numbers in the Americas, Australia, and

TABLE 9.1 REGIONAL POPULATION, 1750-1996

	Million People					
	1750	1800	1850	1900	1950	1996
Asia	480	602	749	937	1,386	3,501
Europe	140	187	266	401	576	728
Africa	95	90	95	120	206	732
North America	1	6	26	81	167	295
South and Central America	11	19	33	63	162	486
Australia and Oceania	2	2	2	6	13	29

Sources: Reinhard et al. 1968:680–1; Population Reference Bureau (1996).

Note: The figures for Africa and to a lesser extent Asia are speculative prior to 1900; figures somewhat lower for Africa and somewhat higher for Asia appear in McEvedy and Jones 1978.

[1]The transition brings birth and death rates down from around 30–35 per thousand (per year) to around 10–12 per thousand.

[2]Boulding 1964:135–6. Boulding (1910–1993), an economist and Quaker, was not normally given to recommending sharp restrictions on freedoms.

Oceania grew much faster, five- or sixfold in 100 years. This reflected both migration patterns and differences in natural increase. After 1950 the locus of fast growth changed. In the ensuing half century, Asian numbers more than doubled, Latin American population tripled, and African population nearly quadrupled. Meanwhile Europe and North America grew much more slowly, having completed the demographic transition by 1950.[3]

By the 1990s humankind accounted for about 0.01 percent of earth's total biomass, and about 5 percent of its animal biomass, ranking with cattle and far outstripping any other mammal.[4] This crowning success of the human species coincided with intense environmental change. Did it cause that change? I will try to answer this with both arithmetic and anecdote. First the arithmetic.

Consider the relationship between global population growth and some global air pollutants. Between 1890 and 1990, world population increased by a factor of 3.5, while emissions of carbon dioxide, the main greenhouse gas, climbed more than 17-fold.[5] A superficial calculation[6] suggests that population then accounted for only about a fifth of that rise. In the same span, global emissions of sulfur dioxide, a major component of acid rain, increased about 13-fold. Population increase accounted for slightly more than a quarter of the growth in sulfur emissions. If one does the arithmetic for the United States alone, population growth accounted for 31 percent of the growth in carbon dioxide emissions; in what is now OECD (that is, the rich countries of) Europe, 41 percent; in Japan and the USSR (and successor states), only 2 percent; in Africa, less than 1 percent.[7] Abandoning this primitive arithmetic for the moment,

[3]The data would look different if presented not by continent but by religion or GNP per capita. However, these indices have changed so much in the 20th century that it is impractical to do so.

[4]Turco 1997:105. Ants outweigh us about 4:1.

[5]These data are from RIVM 1997.

[6]The calculation is superficial for several reasons. It deals with global aggregates, and neglects the possibility that population growth occurred in places where carbon dioxide emissions did not and that the two were unrelated. It assumes a linear relationship between the two, whereas in reality there might well be thresholds below which population increase makes little difference, or above which it does. Regrettably, no precise mathematical relationship between these variables can logically be constructed.

[7]The reason these figures vary so much is that in western Europe and the USA, population growth after 1890 was slow (by world standards) and carbon dioxide emissions were already (comparatively) high. In Japan and Russia, while population growth was also comparatively slow, carbon emissions were very low in 1890 because the countries had just begun to indus-

one may safely suppose that population growth had a minimal role in releasing chlorofluorocarbons (CFCs) into the stratosphere.

For some important forms of air pollution, then, population growth in the twentieth century was a significant but by no means preponderant driving force. This stands to reason: combustion generated most of the air pollution, and the quantity of combustion and the intensity of its pollution were only loosely linked to the number of people. In rich societies, such as the United States and Germany, additional people did substantially raise air pollution levels from 1900 to 1970 or so, because they drove cars, heated with oil or coal, and in general increased total combustion. After 1970, new technology, encouraged by regulation, lawsuits, and a decade of high energy prices, loosened the link: cleaner production and cleaner cars meant additional people caused much smaller increments in air pollution than they had in the 1950s or 1960s. In poor societies additional people had even less impact on air pollution because they caused negligible increases in combustion. Even where population growth and pollution coincided, as in China after 1970, it is hard to conclude that the former caused the latter. Fast and careless industrialization and urbanization probably mattered more than the rate of population growth.[8]

In general, population growth provoked additional pollution of air and water primarily where and when the economy was already industrialized

---

trialize. Harrison (1992) does similar arithmetic for 1961–1988 and concludes that population accounted for 44% of the growth in carbon emissions. Taking these years, when population growth was at a maximum and when emission increase rates declined (after 1975) accounts for the higher figure. Ogawa 1991 and Darmstadter and Fri 1992 examine the years 1973–1987 and rate the role of population growth even higher. In those years, improved energy efficiency and reduced carbon emissions per unit of energy consumption combined to lower carbon emissions by 1% per year, but this was more than counteracted by the impact of population (+1.8% per year) and economic growth (+1% per year). These 15 years were ones of very fast population growth and (by post-1945 standards at least) slow economic growth, so, like Harrison's figures, they are helpful on their own terms but not useful for the century as a whole. Other arithmetic exercises along these lines include Raskin (1995), who finds population growth responsible for 32% of carbon dioxide emission growth for the period 1950–1990; a fine summary of the weaknesses of all such calculations is found in McKellar et al. 1998:120–35.

[8]China's extraordinary demographic trends are summarized in Lee and Feng 1999. Population policies since 1978 kept Chinese population 250 million below what it otherwise would have been in 1998. Growth rates in the 1960s approached 3% per annum, but in the 1970s dropped to under 2%. In 1996, China's annual population growth rate was 1.1%. Smil 1993 considers pollution and population in the recent Chinese context.

and where society (and state) did not value environmental amenities. This was true in the United States, Japan, and western Europe from roughly 1890 to 1970, in Russia from 1960 onward. Population growth in societies without significant industry had much less impact on pollution levels except for human wastes and domestic smoke. In societies undergoing industrialization, such as South Korea (1960–1990) or the USSR, (1930–1960), the rate of population growth mattered much less than the rate and type of industrialization.

The nexus between population and pollution in the twentieth century is hazy enough. But the relationships between population growth and other forms of environmental change are cloaked in still thicker confusion. Population pressure both caused and prevented soil erosion. In places where it drove farmers up steep hillsides, as in Java or northern Morocco, it quickened erosion. Elsewhere it provided enough labor to build and maintain soil conservation schemes, as in the Machakos Hills of Kenya. Moreover, in mountain environments the loss of population sometimes brought on faster soil erosion, as too few people remained to maintain terraces and other soil conservation stratagems. Soil salinization sometimes derived (indirectly via expanded irrigation) from population pressure, but more often from the temptations of commercial or centrally planned agriculture. Population growth and density were only partial determinants in these equations: natural, political, and economic conditions frequently carried greater weight. The best conclusion—a rough one—is that population growth often heightened erosion rates, but dense populations, if stable, could lower them.[9]

Population growth probably accounted for much of the world's increased water use and intensified problems of water scarcity (see Chapters 5 and 6). A superficial calculation suggests 44 percent of it: water use increased ninefold and population fourfold in the years 1900 to 1990, so four-ninths of the century's increment derived from the existence of more people. This, however, is only a rough measure. Changes in water-use efficiency, as well as in pricing and subsidies, blur the picture. In the United States after 1980, as noted in Chapter 5, water use declined while popu-

[9]On Java, see Repetto 1986, who reports a sixfold increase in erosion rates (1911–1983) attributable to population pressure. On northern Morocco, see Maurer 1968; on labor shortage and erosion, see Barker 1995, McNeill 1992b, and Mignon 1981. The tale of the Machakos Hills was given in Chapter 2. Interestingly, in the 1930s, Jacks and Whyte 1939:286–7 argued that a dense population was the best insurance against soil erosion, offering Japan and Java as evidence (and admitting India and China as exceptions)!

lation grew. In almost every society considerable slack existed in the water-use systems, such as inefficiencies and waste, so changes in technology and policy could, and sometimes did, alter matters more sharply than could population growth.[10]

Population growth surely played a large role in driving the manifold changes to the biota in the twentieth century. Food demand drove most of the century's doubling of cropland, helped fuel the Green Revolution, and multiplied the world's fishing effort. Population growth did not produce these changes alone, but in matters directly related to food production it loomed largest.

Some of the important biotic changes, however, had little connection to population or demand for food. Whaling, unlike fishing, did not significantly reflect expanding food needs. Biological invasions had almost nothing to do with human population growth. The vast shifts in human-microbial relations had a lot to do with it, but here the causation was reversed: the environmental shift drove population growth.

Deforestation admirably illustrates the murky conundrum of environment and population. In some cases, such as rural Ethiopia, recent studies conclude that population growth was the main driving force. But historical studies extending back to the nineteenth century show scant forest even when Ethiopia's rural population was only a fraction of what it became and when its growth rate was also much lower. Deforestation, in Ethiopia and around the world, occurred in conditions of population growth, population stagnation, and even population decline (e.g., Russia in the 1990s or Madagascar in the period 1900–1940).[11] A meta-analysis (that is, a statistical study of numerous independent studies) of the relationship between population and deforestation, concluded that

> while population pressure is an important force leading to deforestation,
> it rarely acts alone to produce the outcome. Other determinants appear
> to be necessary as mediation and contingencies for population growth (or

[10]See Falkenmark 1996 for a discussion of contemporary population and water issues. She seems to regard population as a stronger driving force than my rough calculation suggests. The discrepancy, if it is that, may derive from the fact that population growth in the 1990s (Falkenmark's implicit reference point) is greater than the average for the 20th century (my reference point).

[11]On Ethiopia, see Campbell 1991, Grepperud 1996, and McCann 1997. Madagascar lost 4 million ha of forest (1900–1940), while population stagnated or possibly declined. Cash crops, notably coffee, replaced the forest (Jarosz 1993). Kummer 1991:146–9 concludes that population growth played a scant role in deforestation in the Philippines in recent decades.

density) to have a discernible impact. . . . [Q]uantitative analysis
suggested that even if the effects of population growth are statistically
significant, their magnitude is quite modest.[12]

Such a vague conclusion, regrettably, is just about right.

In sum, population growth accounted for a modest share of air pollution–related environmental changes and a larger share of those pertain-
ing to water and the biota, especially those involving food production. Big
environmental change resulted more often from combinations of mutually reinforcing factors than from population growth alone. The latter
probably mattered more late in the century than earlier, mainly because
growth climaxed after 1960.[13]

Migration

Migration often mattered even more than growth, although the two
are often hard to separate because growth sometimes caused, or at least
helped cause, migration. From 1500 to about 1870 most of the world's in-
tercontinental migrants were slaves or "coolies." Then, around 1845 to
1920 the spontaneous migration from Europe to the Americas came to
overshadow other currents around the world. After 1925, international
migration receded for several decades, and when it gathered pace again
by 1960, more diffuse flows prevailed. In all periods, migration often
promoted radical changes in land use and sharp ecological shifts.

From the point of view of environmental change, the most important
migration involved frontier areas. Mass migrations from humid to dry
lands repeatedly provoked desertification.[14] Migrations from flatlands to
sloping lands often led to faster soil erosion. Migrations into forest zones
brought deforestation.

---

[12]Palloni 1994:160.

[13]Repetto and Holmes 1983 attempt to show mathematically the limited role of population in
accounting for natural resource depletion (using mostly 1970s data). The view that population
growth is "probably predominant . . . in environmental problems" is put forward by Myers 1993.

[14]Here desertification means progressive loss of vegetation cover and organic content in the
soils. The process is not necessarily irreversible.

With declining transport costs and tightening integration of markets after 1870, people moved about as never before. They frequently ventured into ecological zones about which they understood little. This, of course, has happened throughout human history. But in the twentieth century more people moved, they had more transformative technologies at hand, and in most cases they had stronger links to markets or national planners, giving them incentives to clear more land, plant more crops, graze more animals, catch more fish, or mine more ore than their own subsistence required.

The scale was enormous. In the period 1830 to 1920, Europe alone sent 55 million to 70 million emigrants to the Americas, Australia, and Siberia.[15] A large minority engaged in pioneer farming, such as Ukrainians on the Canadian prairies or Italians on the coffee frontier of Brazil. (In 1934 nearly half of São Paulo state's coffee farms were owned by immigrants).[16] This epochal migration wound down after its 1913 peak, restricted by World War I, anti-immigration laws in the United States (1924) and elsewhere, anti-emigration laws in the Soviet Union (1926) and elsewhere, and after 1929 by the Great Depression. Simultaneously (1834–1937) about 30 million to 45 million Indians moved, mostly as indentured laborers to plantations in Fiji, Malaya, Burma, Mauritius, Natal, Trinidad, and Guyana. While they took no part in land-use decisions, and while most were sojourners not settlers, their labor converted millions of hectares of forest land to sugar, rubber, and other crops. Large numbers of Chinese labor migrants moved to Southeast Asia, the Caribbean, California, and Peru among other destinations. This "coolie" trade too slowed under the impact of war and depression after 1914.[17]

But migration did not stop. It became less international. Millions upon millions of people after World War I migrated into new rural areas. Fewer and fewer people did so spontaneously. Instead, compulsion and state policy took a larger role after 1920, a return to the pattern that prevailed from 1500 to 1870. In some cases, state policy promoted voluntary migration by offering financial and other incentives. Great Britain paid young people, especially war veterans, to migrate within the Empire to Australia, New Zealand, and Canada. A few hundred thousand signed up, and a few million hectares in the Dominions were consequently broken to the plow. In other cases, state policy forced migrations, as in South

---

[15]Baines 1995; Thumerelle 1996:106–7.
[16]Klein 1995:211.
[17]Zolberg 1997:288.

Africa. There the Land Act (1913) and subsequent laws restricted ownership of good farmland to whites, obliging millions of black South Africans to move to ostensible tribal homelands called bantustans. There they crowded onto South Africa's poorer lands, farming steep slopes and running their cattle on semiarid bushlands, resulting in an archipelago of intense land degradation in that country. Mao's plans after 1949 moved millions of Chinese, accustomed to monsoon agriculture, to dry lands in Inner Mongolia, Xinjiang, and Tibet. Brazil's rulers made settlement of Amazonia a national goal after 1960, calling it "a land without men for men without land." In all these cases, mass migration served as a means to political or social ends. Ensuing environmental changes were sometimes unforeseen side effects; at other times they were expected, accepted, or even desired by the leaders in question. Two of the more determined state efforts, involving varying proportions of compulsion and incentives, took place in the Soviet Union and Indonesia.

From at least the sixteenth century Russians had settled lands beyond the historic Russian heartland. The search for land and furs brought them to Siberia, central Asia, and Alaska. This epic of national expansion continued, except in Alaska, through the end of the tsarist period in 1917. In the final decades of Imperial Russia, the trans-Siberian Railroad brought 3 million to 4 million Russian peasants in search of free land in Siberia. The government, eager to develop Siberia, encouraged this migration and augmented it with a flow of political exiles.[18] Soon after the formation of the USSR (1917–1922), state-sponsored migration quickened. Soviet settlement in Siberia and Central Asia served two purposes. It was both a punishment for counterrevolutionary individuals and nationalities, and part of the brave new future of economic development planned by Soviet leaders. Mass migration began in 1929 with Stalin's collectivization of agriculture. Between 2 million and 3 million Russians and Ukrainians were forcibly removed to Siberia, the Urals, or the Russian far north. Over the whole Stalinist period (1927–1953) more than 10 million people suffered forced migration. More than half were sent beyond the borders of the Russian republic, often to "virgin land." During World War II Stalin deported entire peoples whose loyalty he distrusted. One million ethnic Germans and hundreds of thousands of Chechens, Tatars, and Kalmyks moved to Central Asia and Siberia in the desperate and

[18]Treadgold 1957:159–61. Chesnais 1995:221 gives 5.8 million as the total number of migrants for 1801–1914, from European Russia across the Urals. Of these, 80% were voluntary and 20% were internal exiles.

chaotic days of 1941 to 1944. Millions of Russians took over their lands along the lower Volga, in the North Caucasus, and Crimea.

Stalin was a breaker of nations, and the swirl of migrations he ordered moved millions into unfamiliar environments. Some hacked away at northern forests, others dug canals, built railroads or industrial complexes, while still others tried to till the arid steppe. After Stalin's death in 1953 the Soviet Union used the carrot more than the stick to move another million Russians to plow the Kazakh steppe in Khrushchev's Virgin Lands program (see Chapter 7). All in all, several million Soviet citizens found themselves toiling away in unfamiliar landscapes, radically altering land use and cover, hydrology, soils, and much else. Their government intended that they should do so, for the glory of socialism and the strengthening of the USSR. Given their struggle to survive, it is safe to suppose notions of soil conservation and forest preservation rarely crossed their minds.[19]

Indonesia, like the Soviet Union, incorporated vast expanses that its rulers sought to transform into economically productive areas. It used less compulsion and moved fewer people. The Indonesian scheme was known as "transmigration."

Most Indonesians live on Java, which together with Bali makes up the agriculturally fertile inner core of the country. The far larger outer islands of Borneo, of which the Indonesian portion is called Kalimantan, and Sumatra, together with thousands of smaller islands, have among the poorest soils on earth. Java and Bali have long hosted dense populations of rice farmers, who over the centuries carved their islands' rugged terrain into mosaics of irrigated terraces.[20] The outer islands, in contrast, supported scant population, remained mostly under forest, and contributed little to the wealth and power of the state. That state for centuries was a colonial one, run by the Dutch.

The Dutch began the transmigration project in 1905, acting on ideas in circulation since at least the eighteenth century. Dutch colonials had long thought that the population of Borneo and Sumatra needed augmenting so that the resources of these islands, notably timber and gold, could be brought to market.[21] In 1905 the Dutch East Indies had about 37 million

---

[19]The migration data come from Polyakov and Ushkalov 1995. Zemskov 1991 reports the forcible migration of 1.8 million kulaks in 1930–1931 alone. A zestful account of logging's ravages in the 1930s is reported in Andreev-Khomiakov 1997:27–38 et passim.

[20]See Geertz 1963.

[21]Knapen 1998.

people, of whom 30 million lived on Java. By awarding nearly a hectare of land to any family that would move, the Dutch induced about 200,000 people to migrate, mostly to southern Sumatra, by 1941. This did next to nothing to relieve population pressure, a stated goal. In its peak year (1941) Dutch transmigration involved 36,000 people. Java acquired this many people every two to three weeks from natural population increase. Transmigration stopped while Japan occupied the islands (1942–1945) and resumed when Indonesia became independent in 1949. Indonesian planners hoped for an outflow from Java of 2 million people per year, and intended that by 1987 no less than 50 million Javanese would populate the outer islands. This would alleviate pressures on Java, turn the outer islands into revenue-producing areas, and conveniently swamp their local populations, whose devotion to the Indonesian state at best wavered. This grand scheme enjoyed the support of diverse backers, from the Indonesian Communist Party to the World Bank, which chipped in half a billion dollars. But it foundered on—among other things—ecological ignorance.

Reality proved triply disappointing. Too few took the lure, most who did stayed poor, and the resentments created by transmigration threatened state security more than the process of swamping indigenous people with loyal Javanese helped. By 1987 only 4.6 million people had moved, equivalent to about three years' increment in Java's population growth at 1970s rates. The lure had grown to 4 to 5 hectares of land per family. But still the vast majority of Javanese preferred the crowded conditions of rural Java, the squalor of Djakarta's shantytowns, or labor migration to Malaysia to the hardships of pioneer life on the outer islands. Java's peasants were experts at growing paddy rice, but their skills applied in only a few places in the outer islands. Their heroic efforts to clear the forest too often led to poor harvests, dashed hopes, land impoverishment, and abandonment. On the outer islands luxuriant forest growth derived from constant nutrient cycling between trees and the organic litter on the forest floor. The soils contained few nutrients, as is widely the case in the tropics. Farming successfully in such environments required different skills, normally swidden techniques.[22] Some transmigrants adopted appropriate techniques, some did not. Some enjoyed strong state support in the form of

[22]Swidden or slash-and-burn is a technique in which farmers burn forest, plant crops among the ashes, harvest their crops for one to three years, and then move on to repeat the cycle elsewhere. They (or others) can return to the original site and use it again, but only after 10–30 years, enough time to allow sufficient vegetation to grow that, when burned, will release enough nutrients for farming.

schools, health clinics, and so on, but others did not. In many cases migrants gave up on eroding lands, on fields invaded by tough and useless *imperata* grass and other pests. Many found themselves in conflict with local populations on the outer islands, who wanted to farm the best lands themselves. About 20 percent of the migrants improved their standard of living; many of the rest felt swindled. By the late 1980s the transmigration scheme had ground nearly to a halt.[23] While it lasted, it provoked considerable environmental change; had it succeeded as envisioned by its promoters, it would have changed the outer islands utterly.

Officially sponsored migration helped decimate two of the twentieth century's three great rainforests, in Indonesia and Brazil. Africa's rainforests avoided this fate in part because of the continent's political fragmentation: impoverished Sahelian farmers and herders were not encouraged to migrate to Zaire's forests. Had Africa been unified, as pan-Africanists had hoped, the life expectancy of Africa's tropical forests would surely have been shortened.[24]

Around the world, spontaneous and officially sponsored migrations moved tens of millions of people into ecologically unfamiliar areas in the twentieth century. A hefty proportion of the world's environmental changes resulted. This was especially true in those categories strongly affected by frontier farming: vegetation cover, biodiversity, soil condition, and, in dry lands, water use.

The Footprints and Metabolisms of Cities

Twentieth-century urbanization affected almost everything in human affairs and constituted a vast break with past centuries. Nowhere had humankind altered the environment more than in cities, but their im-

[23]Abdoellah 1996; Hardjono 1977, 1988; Levang and Sevin 1989. The World Bank's role is treated in Rich 1994:34–8.

[24]Note the limits of this argument. Tropical deforestation in Brazil and Indonesia was caused by other things as well as in-migration (see Chapter 8); in-migration in these countries took place spontaneously as well as with official help; and international migration within Africa, both during and after the colonial period, did (to a small degree) transform forests into croplands, most notably in West Africa. On the migrations to the West African forest zone, see Cordell et al. 1996.

pact reached far beyond their boundaries. The growth of cities was a crucial source of environmental change.

First, consider the scale and tempo of the process. A millennium ago China and the Islamic Middle East had the world's most urbanized populations, but even in these lands 90 to 95 percent of people lived outside cities. City size was strongly constrained by the nearly stagnant agricultural productivity of their hinterlands and the sharply fluctuating ability of cities to dominate these hinterlands politically. In 1700, only five cities in the world, all political capitals, had more than half a million people: Istanbul, Tokyo, Beijing, Paris, and London. By 1800 there were still only six. (Canton/Guangdong had joined the list.) The hazards of city life, mainly endemic childhood diseases but also epidemics, killed people faster than others were being born. London, for example, in 1650 required 6,000 in-migrants per year to stay even. In 1750 London's lethality counteracted half the natural increase of England as a whole.[25] But the constraints on urban growth soon relaxed: farms were growing slightly more productive; cities, at least capital cities, were more able to dominate hinterlands; fossil fuel transport extended the reach of cities farther into the countryside; and public health programs began to take effect. By the late 1880s, city folk in Austria and Bavaria had longer life expectancies than their country cousins. By the 1920s, urban Chinese outlived Chinese peasants.[26] For 8,000 years cities had been demographic black holes. In the span of one human generation they stopped checking population growth and started adding to it: a great turning point in the human condition.

By 1900 there were 43 cities of over half a million, mostly in western Europe, eastern North America, and on the seacoasts of export-oriented economies, often in European colonies. By 1990 about 800 cities, scattered all over the world, surpassed the half million mark. Some 270 had more than a million residents, and 14 topped 10 million. It remains to be seen what constraints on city size now exist.[27]

The first country in which urban dwellers accounted for a tenth of the population was the seventeenth-century Netherlands, propelled by the

[25]Macfarlane 1997:22.

[26]On Austria and Bavaria, see Hohenberg and Lees 1985:259, and Munch 1993. On China, see Lee and Feng 1999. This transition may have begun first in Japan (Hanley 1987). One of the few exceptions to the greater longevity of urban populations in the 20th century was 1980s Poland, where acute pollution contributed to shorter lives among city folk.

[27]Berry 1990:104–6; WRI 1997:8–9.

twin engines of global commerce and unusually rich local agriculture. The first country in which half the population lived in cities was industrializing England, in about 1850. The United States reached this level at about 1920, Japan about 1935, the USSR and Mexico about 1960, South Korea about 1975, South Africa about 1985. In 1998 the world as a whole approached it.[28] In very general terms, rapid urbanization happened in Japan, western Europe, and eastern North America in the nineteenth and early twentieth century, in the Soviet Union and Latin America in the mid-twentieth century, and almost everywhere else after 1960. In national terms, by far the fastest urbanizations occurred in the USSR in the 1930s and in China in the 1980s. Each of these reflected burgeoning industrialization, in the Soviet case while building communism (and starting cities from scratch) and in the Chinese case while dismantling it. Chinese policy under Mao had successfully stalled urban growth, but after his successors abandoned Mao's restrictions in the late 1970s, cities grew with pent-up force. Table 9.2

TABLE 9.2 URBAN PROPORTIONS, 1890-1990

Region	Percent of Total Population					
	1890	1910	1930	1950	1970	1990
USA	35	46	56	64	70	75
Japan	30	40	48	56	71	77
Western Europe	35	45	55	63	72	78
Latin America	5	7	17	41	57	71
USSR	12	14	18	39	57	66
Africa	5	5	7	15	23	34
China	5	5	6	11	17	33
South Asia	5	8	12	16	21	28
World	14	18	23	29	37	43

Source: RIVM 1997:20. Somewhat different figures, especially for early in the century, appear in Thumerelle 1996:75.

Note: These data ultimately derive from national censuses (and estimates) that define "city" and "urban" very differently. In Japan a community normally had to have 20,000 people to qualify as urban; in Turkey, 10,000; in the USA, only 2,500.

[28]By one calculation (Clark 1998), the world was already half urban by 1996.

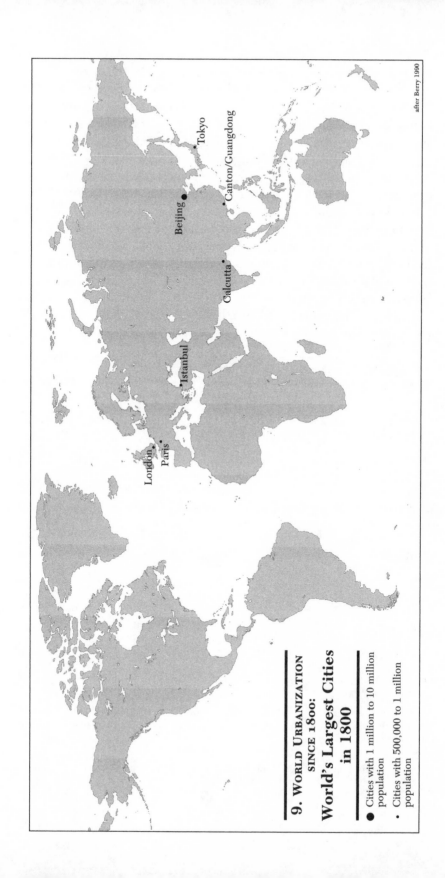

9. WORLD URBANIZATION
SINCE 1800:
**World's Largest Cities
in 1800**

● Cities with 1 million to 10 million population

· Cities with 500,000 to 1 million population

London
Paris
Istanbul
Beijing
Calcutta
Canton/Guangdong
Tokyo

after Berry 1990

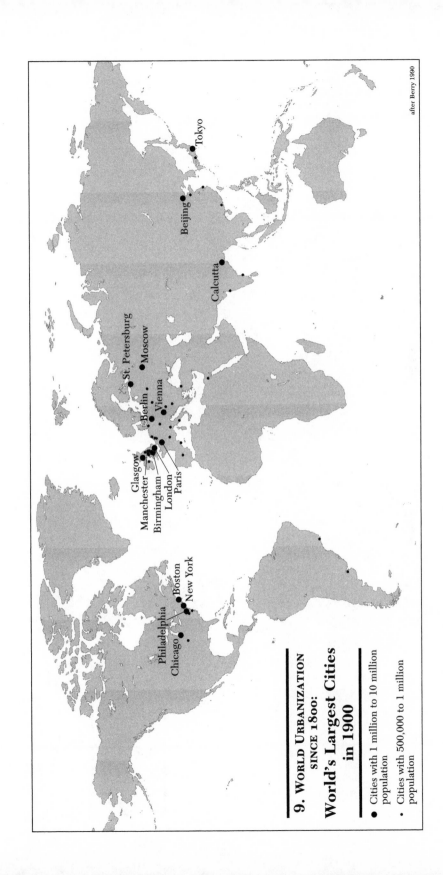

after Berry 1990

9. WORLD URBANIZATION
SINCE 1800:
World's Largest Cities
in 1900

- Cities with 1 million to 10 million population
- Cities with 500,000 to 1 million population

Tokyo

Beijing

Calcutta

St. Petersburg
Moscow
Berlin
Vienna
Glasgow
Manchester
Birmingham
London
Paris

Philadelphia
Chicago
Boston
New York

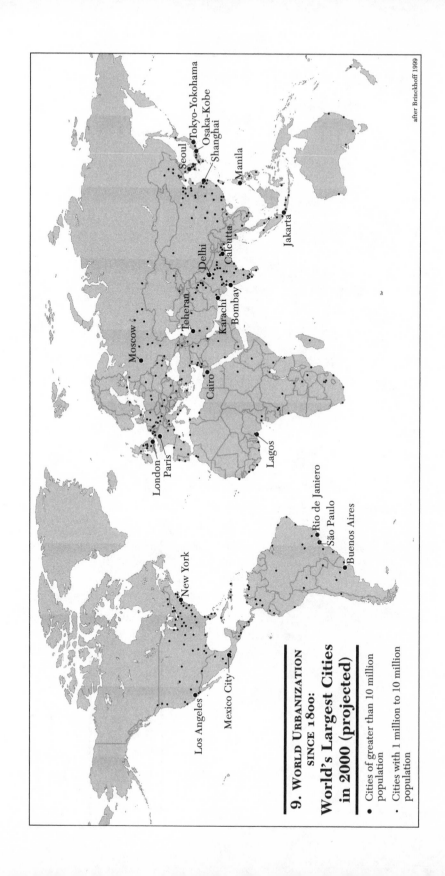

after Brinckhoff 1999

Tokyo-Yokohama
Osaka-Kobe
Seoul
Shanghai
Manila
Jakarta
Delhi
Calcutta
Teheran
Karachi
Bombay
Moscow
Cairo
Lagos
London
Paris
Rio de Janiero
São Paulo
Buenos Aires
New York
Mexico City
Los Angeles

9. WORLD URBANIZATION
SINCE 1800:

World's Largest Cities
in 2000 (projected)

● Cities of greater than 10 million
   population

· Cities with 1 million to 10 million
   population

summarizes the recent history of urbanization for some large regions of the world. As the table shows, the world's urban percentage tripled in the twentieth century. The total number of urban dwellers rose from about 225 million in 1900 (most of them in Europe and North America) to 2.8 billion in 1998, a 13-fold rise.

The environmental meaning of the world's tumultuous urbanization was vast and variable. Urban impacts extended beyond city limits to hinterlands, to downwind and downstream communities, and in some respects to the whole globe. Cities absorbed ever larger quantities of water, energy, and materials from near and far. In exchange they pumped out goods and services—as well as pollutants, garbage, and solid wastes. Broadly speaking, this process of urban metabolism generated two categories of environmental change: pollution effects and land use effects.[29]

GARBAGE AND POLLUTION. Rapid urbanization normally generated severe pollution stress. Infrastructure rarely kept pace with heady urban growth. Many people in 1900 and about 800 million in 1990 lived without piped water, sewage systems, gas, or electricity.[30] Consequently they lived amid their own wastes and pollution (see Chapters 5 and 6). No novelty in that: most city dwellers from ancient times had lived—and died— that way. But far more participated in this grim existence in the twentieth century, often in ramshackle shantytowns built around urban cores. Shantytown inhabitants usually lacked formal title to their homes, could be evicted at any time, and consequently invested little time or money in neighborhood improvements. City governments occasionally did so, as in Curitiba (Brazil) during Mayor Jaime Lerner's tenure in the 1970s and 1980s, when recycling and public transport attained unusual efficiency. But often they did not, because of lack of funds or lack of concern. The environmental conditions created by speedy urbanization were among the bleakest humans experienced. Slimy water and gritty air were the worst but not the end of it.

Consider garbage and solid waste. In the nineteenth century, cities everywhere—except maybe Japan—reeked of garbage.[31] But when addressing water supply and sewage issues after 1870 or so, most municipalities in the Western world also organized garbage collection and

---

[29]General treatments with some historical dimension include Berry 1990, Douglas 1994, and Gugler 1996.

[30]WRI 1997:152–3 has a table with these data for 1980–1995.

[31]Hanley 1987 says Japanese cities organized garbage collection in the seventeenth century.

disposal. Early in the twentieth century, New York had garbage barges that dumped their contents in the waters outside the city's harbor. Other cities did things differently, but almost all shunted garbage onto neighboring land or into neighboring bodies of water, improving urban health. By midcentury the scale of garbage and solid-waste problems began to grow, overwhelming the capacity of many cities to cope. Mexico City in 1950 produced about 3,000 tons of garbage per day. By the late 1990s it generated three times as much. At least a quarter of it piled up in streets, ditches, and ravines. Garbage hills and ridges sprouted in almost every fast-growing city, from Manila to Maputo and Quito to Karachi. Streets and alleys acquired a rancid and rotting patina of household and other wastes. In Surat, a city of 2.2 million in India, a fifth of the city's garbage went uncollected in the early 1990s. This created ideal rat habitat, and helped launch an outbreak of bubonic plague in 1994. Prompt official response and tetracycline limited the toll to 56 deaths. Galvanized by the plague, Surat managed to clean its streets and collect its rubbish so efficiently that by 1997 it was deemed India's second-cleanest city.[32] Few other cities, however, matched this civic miracle. Fortunately, highly unsanitary conditions have to date produced no sweeping epidemics of the sort that culled urban population in earlier centuries, testimony perhaps to the power of modern medicine.

Where infrastructure caught up to urbanization, solid waste was collected and kept away from human population. After World War II, New York City consigned its trash to Staten Island, where the world's largest landfill opened in 1948, now a towering monument to wastrel ways. By the late 1990s, New York sent caravans of garbage rolling toward states such as Virginia where landfill charges were cheaper. Tokyo, which in the 1990s generated three times as much solid waste as Mexico City, had virtually no health problems related to garbage. Indeed, the Japanese had enough money and ingenuity to find good uses for garbage, converting some into construction materials.[33]

Considering all urban pollution and waste problems together, I hazard the following chronology and conclusions. Cities in the richer countries suffered serious sanitation problems as they grew rapidly. But by 1940 they had addressed these difficulties, through garbage collection, sewage

[32]On Surat, see WRI 1997:42–3.

[33]On New York's landfill (25 times the size of the great pyramid at Giza in 1994), see Trefil 1994:23. Calcutta's garbage habits are described in Basu 1992. On American practices, see Melosi 1981.

systems, and water treatment plants. This left them with pollution derived from industry and transport, the scale of which continued to grow. After about 1970 (earlier in the case of smoke and soot), cities faced these problems and often reduced their intensity. To some extent, rich cities "solved" their problems by shunting them off on downstream or downwind neighbors, but eventually this proved less practical as neighbors learned to use politics and the courts to prevent it. By and large, cities did not address pollution that threatened only diffuse, disorganized, or powerless communities. Cities in wealthy countries took about a century to organize partially effective responses to the pollution effects of urbanization.

Cities in poorer countries did not follow the same trajectory. In many cases they grew too fast, with too little accompanying economic growth, to afford clean water and garbage collection. The sanitary revolution that came to western Europe, Japan, and North America (c.1870–1920) came late and as yet incompletely to Brazil and South Africa, and scarcely at all to Bangladesh and Papua New Guinea, where urbanization rates were among the fastest in the world in the 1990s. And it scarcely came to the world's shantytowns, which spread exuberantly after 1950.[34] Hence, in the latter half of the twentieth century, pollution derived from household wastes continued to grow in poor cities. In addition, poor cities rapidly acquired the pollution problems derived from industry and from fleets of cars, trucks, and buses. Djakarta in 1980 and Bombay in 1990 suffered a double burden of pollution almost never experienced in the rich world, except perhaps in the first industrial cities during the period 1820 to 1860. Only a few societies managed to accumulate capital for investment in pollution abatement at rates faster than pollution itself intensified. And even in those that did, ruling elites usually found it easier to insulate themselves from pollution rather than to reduce it. So cities remained concentrated nodes of pollution—far larger yet far less lethal than before, thanks to vaccination, antibiotics, and other public health measures.[35]

THE ECOLOGICAL FOOTPRINTS OF CITIES. Cities themselves covered perhaps 0.1 percent of the earth's land surface in 1900 and about 1 percent in 1990.[36] Their spatial growth, however, was only a small fraction

[34]The growth and morphology of shantytowns (*favelas*) in Rio de Janeiro is traced in Abreu 1988:106, 125–6; see also Pineo and Baer 1998.

[35]A survey of early 1990s conditions is given by Hardoy et al. 1993.

[36]Douglas 1994 offers the figure used for 1990. I calculated the 1900 figure on the basis of a world population one-third as urban, one-fourth as large, and with a slightly higher density in 1900 than in 1990.

of their environmental impact, for they spread their tentacles far and wide, drawing in food, water, and energy sometimes from other continents. The space needed to support a city, and to absorb its wastes is, metaphorically speaking, its ecological footprint.[37]

For millennia cities were biological oddities. They made good habitat for cats, rats, pigeons, and a small array of weeds, but they were shorn of most other animal and plant life. This changed little in the twentieth century. For each city that added a "green belt" as London did in 1936, or planned green spaces as Ankara did in the 1930s, several lost what green space they had, as Mexico City did between 1950 and 1990.[38] Perhaps the greatest biological change came with the gradual disappearance from cities of horses, camels, donkeys, and other beasts of burden.

In physical as opposed to biological respects, cities often changed fundamentally. In the wealthy world, cheaper and stronger steel allowed the construction of skyscrapers from early in the century. Electrification brought a myriad of changes, not least the electric trolley (invented in 1887 in Richmond, Virginia), which encouraged suburbanization around the turn of the century. Then cars replaced horses and buses all but eliminated trolleys. Electrification, sewage systems, and piped water converted the cities' substrate into a warren of tunnels, mains, and pipes. European and North American cities sprouted toward the sky, burrowed underground, and crept across the earth's surface simultaneously. In 1870, most cities were held together by muscle and bone: people and horses carried or pulled all the food, water, goods, wastes, and information that circulated. By 1920, cities in the wealthy parts of the world (and a few elsewhere) were immensely complex systems of interlocking technical systems. Not coincidentally, they were the powerhouses of the modern economy and the hothouses of modernism in art and literature.[39]

Such changes in cities' built environments came between 1880 and 1940 in the rich world. They came only a few years or decades later in colonial cities such as Bombay or Tunis. Many colonial cities acquired modern infrastructure—ports, warehouses, railroads, roads—to meet the needs of

[37]See Rees 1992.

[38]The *Wall Street Journal,* 4 March 1993, reported that Mexico City had 21% tree cover in the 1950s and 2% in 1993. On Ankara, see Türkiye Çevre Sorunları Vakfı 1991:50.

[39]In 1939, New York City had 72,000 kilometers of underground mains, pipes, and ducts (Konvitz 1985:139). On environmental change and urban morphology in U.S. and European cities, see Ausubel and Herman 1988, Gugler 1996; Hurley 1997; Melosi 1990, 1993; Platt 1991; Relph 1987; St. Clair 1986; Vance 1990; and Whitehand 1987, 1992. On Latin American cities, see Pineo and Baer 1998.

an export economy. Some acquired sanitation infrastructure, usually only for parts of the city where Europeans or Japanese settled. British Singapore conformed to this pattern. The discrepancies between the European and the Chinese or Malay parts of Singapore rankled, probably helping to motivate independent Singapore's singularly determined campaigns to achieve a thoroughly orderly and sanitary environment after 1960. Seoul, which fell under Japanese rule in 1910, appalled its new masters with its filth and pollution, which they soon set about controlling.[40]

After 1950, cities in all parts of the world increasingly spilled out over neighboring territory. In the United States, suburbanization became the dominant trend in urban life after 1945. By 1980, roughly two-thirds of the population in the 15 biggest metropolitan areas lived in far-flung, low-density, automobile-dependent suburbs. Cheap land, cheap cars, cheap gasoline, and mortgages made cheaper by the federal tax code contributed to the unusual American pattern.[41] Elsewhere suburban sprawl more often featured government-financed apartment blocks or, more often still, shantytowns. Delhi covered 13 times as much land in 1990 as in 1900, engulfing a hundred villages and countless peasant plots in the process. Beijing doubled in extent in the 1990s alone.[42] Suburban sprawl, whether in the form of genteel lawns and shopping malls or jerrybuilt shacks, typically came at the expense of farmland. But while hundreds of millions of people took part in these socially momentous changes, the land directly subsumed came to only a few million hectares, equivalent roughly to the area of Costa Rica or West Virginia.

Urban growth nonetheless had tumultuous effects on water, land, and life—because cities have metabolisms. They take in water, food, oxygen (and more) and discard sewage, garbage, and carbon dioxide (and more). Fast-growing cities, like teenagers, have higher metabolisms than those that have stopped growing. Consider the daily diet of one leviathan, Hong Kong. In 1830 Hong Kong was an overgrown village. The collision between China and Great Britain in the First Opium War (1839–1842) resulted in over 150 years of British rule, during which Hong Kong prospered as one of the few conduits between China and the rest of the world. In 1900, Hong Kong had a quarter million people, in 1950 nearly

[40]On Singapore, see Ho 1997 and Yeoh 1993; on Seoul, see Duus 1995. Other useful studies with reference to colonial cities are Abreu 1988 (on Rio de Janeiro), Coward 1988 (on Sydney), Kosambi 1986 (on Bombay), and Low and Yip 1984 (on Kuala Lumpur).

[41]Jackson 1985.

[42]Hardoy et al. 1993:115.

2 million. By 1971, when Hong Kong had 4 million people, its daily urban metabolism redirected prodigious flows (see Table 9.3). Seven-eighths of Hong Kong's food came from outside the Hong Kong Territory. A quarter of its fresh water was piped in from China, which in exchange took 40 tons of human excreta per day (for fertilizer). It required major administrative and engineering feats to maintain Hong Kong—and greater ones in larger and more industrial cities like Beijing, Tokyo, or São Paulo.[43]

Hong Kong gets plenty of rain. Assuring other cities adequate water required stronger measures, all the more so because of the physical character of modern cities. Cities' roofs and roads prevented water from

TABLE 9.3 DAILY METABOLISM OF HONG KONG, 1971

Inflows (thousand tons)		Outflows (thousand tons)
27	Oxygen	0
0	Carbon dioxide	26.5
0	Carbon monoxide	0.16
0	Sulfur oxides	0.31
0	Nitrogen oxides	0.12
0	Dust	0.04
1,068	Fresh water	819
0	Sewage water	819
6.3	Food	0
11.7	Petroleum	0
0	Food waste	0.8
0	Solids in sewage	6.3
18	Miscellaneous cargo	8.1
0.53 (8,827 persons)[a]	People	0.52 (8,632 persons)[a]

Source: Boyden et al. 1981:116–7.

[a]Boyden's number of persons was converted to tons assuming a 60-kg average weight.

[43]Boyden et al. 1981. On Beijing, see Sit 1995.

percolating into the earth and thereby increased surface runoff. A big city like Chicago, built on what in 1850 was a sodden prairie, changed the hydrology of surrounding waterways, especially their flood regimes. Ambitious engineering works were often needed to cope.[44] Mexico City used 30 to 35 times more water in 1990 as in 1900. It was already subsiding slowly in 1900 as its aquifer drained; between 1940 and 1985 the city dropped (unevenly) up to 7 meters in elevation, damaging building foundations and the city's sewer system. Children amused themselves by marking their height on well casings to see whether the ground was sinking faster than they were growing. By 1960, Mexico City had to commandeer surface water from outside the valley of Mexico, subsequently lowering the level of Mexico's largest freshwater lake. Hundreds of cities around the world used their political clout to absorb water from miles around, often at the expense of people, livestock, crops, and fish living in their hinterlands.[45]

Growing cities also needed timber, cement, brick, food, and fuel. Prior to the age of rail, all this came by ship or came from nearby. With railroads and trucks it could come from much further afield, dispersing the environmental effects across broader hinterlands, thus enlarging the ecological footprint of cities. Chicago by 1900 exerted a gravitational pull on timber, livestock, grain, and other fruits of the land from a huge region in the heart of North America. When the poet Carl Sandburg christened Chicago "hog butcher to the world, stacker of wheat" in 1916, some 15 million to 20 million hogs and cattle per year entered its stockyard gates on foot and left as meat. Broad swaths of North America, from Texas to Ohio to Montana, had been converted to livestock habitat because Chicago's packers could find outlets for all that meat. Similarly, Chicago's grain traders helped drive the conversion of prairie grassland into corn and wheat belts, while Chicago's timber merchants encouraged the great cutover of the northern woods of the Great Lakes basin. Few cities matched Chicago's role in organizing and speeding the transformation of

[44]NRC 1992:22. Chicago in 1990 was by world standards a leafy city, with only 45% of its surface impervious to water. In Madrid or Damascus the percentage was (at a guess) nearly twice as high.

[45]For Mexico City, see Ezcurra and Mazari-Hiriart 1996, Pick and Butler 1997, and WRI 1997:64–5. Vivid details are given by Simon 1997:60–90. A crucial book I found only after writing these words is Romero Lankao 1999. The urban imposition on rural water was most serious in arid Asia and Africa. Modern growth often outstripped the reliable water from Mauretania to Mongolia. On Tunis's effect on rural Tunisian water supplies, see Omrane 1991. On sub-Saharan Africa, see Vennetier 1988.

The ocean of animal flesh known as the Great Union Stock Yards in Chicago, around 1909. The wastes and offal from the stockyards created special pollution problems for Chicago after the stockyards and slaughterhouses were centralized in one spot in 1865. The South Branch of the Chicago River carried stockyard wastes into Lake Michigan until 1900, when Chicagoans reversed the flow of the river and sent their sewage to the Mississippi. Philip Armour owned one of the great meatpacking houses.

rural ecologies, but every city had major impacts on the surrounding environment. Delhi in the 1980s imported fuelwood from forests 700 kilometers away. Vancouver around 1990 needed more than 20 times its own area for its food and fuel needs. Most cities after 1960 imported petroleum from thousands of kilometers away: the Persian Gulf became a crucial hinterland for thousands of cities.[46]

[46]Chicago's role is detailed nicely in Cronon 1991. Chicago's stockyards declined in the 1930s and finally closed in 1970. But from 1865 to 1960 (when the big meatpackers left Chicago) the yards created great wealth for a few, jobs for many, and a unique pollution problem for the south side of the city. See Roberts 1994:318 (on Delhi), and Rees 1992 (on Vancouver).

Conclusion

Among the many engines of environmental change in the twentieth century, urbanization, together with population growth and migration, were among the most powerful. The choices, both individual and sometimes political, that affected the reproductive behavior and geographic movement of billions of people over the century helped give environmental change its impetus and directions. Almost none of the choices took environmental considerations consciously into account.

The growth of cities, and their transformation into demographically self-sustaining entities, marked a turning point in human history and in environmental history alike. Cities had for many centuries dominated political life and high culture, but in the twentieth century they became common habitat for the human species. This shift recast the cities themselves, which grew spatially and evolved into new combinations of materials, energy, and wastes. It also reshaped much of the rural world, a larger share of which was converted to serving the needs of city populations.

The expansion of the cities derived both from migration and from population growth. Rural migrants not only drifted to the cities, but also to new rural terrain. Often they were encouraged or compelled by political programs to find new homes in new lands. They brought their own knowledge and practices with them and normally applied them to their new surroundings, with unusually powerful environmental consequences. Population growth, often reckoned to be the main cause of environmental disruptions, probably fit this description only in certain select circumstances. Precise evaluation of its role is quite impossible, however, especially considering its indirect impact in promoting migration, urbanization, technological change, political initiatives, and much else.

10

Fuels, Tools, and Economics

Men are too eager to tread underfoot what they have once too
much feared.
                    —Lucretius, *De Rerum Natura*

Over the course of the twentieth century, more and more people acquired greater and greater leverage over the environment, through new
energy sources, new tools, and new market connections. Energy, technology, and economic systems were tightly interlocked. They coevolved,
each one influencing the paths of the others. At times, new combinations of energy sources, machines, and ways of organizing production
came together, meshed well, and reoriented society and economy. Borrowing from the vocabulary of the history of technology, I will call these
combinations of simultaneous technical, organizational, and social innovations "clusters." Early industrial clusters were built around water-
powered textile mills and then factories and steam engines. After the
mid-nineteenth century the dominant cluster emerged as coal, iron, steel,
and railroads: heavy-engineering industries centered in smokestack cities.
Call it the "coketown cluster" in honor of Charles Dickens's Coketown in

his novel *Hard Times* (1854). The next cluster coalesced in the 1920s and 1930s and predominated from the 1940s (helped along by World War II) until the 1990s: assembly lines, oil, electricity, automobiles and aircraft, chemicals, plastics, and fertilizers—all organized by big corporations. I will dub that the "motown cluster" in honor of Detroit, the world center of motor vehicle manufacture. The coketown cluster and the motown cluster each spurred the emergence of giant corporations in North America, Europe, and Japan, and the relative efficiency and returns to scale enjoyed by these corporations in turn helped to advance each cluster; technological systems and business structures coevolved.

These clusters, and the rapid changes to society, economics, and environment that came with them, affected the whole world, but unevenly. The dominant innovations came disproportionately from the United States, Europe, and Japan, and the wealth and power they helped to create were concentrated there. But the ecological ramifications of these clusters were felt everywhere, if not in the same ways.

Energy Regimes and the Environment

Every society has its "energy regime," the collection of arrangements whereby energy is harvested from the sun (or uranium atoms), directed, stored, bought, sold, used for work or wasted, and ultimately dissipated. Most twentieth-century societies had complex energy regimes involving several different energy sources, modes of conversion, storage, and use. Oil, hydroelectricity, and nuclear fission joined coal, wind, and muscles in powering the twentieth century.

For the most part, the twentieth-century world ran on fossil fuels, mainly coal and oil. Both lay scattered unevenly around the world, so a huge business emerged to extract, transport, process, and deliver fossil fuels to final users. Extraction of both coal and oil were dirty affairs. Transport of oil may have been messier than coal transport. In its final use—combustion—coal was much the grimier of the two fuels. Coal mining, combustion, and disposal of slag and cinder[1] had pronounced effects on land, air, and water. But because oil had so many more applications

---

[1]Bourgeois et al. 1996 in a work of garbage archeology show that in Ghent (Belgium) coal scraps and cinders formed the largest share of urban trash through the 1950s.

and could be distributed cost-effectively more widely, it spread pollution more broadly around the globe, whereas coal had concentrated it around a few thousand mines, furnaces, and steam engines. The pollution derived from fossil fuel burning is treated in Chapters 3 and 4. Here I will treat only the extraction and transport of one fossil fuel: oil.

After 1820 the world's economy became increasingly based on work done by nonmuscular energy. By 1950 any society that did not deploy copious energy was doomed to poverty. The scale of energy use grew so vast that the choice of energy regime became a prime determinant of the world's environmental condition. After 1820 an energy transition to fossil fuels took place. Within that, a transition within a transition, from coal to oil, occurred. By 1930, oil replaced coal as the world's main fuel in transport; by the late 1950s it usurped king coal's position in industry. The United States pioneered this energy path between 1901 and 1925. For world environmental history, few if any things mattered more than the triumph of oil.

OIL EXTRACTION. At the turn of the century, oil scarcely mattered at all. Its main use was as kerosene for lamps. But soon cars, ships, and eventually airplanes and trains came to run on oil products. A goodly share of heating fuel came from oil, as later still did the feedstocks for plastics, synthetic fibers, and chemicals. By and large, the United States shifted to oil first, between 1910 and 1950. Western Europe and Japan, which had stronger political attachments to coal, followed in about 1950 to 1970.[2] High prices from 1973 to 1984 made oil extraction especially attractive. All this provoked a determined search for oil deposits around the world and the construction of a vast network of wells, pipelines, tankers, and refineries designed to carry and process crude oil.

Although hard-rock drilling for oil began in Pennsylvania, the first big gushers came in the 1870s around Baku on the Caspian Sea. The Russian Empire led the world in oil production at the turn of the century. Derricks also sprouted in Rumania and the Dutch East Indies (Sumatra). Then, on the tenth day of the new century, at Spindletop in east Texas, came the first big American oil strike. American oilmen drilled for oil widely, first in Texas, Oklahoma, and California and then all over the world. A new age slowly dawned.

[2]On Japan, see Hein 1990. By 1958, oil was cheaper than coal in Japan. In the USA, oil use surpassed coal around 1948; in 1900 the country used 20 times as much coal as oil (Adams 1995:181).

Spindletop, Texas, in 1901 was the site of the first large gusher in American oil-drilling history, opening a new era in world environmental history. Cheap energy vastly enhanced the human capacity to alter environments, and getting that cheap energy, at Spindletop and elsewhere, required altering local environments thoroughly.

Drilling was a dirty business in those days, not least in Mexico. Oil lay under the rainforests of Veracruz along the shores of the Gulf of Mexico.[3] Here the capital came from American and British firms, the equipment often secondhand from Texas, and the labor from Texas and from the local indigenous population, Huastec and Totonac Indians. To oilmen and to successive Mexican governments, their rainforest ways seemed backward and pointless, an anachronism in the new century. Widespread drilling began in 1906.

With the Mexican Revolution (1910–1920), ambitious new leaders saw in oil a way to propel Mexico forward. Boosters thought that northern Veracruz could support 40 million people, if only trees and Indians would make way for oil and oilmen. World War I helped in the regard Lord Curzon noted: "The Allies floated to victory on a wave of oil."[4] Much of it flowed from Tampico. Mexico stood third in world oil production by 1915, and second by 1919, thanks to the wartime boom and to revolution and chaos in Russia. The Mexican boom peaked in 1921.

Oil recast both ecology and society in northern Veracruz, and almost overnight. Tampico was a sleepy, swampy port in 1897. By 1921 it had 58

[3]The following discussion draws on Santiago 1997 and 1998.
[4]Quoted in Melosi 1985:97.

Oilmen liked nothing better than a scene like this one, a gusher in southern California in about 1910. With the shift to oil as a major fuel of the twentieth century gathering pace by 1910, fortunes were made overnight. In such a heady atmosphere, few oilmen paused to worry about what oil drilling did to landscapes. Those who objected, like the Huastec and Totonac Indians in Mexico, had little power to prevent drilling. The early oil industry was a messy affair in California, Mexico, Russia, and elsewhere, but oil when burned generated much less air pollution than did coal.

oil companies, 16 refineries, 24 law firms, 6 bakeries, 77 liquor stores, and nearly 100,000 people.[5] The surrounding region had (by 1924) thousands of oil wells and pools, and enough pipeline (4,000 km) to stretch to Hudson Bay or Chile. Spills, leaks, blowouts, and fires, while wrenching for the Huastec and Totonac, were a necessary cost of doing business for oilmen. Indeed, they positively rejoiced in a gusher. But according to the Minister of the Interior, the oil business "ruined" the land.[6]

The bonanza subsided in the early 1920s. Salt water seeped into the oil fields, complicating production. Then United States and Venezuelan oil fields began producing more than the market could absorb. The Mexican government nationalized the oil industry in 1938 and forbade exports of crude. Foreign companies, indignant at expropriation, boycotted Mexican oil anyway. Production plummeted and low forest slowly recolonized much of the oil fields. Old derricks stood out here and there like Mayan pyramids amid the jungle.

[5]Santiago 1997. The refineries datum is actually for 1924.
[6]Santiago 1998:182, quoting (later president) Plutarco Elías Calles in 1921.

Variations on this pattern of boom and bust played out around the world. The oil patches of the United States polluted land and water in Texas and Oklahoma merrily, impeded only slightly by an antipollution law of 1924.[7] Venezuela became the world's number two oil producer in 1928 and leading exporter by 1946. Lake Maracaibo, the largest lake in South America and the focus of Venezuelan production and refining after 1918, became an oily morass.[8] The early Russian fields around Baku (Azerbaijan) eventually became an oily backwater when the Soviet Union tapped its huge Siberian oil reserves. Northern Azerbaijan was left with a grimy residue of polluted water and abandoned derricks. But the world's mother lode of oil lay in the vicinity of the Persian Gulf. Here extraction took place in environments with few people and indeed comparatively little life of any sort, which helped lower the ecological costs of spills and leaks.

In 1973 the Organization of Petroleum Exporting Countries (OPEC) restricted production, inadvertently opening a new era in oil exploration and extraction. High crude prices, pushed higher by the Iranian Revolution of 1979, quickened exploration and production in nonmember countries and encouraged oil operations in Alaska, Alberta, the Gulf of Mexico, the North Sea, Angola, Ecuador, and on a gigantic scale in western Siberia.[9] OPEC's high prices shaped world economic history from 1973 to 1985, weakening oil-importing industrial economies and strengthening some—not all—oil exporters. High oil and natural gas prices helped prop up the Soviet Union. But OPEC's initiative also shaped environmental history. First, it encouraged energy conservation in industrial economies, notably in Japan. Furthermore, after the 1970s, the environmental impacts of oil production—with its construction projects, its pipelines and refineries, and its leaks, spills, and fires—spread much more widely around the world. High prices also tempted those, like Nigeria, which flouted cartel rules.

Nigeria's oil lay in the Niger delta, home to about 6 million people in 1990. Royal Dutch Shell and British Petroleum (Shell-BP), which had been granted exploration licenses by the British colonial government, struck oil in 1956. Production began in the 1960s. Shell-BP built a refinery at Port Harcourt in 1965, stimulating production. Shell-BP prudently

[7]Melosi 1985:151–2.

[8]Vitale 1983:95. For statements on production rankings I relied on Etemad and Luciani 1991.

[9]The big oil fields in the USSR were tapped in 1948 in Tatarstan and in the 1960s in the middle Ob swamp-forest of western Siberia. In 1961 the USSR became the world's second oil producer (Dienes and Shabad 1979:50–61).

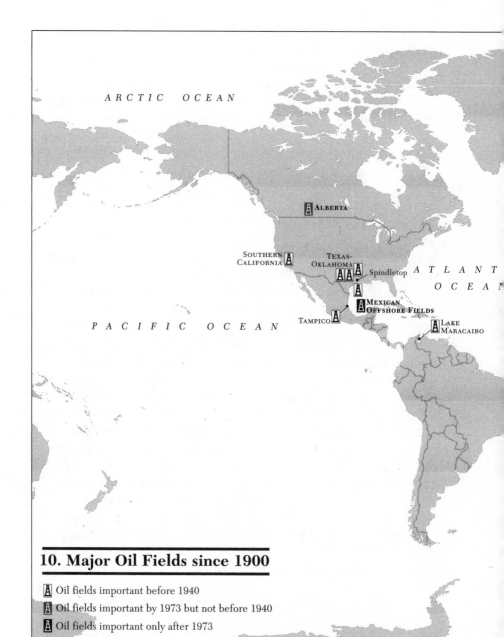

ARCTIC OCEAN

ALBERTA

SOUTHERN
CALIFORNIA

TEXAS-
OKLAHOMA

Spindletop

ATLANT
OCEAN

PACIFIC OCEAN

MEXICAN
OFFSHORE FIELDS

TAMPICO

LAKE
MARACAIBO

10. Major Oil Fields since 1900

Oil fields important before 1940

Oil fields important by 1973 but not before 1940

Oil fields important only after 1973

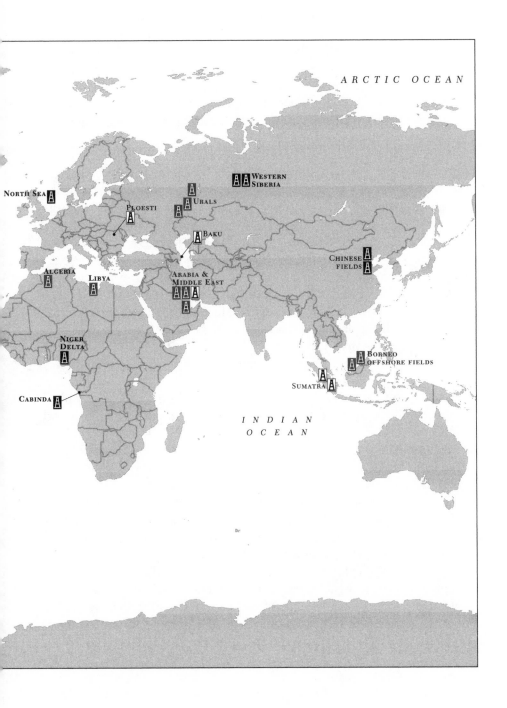

ARCTIC OCEAN

NORTH SEA

WESTERN
SIBERIA

PLOESTI

URALS

BAKU

CHINESE
FIELDS

ALGERIA

LIBYA

ARABIA &
MIDDLE EAST

NIGER
DELTA

BORNEO
OFFSHORE FIELDS

CABINDA

SUMATRA

INDIAN
OCEAN

backed the victorious central government in the civil war of 1967 to 1970, in which southeastern Nigeria (Biafra) attempted to secede and take the oil revenues with it. After the price hikes of the 1970s, Shell-BP pumped out oil while Nigeria pretended to comply with the cartel's rules. Oil royalties splashed through the state's coffers, greasing the wheels of the corruption for which Nigeria became justly famous in the 1980s. Leaks, spills, and perhaps sabotage splashed oil throughout the delta, fouling the fisheries and farms of local peoples, notably the half-million-strong Ogoni.[10] Their protests and rebellions, which featured environmental grievances prominently, were met with intimidation, force, show trials, and executions of prominent Ogoni. Nigeria's military government by the 1990s derived some 80 to 90 percent of its revenue from oil, and the rulers skimmed their personal riches from it. They brooked no challenges, least of all from fisherfolk, farmers, and small ethnic minorities. In 1992 the United Nations declared the Niger to be the world's most ecologically endangered delta. Shell-BP came under unwanted scrutiny and international pressure, and in 1995 began to address environmental and other complaints.[11] Nonetheless, the Niger delta at the end of the century, like Tampico at the beginning, became a zone of sacrifice. The Ogoni, like the Huastec and Totonac, lacked the power to resist the coalition of forces that created and maintained the twentieth century's energy regime.

OIL TRANSPORT. The energy regime of the twentieth century implied massive oil transport, especially after petroleum-poor European and Japanese economies converted to oil. At any given time after 1970, about 5 gallons of oil were in transit at sea for every man, woman, and child on the face of the earth. Most of it got safely to its destination. A small fraction did not.

In the first six months of the Battle of the Atlantic (January–June 1942), German U-boats sank American tankers and spilled about 600,000 tons of crude into the sea. Tankers grew in size 30-fold between 1945 and 1977, so a single spill could do serious damage, equivalent to a month's work by the U-boats. Big tanker spills on the world's seaways became commonplace after the *Torrey Canyon* broke up off of Cornwall, England, in 1967, spilling 120,000 tons of oil into the English Channel. As tankers got safer in the 1980s, the frequency of big spills abated. New

---

[10]Salau 1996:257–8 writes of "devastation" and notes that in 1976–1988, recorded oil spills in the Niger delta averaged about 200 per year.

[11]On Nigeria, see Lowman and Gardner 1996, and Osaghae 1995.

rules restricted tankers from cleaning out their tanks at sea. The total human contribution to oil in the seas consequently declined sharply by 1990. Oil cleanup techniques also improved with practice. But smaller spills like that along Alaska's coast in 1989, when the Exxon *Valdez* spilled 34,000 tons of crude, occurred about once a year in the 1990s. Moreover, most oil in the seas came not from accidents but from routine dumping and tank cleaning, which, although legally regulated, was difficult to control. All told, human action by 1990 put 10 times as much oil into the seas as did natural seeps.[12]

Tanker accidents damaged marine life for months and years. Their residual ecological effects lasted for decades in the worst cases. The same was true of offshore blowouts, the worst of which, at Ixtoc I off the Tabasco coast of Mexico in 1979, spewed 600,000 tons of oil into the Gulf of Mexico and sent an oil slick nearly the size of Connecticut drifting toward Texas. But blowouts and tanker spills were one-time events. The sun and sea eventually covered their tracks, evaporating, breaking up and dispersing the oil to inconsequential concentrations. Ongoing production, as in Veracruz or the Niger delta, or persistent pipeline leaks on land, as in Siberia, led to more enduring environmental effects—as well as more social and political frictions.[13]

Oil left a larger mark on the environment than the stains arising from drilling and transport. Petrochemicals derived from oil created new species of materials, notably plastics, that replaced wood in many uses but added to the tonnage of durable wastes. Many petrochemicals proved to be toxic pollutants themselves. Oil also gave us the car as we know it, with all its implications. It made tractors and farm mechanization possible. And because oil's price fell, especially during the interwar years (1919–1939), and again from 1948 to 1973 but also after 1984, it strongly encouraged more and more applications of energy, in various technological forms, from lawn mowers to power plants, all of which affected ecology to some degree. This energy regime allowed wealth and ease on scales quite impossible in earlier centuries for a billion or two people.[14] It had enormous social, economic, and geopolitical consequences for the

---

[12]This estimate is from Holdren 1991:124. See also Buxton 1993, Earle 1995:261–91, Gorman 1993:119, and ReVelle and ReVelle 1992:408–11.

[13]Burger 1997 summarizes the history of oil spills nicely.

[14]In 1992 the U.N. Development Program (UNDP) calculated the world's richest billion people were 150 times richer than the world's poorest billion (reported in the *Economist*, 25 April 1992:48.

twentieth century. It also polluted air and water and changed environments generally on scales equally impossible in earlier centuries.[15] Oil, on one reckoning at least, was the single most important factor in shaping environmental history after the 1950s.[16]

Technological Change and the Environment

A century ago Oscar Wilde wrote that

civilization requires slaves. The Greeks were quite right there. Unless there are slaves to do the ugly, horrible, uninteresting work, culture and contemplation become almost impossible. Human slavery is wrong, insecure and demoralizing. On mechanical slavery, the slavery of the machine, the future of the world depends.[17]

Wilde was quite right too: the course of the twentieth century did depend on machines. The technologies of the twentieth century, intertwined with related changes in energy and economy, powerfully determined the rates and kinds of environmental changes.

As with energy paths, different technological trajectories implied different environmental outcomes. The coketown cluster meant, in particular, urban air pollution. The motown cluster meant far more because it spread so widely, so quickly, and involved such energy intensity. A given technology could magnify or minimize ecological impacts, but alone it merely modified the consequences of social forces. A technological cluster, on the other hand, could exert an influence at least as great as population or politics. Consider three technologies, one prosaic by twentieth century standards (chainsaws), one emblematic of the course of the twentieth century and the centerpiece of the motown cluster (automobiles),

---

[15]Which is not to say other energy regimes would have been environmentally neutral. One based on muscle and biomass (the Haitian model) would have stripped the world of combustible vegetation. One based on coal (the Polish model) would have markedly increased air pollution. One based on nuclear energy (anyway not available before 1954) would have run greater risks of meltdowns and committed the world to millennia of management of more lethal wastes. As for one based on photovoltaics, wind power, and fuel cells—we don't know (yet) what that might entail.

[16]The reckoning is Pfister's (1995).

[17]Quoted in Silversides 1997:1.

and one for which future millennia will remember the twentieth century (nuclear reactors).

CHAINSAWS. Before the invention of practical chainsaws, the bottleneck in logging consisted of the enormous labor demands of felling timber. In North America, armies of men filled the woods in fall and winter, swinging broadaxes and pulling crosscut saws. The lumber camps came alive only seasonally because hauling felled timber was much easier over snow and ice, and because, in eastern North America, the men, horses, and oxen mostly came from farms: after the harvest there was a slack season and labor could be spared. In parts of the world where abundant labor was harder to find, forests often survived.

The fundamental constraint was one of energy. Human muscle had its limits. The chainsaw changed social and ecological landscapes, in North America and beyond, by unleashing the energy of fossil fuels in the forests. In an eon-straddling irony, the new machines allowed loggers to use the energy derived from ancient vegetation (the source of oil) against modern forests.

While the first chainsaw patent dates to 1858 and its first manufacture to 1917, its real impact came after World War II. The war brought vastly improved air-cooled engines and light metals (aluminum), which together allowed a practical, gasoline-powered chainsaw. Between 1950 and 1955, chainsaws revolutionized logging and pulping in North America. In eastern Canada, for example, bucksaws and axes still cut all pulpwood in 1950. By 1955, chainsaws accounted for half the total, and by 1958 all of it. Lumber and pulpwood firms had to mechanize by the 1950s, because farms had mechanized and there was no longer an available army of seasonal labor (and horses). Soon far bigger machines that looked like "giant insects from another planet" and could snip trees off at the base, took over the lumber and pulpwood business in North America. The age of the lumberjack, a distinctive figure in the cultural landscape of North America, closed.[18]

Elsewhere the chainsaw remained cutting-edge technology. It allowed men to cut trees 100 or 1,000 times faster than with axes. Without the chainsaw, the great clearance of tropical forests (see Chapter 8) would either not have happened, have happened much more slowly, or have re-

---

[18]Ibid. The quotation (from Silversides 1997:107) is by a Canadian lumberjack, Armst Kurelek.

Without chainsaws or machines of any kind, it took these men five days to fell this kapok tree in Cameroon in about 1950. It took them three more days to chop it up and burn it. They cleared this patch of African rainforest because French geologists hoped to find diamonds beneath it. Machines made tree cutting much easier and much faster in the second half of the twentieth century. This photograph also captures some of the social distinctions in colonial Africa.

quired 100 or 1,000 times as many laborers. Hundreds of small-scale technologies, equally as prosaic as the chainsaw, altered twentieth-century environmental history in small and not-so-small ways.

FROM RAILROADS TO CARS. Transport technology made even larger differences. At the end of the nineteenth century most societies depended on combinations of railroads and animal- or human-drawn carts and carriages. Such a transport regime had its environmental consequences. American railroads, for example, demolished forests. They usually burned wood in their boilers. Boxcars were made of wood and some rails were too. Crossties, which had to be replaced every few years, consumed the most wood of all. The locomotive may have been the iron horse, but the railroad was mainly a wooden system. When the system

With heavy equipment provided to France by the U.S. Marshall Plan, logging in colonial Africa suddenly became more efficient. This photograph was taken at Eseka, Cameroon, around 1950, where a French firm undertook to export timber. Mechanized logging, together with clearing for agriculture, made short work of the West African rainforests between 1950 and 1990, part of the great clearances of tropical forests in the late twentieth century.

was growing fastest (1890s), it threatened to gobble up American forests. Fears of a timber famine arose, and Theodore Roosevelt decided to create a national forest service to rationalize the use of the country's remaining timber. The same fears helped breathe life into the American conservation movement at the turn of the century and generate political support for the system of national parks. Railroad technology put enormous strains on American forests, provoking social and political responses. But soon Americans launched new technologies, in turn provoking new responses.

Two new technologies rescued American forests from the iron horse: creosote oil and cars. By 1920, creosote oil, a wood preservative derived from coal tar, coated half of American crossties, reducing the need for new ones. Then the railroad network in the United States stopped growing in the 1920s—because of the automobile.[19]

Before it replaced the railroad in intercity travel, the automobile displaced the horse within cities. Horses, like the railroad, brought environmental problems of their own. It took about 2 hectares of land to feed a horse, as much as was needed by eight people. So in Australia, which in 1900 had one horse for every two people, much of the country's grain land went to sustain horses. In 1920 a quarter of American farmland was planted to oats, the energy source of horse-based transport. Supplying inputs was only part of the horse problem. Horses deposited thousands of tons of dung on the streets, making cities pungent, fly-ridden, filthy, and diseased. A big city had to clear 10,000 to 15,000 horse carcasses from the streets every year. Part of the automobile's manifold appeal in 1910 was its modest emissions and the liberation it promised from the urban environmental problems associated with horses. By 1930 the urban horse was on the road to extinction.[20]

The automobile is a strong candidate for the title of most socially and environmentally consequential technology of the twentieth century. Cars in 1896 were such a curiosity that they performed in circuses along with dancing bears; by 1995 the world had half a billion cars. The history of their adoption and of the air pollution consequences appears in Chapters 3 and 4. Their total ecological impact was much greater, however. Their fuel needs helped propel the oil industry. Their country cousins—tractors and small trucks—helped revolutionize agriculture (Chapter 7). Cars and car culture had many requirements and impacts.

Making a car took a lot of energy and materials. In Germany in the

[19]See Ausubel 1989 (on railroads, crossties, and creosote), and Grübler and Nakičenovič 1991 (on the evolution of transport technology generally).

[20]On horse problems, see Grübler and Nakičenovič 1991:56–7, and Lay 1992:131–3. Britain in 1900 had one horse for every 10 people; in the United States there was one horse for every four people.

1990s, the process generated about 29 tons of waste for every ton of car. Making a car emitted as much air pollution as did driving a car for 10 years. American motor vehicles (c.1990) required about 10 to 30 percent of the metals—mainly steel, iron, and aluminum—used in the American economy. Half to two-thirds of the world's rubber went into autos. This requirement alone led to the creation of rubber plantations in Sumatra and Malaya on a large scale; in Sri Lanka, Thailand, Cambodia, and Liberia on smaller scales; as well as failures on the grand scale in Amazonia.[21]

Making room for cars took a lot of space. The United States built a road network from a very modest start in 1900 to 5.5 million kilometers of surfaced roads by 1990, exceeding the length of railroads at their maximum by 10 or 15 times. Most of that road-building spree happened from 1920 to 1980, partly because the federal government subsidized road building from 1916 onward. In the 1930s, Franklin Roosevelt's New Deal put thousands of unemployed Americans to work on road construction. The fastest growth in the road network occurred in the late 1940s. The interstate system that now crisscrosses America dates from 1956. All these roads, especially the interstates, attracted people, settlement, and businesses like iron filings to a magnet, reorganizing America's broad spaces into new patterns, which, in turn, made car ownership almost essential for most adults. No other country achieved the same automobile saturation as the United States, although some small countries got far higher road densities. All in all, in North America, Europe, and Japan, auto space took about 5 to 10 percent of the land surface by 1990.[22] Worldwide it took perhaps 1 to 2 percent, matching the space taken by cities (and overlapping with it).

Cars also killed a lot of people. In the United States the toll ranged from 25,000 to 50,000 per year after 1925, totaling perhaps 2 million to 3 million over the century—roughly five or six times the American war dead of the twentieth century. Worldwide, auto accidents killed about 400,000 people annually by the end of the century. Nonetheless, cars were convenient and conveyed social status, so they remained irresistibly popular.[23]

---

[21]Tucker forthcoming; his Chapter 5 deals with the global rubber industry and its ecological impacts. Dean 1987 details the Brazilian experience, in which first Henry Ford and then Daniel Ludwig tried to carve rubber plantations from Amazonia's forests.

[22]This space included roads, parking lots, gas stations, car junkyards, and so forth. The data are from Freund and Martin 1993.

[23]On cars and their impacts, see Freund and Martin 1993; Jackson 1985:157–71, 246–71; Kay 1997; and Melosi 1985:105–12. Cars, incidentally, were much safer (measured by deaths per mile traveled) than horses by 1925; by 1985 they were 10 times safer than in 1925 and 15 times safer than travel by horse. The USA data are from Lay 1992:176.

THE STRANGE CAREER OF NUCLEAR POWER. Nuclear power was an unpopular and uneconomic innovation, less lethal than cars, but with mind-boggling ecological implications. Like cars, atomic power had its origins in European science, reached maturity in the United States, and subsequently spread (unevenly) around the world. Humankind's first self-sustaining nuclear reaction took place in 1942, in a squash court at the University of Chicago, amid the hectic U.S. drive to build an atomic bomb. Civilian nuclear power started up in 1954 in the USSR, 1955 in the United Kingdom, and a year later in the United States. Nuclear power held some of the same political attraction as dam building: it signified vigor and modernity. Admiral Lewis Strauss, head of the American Atomic Energy Commission, predicted in the 1950s that by the 1970s nuclear power would be too cheap to meter. Such optimism helped inspire governments, especially in the United States, USSR, Japan, and France, to invest in civilian reactors or assist private utilities in doing so. By 1998, 29 countries operated some 437 nuclear power plants.[24] But no nuclear power plant anywhere made commercial sense: they all survived on an "insane" economics of massive subsidy.[25] In Britain, which privatized the electricity industry in the late 1980s, there were no takers for nuclear power plants. Closing down old or dangerous nuclear plants proved horribly expensive. Nervousness about accidents accounted for many closings.

Scores of mishaps beginning in 1957 (Windscale, U.K.) climaxed at Chernobyl (in Soviet Ukraine) in 1986, by far the most serious civilian nuclear accident. There, human error led to an electrical fire and explosions that nearly destroyed one reactor. Thirty-one people died quickly. Untold numbers died (and will yet die) from Chernobyl-related cancers, primarily among the 800,000 workers and soldiers dragooned into cleanup operations, but also among local children whose thyroid glands absorbed excessive radiation. About 135,000 people had to leave their homes indefinitely, although some desperate souls eventually returned. The total release of radiation, officially put at 90 million curies, was hundreds of times greater than that given off by the bombs at Hiroshima and Nagasaki, which continued to cause health problems for decades after detonation. Everyone in the Northern Hemisphere received at least a tiny dose of Chernobyl's radiation.

The accident and initial denial and cover-up knocked one of the last props out from under the Soviet Union. It completely changed the pub-

---

[24]Berkhout 1994:324. Some 94% of civilian reactors were in industrial countries.
[25]*Economist*, 28 March 1998:63.

lic perception of nuclear power plants around the world, but especially in Europe, making it politically unpalatable except in a few countries (such as France and Belgium). Outside of Europe, only Japan, South Korea, and Taiwan showed much interest in nuclear power after Chernobyl. When none but historians remember the USSR, the environmental imprint of its nuclear power will remain. Some nuclear wastes and part of Chernobyl's fallout will be lethal for 24,000 years—easily the most lasting human insignia of the twentieth century and the longest lien on the future that any generation of humanity has yet imposed.[26]

Nuclear power did not replace other forms of energy production, as the car did the horse. It did not find companion innovations, technical and social, to form a new cluster that would remake the world, the way oil and internal combustion engines had done. Instead, nuclear power complemented fossil fuels; it never accounted for more than 5 percent of the world's energy supply. But it did slightly reduce air pollution by providing an alternative to fossil fuel combustion. It created a different set of environmental consequences and risks. All significant technologies carried their specific packages of environmental implications, aggravating some problems while mitigating others. No single technology, not even nuclear power, matched the motown cluster in its capacity to alter both society and nature.

ENGINEERING GENES AND BYTES—A NEW CLUSTER? A new technology cluster may be emerging, one that may also succeed in revolutionizing human life and the globe's environment. Since 1750, new clusters have come at 50- to 55-year intervals, and another was "due" in the 1990s. Genetic manipulation and information technology may be at the center of it. At the century's end, momentous changes were afoot in biotechnology, especially the fervent efforts to turn new knowledge about genes to good use (and profit). For millions of years, genetic selection dominated evolution; then, with human society, cultural evolution slowly

[26]Savchenko 1995:2 says that estimates for the eventual cancer toll range from 14,000 to 475,000; see also his pp. 78–84 (on further health effects as of 1990) and pp. 128–30 (on genetic mutations). Contamination at Chernobyl required the retirement from human use of fields and forests over an area of about 640,000 ha, or two-thirds the size of Lebanon (p. 142). The Nuclear Energy Agency—the international "club" of the nuclear industry in OECD countries—suggests that only a few thousand people (4 per million in the Northern Hemisphere) will contract cancers as a result of Chernobyl, mostly Europeans who were young children in 1986 (OECD Nuclear Energy Agency 1995). A Ukrainian doctor (Shchebrak 1996) believes 32,000 already died within the first 10 years after Chernobyl. Eisler 1995 details Chernobyl's biological consequences.

emerged as a rival force. From the 1990s, the two began to merge as science acquired the capacity to intervene directly in the selection and propagation of genes. Genetically engineered creatures, especially tiny ones, appeared set to change procedures in pest control, fertilizers, mining, recycling, sewage treatment, and other realms with direct connections to the environment. Scientists in Scotland cloned sheep; colleagues in Japan cloned cattle. Brave new worlds loomed, or beckoned.

Once expected to save paper, minimize commuting, and so forth, computers by 1999 had negligible environmental consequences but, like genetic manipulation, limitless possibilities. The Internet, still in its infancy, promised untold changes wherever electricity and computers reached. Some of these presumably would yield unpredictable environmental consequences. But the eventual impact of information technology and the new cluster (if it is that) remained opaque.

Technologies, energy regimes, and economic systems coevolved, occasionally forming revolutionary clusters, but this was only part of the picture. These clusters in turn coevolved with society and the environment in the twentieth century, as at all times. Successful, widely adopted clusters must fit with contemporaneous conditions and trends in society and environment. At the same time, society and environment were affected by and adjusted to successful clusters. Thus, while all three codetermined one another, their relative roles changed. In prior centuries, the environment played a stronger role in influencing society and technologies, whereas in the twentieth century, technology's role, especially within the motown cluster, expanded and shaped society and environment more than in the past. But if certain environmental perturbations—such as significant global warming or biodiversity loss—prove fundamental, then the equation will be revised again in the direction of a stronger determinative role for the (new) environment. Paradoxically, if humanity is to escape projected environmental crises, then technology, which helped bring them on, will be asked to lead us out. A new cluster of related technologies, with or without a new energy system and economic order, could lead almost anywhere.

Economic Changes and the Environment

The three dominant features of twentieth-century economic history were industrialization, "Fordism," and economic integration. They were all intertwined, and together intermingled with the spread of fossil fuels and technological change. They, too, helped spread disruption and prosperity, foment the economic miracles of the twentieth century, and provoke massive environmental change.

INDUSTRIALIZATION. In the late eighteenth century in Britain, industrialization took off, quickly reaching an intensity never before approached, not even in Song China. From there it spread by leaps and bounds, intensified further, and changed form several times. Industrial labor efficiency increased about 200-fold between 1750 and 1990, so that modern workers produce as much in a week as their eighteenth-century forbears did in four years. In the twentieth century alone, global industrial output grew 40-fold.[27]

The coketown cluster centered in the United States and northwestern Europe. It spread to Japan early in the century, the USSR in the 1930s, and to Soviet satellite countries in the 1950s, some of which already had pockets of it, in Bohemia and Silesia. The motown cluster first took shape in the United States but quickly crystallized in Canada, western Europe, Japan, Australia, and New Zealand. It extended only partially to the USSR (where innovations often met stern resistance after Stalin's consolidation of power) and to Latin America, and scarcely at all to Africa and southern Asia. China in effect attempted to create a version of the coketown cluster overnight with the backyard steel furnaces of the Great Leap Forward of 1958 to 1960. Despite the geographic spread of industrialization, since the 1920s about two-thirds of industrial production (by value) occurred in the core areas: the United States, Canada, Japan, and western Europe.

Like urbanization, industrialization changed the structure and pace of energy and material flows. Industry too has metabolisms. Here I will pass over specifics and examples and offer only two generalizations, one obvious and one hidden. First, industrialization everywhere and at all times increased resource use and pollution. The coketown cluster was especially dirty, even in some of its late-twentieth century incarnations (such as Silesia). The 40-fold increase in industrial output in the twentieth century implied a vast rise in raw material use and industrial pollution.[28] Vast but not 40-fold.

[27]Grübler 1994:44.

[28]Ayres 1989 estimates that in the 1980s, U.S. industry mobilized 10 tons of minerals and biomass per capita per year, 94% of which was instantly waste (e.g. slag, chaff, husks). Most of that waste was in mining, where, for example, to extract an ounce of gold, miners chewed through 25 tons of rock.

Second and less obviously, industries over time grew less dirty and less demanding. Their energy efficiency improved, and so they emitted less carbon into the atmosphere per unit of production, allowing industrial economies to "decarbonize." Industries also learned to use less raw material per unit of output, permitting "dematerialization." The energy intensity (ratio of energy use to GDP) of the British economy peaked around 1850 to 1880; it was probably the most inefficient, energy-guzzling economy in world history.[29] Energy intensity in Canada declined after about 1910, in the United States and Germany after about 1918, in Japan after 1970, in China 1980, and Brazil 1985. The United States used half as much energy and emitted less than half as much carbon per (constant) dollar of industrial output in 1988 as in 1958. South Korea achieved the same efficiency gains in half the time, between 1972 and 1986. In the world as a whole, energy intensity peaked around 1925 and by 1990 had fallen by nearly half. This meant far less pollution (and resource use) than would otherwise have been the case in the twentieth century. But this happy trend was masked by the strong overall expansion of the scale of industry.[30]

FORDISM AND MASS CONSUMPTION. Fordism here refers to both assembly-line production and the historic compromise of the twentieth century between industrial workers and employers. As a result of a myriad of managerial developments, including Henry Ford's electrified assembly line, inaugurated in 1912, and Taylorism, the so-called scientific management involving the choreography of each laborer's motions, industrial economies achieved enormous productivity gains in the early twentieth century. Henry Ford saw that sharing these gains with his workforce suited his own interests, and from January 5, 1914—the birthday of consumer society—he paid laborers enough that they could hope to buy a Model-T. In 1923 his workers could buy one with 58 days' wages.[31] Mil-

[29]Depending on how one estimates Soviet and Soviet-bloc GDPs for c.1960–1990, these economies may have been even more energy-intensive than Britain's a century before.

[30]This section draws heavily on Ausubel 1989, 1996; Grübler 1994; Herman et al. 1989; and Nakicenovic 1996. See also Adams 1995 and Schurr 1984. The energy-intensity data are from Dessus and Pharabod 1990:292, Grübler 1994, and Smil 1994:205–7. Slightly different data appear in Reddy and Goldemberg 1991. Dematerialization, while not as pronounced as decarbonization, is also a long-term trend, but is also masked by overall growth. See Wernick et al. 1996.

[31]Adams 1995:187–8. While consumer society may have been born in 1914, it reached maturity, after a troubled adolescence, only in the 1950s. See Pfister 1995 for a European perspective.

One of the earliest assembly lines of the Ford Motor Company in 1913 at Highland Park outside the city of Detroit, the world capital of automobile production. Henry Ford's assembly line, and the high wages it allowed him to offer, revolutionized both production and consumption patterns in the United States and eventually in much of the world, with countless social, political, and environmental consequences.

lions of Americans did buy cars, radios, phonographs, then refrigerators and washing machines. They enjoyed an affluence and leisure that in the nineteenth century would have required an army of household servants. Fordism amounted to a renegotiation of the social contract of industrial society.

The production systems pioneered by Ford in the United States spread to Canada, Europe, Japan, the Soviet Union, and outposts elsewhere. The social compromise that converted the gains of mass production into mass consumption took many different forms and took varying lengths of time. In Europe the state brokered Fordism, striking bargains between unions and employers. (In France and Italy the employer was often the state itself.) In Japan, Fordism made rapid headway after 1945, when both factories and society were reconfigured under the American occupation (1945–1952). Mass consumption arrived in the 1960s. In the USSR the state was in effect the sole employer, and its rulers from the 1920s were much smitten by American factory efficiency. The Soviet Union committed itself ideologically to sharing its gains with industrial workers, but it did so in the form of secure employment, not mass consumption, creating a variant of Fordism in which industry produced mainly for the state, not the citizen. Whatever their prevailing ideology and political economy, industrialized societies spread the wealth sufficiently to keep the machines humming, the workers working—and usually buying. In short, outside the Soviet sphere, the enormous revolutions in production permitted—and required—enormous revolutions in consumption.

Social arrangements, from family relations to class structure, changed accordingly. Intergenerational and gender relations had to change with the demands of mass production and the delights of mass consumption. Young people tolerated the clanking inferno of early assembly lines better than did their elders, whose venerable skills no longer counted for much. For tasks that emphasized precision and endurance, employers often preferred women to men. Affordable household appliances changed the lives of millions of wives and daughters. Old, usually unspoken, social contracts were abrogated and renegotiated within families and within societies, sometimes bitterly. Fordism's social impact was felt first in the United States around 1912 to 1945, in western Europe about 1925 to 1960, in Japan around 1950 to 1970, and in South Korea and Taiwan after 1980. The social changes involved a fair amount of strife because some people enjoyed the benefits of Fordism while others felt left out in the cold, buffeted by the brutally efficient competition of assembly-line production. Small businessmen and artisans in Germany, for instance, crushed by the efficiency of assembly lines, often turned to radical politics in the late 1920s, especially Nazism because Hitler explained their troubles comfortingly and expressed their anger compellingly. Revolutions of production and consumption can tear apart societies as completely, if not as quickly, as political revolutions.

Ecological arrangements had to change too. To sustain the new social arrangements, fields, factories, and offices needed more fuels, fertilizers, water, wood, paper, cement, ores—more of almost everything except horses, oats, whalebone, and a handful of other raw materials consigned to the dustbin of history. All these inputs were converted into energy, food, goods, pollution, and garbage. Without Fordism, without mass consumption, the environmental history of the twentieth century would have been much calmer.

Fordism extended few tendrils into Africa, Latin America, or South Asia in the twentieth century. Some isolated pockets of Fordist production techniques developed, such as the Tata family's iron and steel mill near Calcutta, long the largest one in the British Empire. But nowhere did these translate into the social compromise that yielded mass consumption. Had they done so, the environmental history of the twentieth century would have been even more tumultuous than it was. Should mass consumption society emerge in China, India, Nigeria, and Brazil in the next century—a prospect that remains uncertain—further excitement lies in store.

ECONOMIC INTEGRATION. Now often called globalization, economic integration has a long history of fits and starts. In modern times it got boosts from the explorations and trade links pioneered mainly by Europeans and Chinese between 1405 and 1779,[32] then from colonialism, and new transport and communications technologies—railroad, steamship, telegraph—in the nineteenth century. Indeed the era 1870 to 1914 was one of great integration and consolidation in the world economy, observable in flows of trade, migration, and capital. World War I and the Russian Revolution stopped this trend. Soviet Russia veered toward an ideal of autarky, followed by fascist Italy in the 1920s. Then international trade and investment flows plummeted during the Great Depression and World War II.

That disastrous experience weighed on the minds of the architects of the postwar economic order. They understood that prosperity depended on trade, and fashioned a new regime of monetary and trade agreements under American leadership. This regime promoted and achieved rapid integration of western Europe and North America from the late 1940s, Japan from the time of the Korean War (1950–1953), the big oil exporters of the Middle East by the mid-1950s, South Korea and Taiwan from about 1970, and, less thoroughly and less quickly, Latin America, Africa, and South Asia. Meanwhile the USSR organized a much smaller and less well integrated rival system including eastern Europe and, briefly, China. In the 1970s the two blocs began to integrate as Moscow weakened its commitments to economic autarky, seizing the opportunity to sell oil and gas, and facing the necessity to buy grain.

---

[32]The dates here represent the beginning of oceanic voyaging as organized by Henry the Navigator in Portugal and the Ming dynasty in China (1405), and the completion of Captain Cook's voyages (1779), which integrated the last significant lands and populations into a single system.

Despite slumps and setbacks the momentum of integration continued. Indeed, it accelerated in the 1980s and 1990s, propelled by falling transport costs,[33] instant electronic communications, by assertive privatization and deregulation of financial markets and major industries, and most importantly by the ideological collapse of autarkic socialism in China (after 1978) and its political collapse in eastern Europe and the USSR (1989–1991). In terms of both the prevailing economic ideology and the prominence of international trade and finance, the post-1980 era resembled that of pre-1914.

All this, while socially disruptive everywhere and grating for those who did not enjoy American leadership, served with industrialization and Fordism as impetus for the world's amazing economic growth after 1945. The environmental consequences of this surging economic integration extended beyond those of mass consumption.

Economic integration often commodified nature suddenly. When groups of consumers, through the magic of markets, were presented with the opportunity to buy something hitherto unavailable, they often did so. If that thing was elephant ivory, rhinoceros horn, giant panda skin, alligator hide, ostrich feathers, beaver fur, tortoiseshell, whale oil, teak, or the like, then the linkup between consumer and source of supply changed ecology in the zone of supply—often drastically. This was because supply was governed by rhythms of reproduction not subject to rapid acceleration. Rhinos will not procreate on demand. After 1970 the market for rhino horn in East Asia (for medicines) and in North Yemen (for dagger handles) overwhelmed the rate of rhino reproduction. By 1997 their numbers fell 90 percent, to about 5,000 to 7,000 worldwide.[34]

Economic integration focused the dispersed demand of millions upon limited zones of supply. These zones were often sparsely populated frontiers, where the human touch had hitherto been light and where social restraints upon rapacity were few. The result was rapid exhaustion of commodities and transformation of ecologies. While this effect was most dramatic in the case of wildlife, it extended to valuable plants and trees, such as mahogany and cedar, and to landscapes under which valuable minerals lay. The impact of nickel mining on New Caledonia (see Chapter 2) would have been negligible had New Caledonia not been inte-

---

[33]A big part of this was due to the rise of container shipping, invented in 1955 but a worldwide practice only from the 1980s.

[34]The numbers about rhinos are taken from *http://www.mamba.bio.uci.edu* (2 October 1997), a web site maintained by Peter J. Bryant of the University of California-Irvine.

grated into an international trading system. New Zealand and Argentina converted suitable land into pasture to meet the overseas demand for meat, butter, and cheese after transport changes—in particular, refrigerated shipping, invented in the 1880s—connected these landscapes to distant urban markets. From the 1950s, Central American forests became cattle ranches to meet North American demand for beef. Malayan forests became rubber plantations, Brazilian ones coffee plantations, and Ghanaian ones cocoa plantations—all because of market integration. Even illegal trades, in cocaine or marijuana, drove ecological change in places such as Peru, Bolivia, and northern Morocco. After 1965 the borderlands of northern Mexico—a different kind of frontier—industrialized rapidly and dirtily because Mexican manufactures achieved greater access to the U.S. market. Economic integration, especially modern globalization, left fewer and fewer landscapes, seascapes, or habitats untouched by the effects of "frontier economics."[35]

Economic integration, at least when sudden, also disrupted common property regimes that checked environmental change. Around the world, fisheries, forests, pastures, aquifers, and other resources were (and are) often governed by rules of access that allow many to use the resource but none to exhaust it. Some such arrangements were old, such as those safeguarding subterranean waters in Valencia, Spain, or rotations among herder groups in the Sahel of southern Niger, whereby different groups took turns exploiting different grazing lands. Others were new, like the lottery system organized in the 1960s by fishermen in Alanya, Turkey; to prevent the depletion of their fish stocks, the fishermen rationed access to fishing grounds and let chance determine who fished where and when. The buffeting winds of globalization brought new shocks to these small-scale social systems. In fishing, for example, bigger operators tapping distant markets introduced trawlers and overwhelmed artisanal fishermen, whose common property regimes often collapsed. Free-for-alls ensued, and the fisheries collapsed too. Such regimes easily gave way to "tragedies of the commons" when strangers selling to distant markets—and thus operating outside the usual system of sanctions for miscreants—got involved.[36]

---

[35]That phrase, I believe, originated with Kenneth Boulding. On Central America, see Augelli 1989; on Brazil, see Dean 1995 and McNeill 1988; on Ghana, where the process led to fleeting riches and durable impoverishment, see Amanor 1994. On cocaine, see Dourojeanni 1989; on marijuana, see McNeill 1992b. On Mexico's *maquiladoras*, see Nuccio and Ornelas 1987. See also the aptly titled work of Mander and Goldsmith (1996).

[36]In the copious literature about environment and property regimes see especially Berkes 1992 and Ostrom 1992.

Similar environmental effects frequently derived from insecurity of property, even without the collapse of systems of regulated access. Wherever landowners, fishermen, herders, hunters, or miners feared that access to the resources that underwrote their livings (or their fortunes) might be lost tomorrow, they had every incentive to get as much as possible out today. Such fears, while commonplace throughout history, may well have grown with the rapid ebbs and flows of colonial empires, of communist revolutions, and other political shifts that rewrote the rules of property and access to resources. Ethiopians after 1935, to take one example, faced a chain of events featuring war, colonial occupation and expropriations, revolution, and civil war. Chinese and Russians found themselves in situations of nearly equal uncertainty. People operating on the fringes of the world economy—where links to distant markets made land and resources valuable, but where property rights, and rule of law generally, were hard to enforce—faced a similar logic. In backwoods Brazil and similar "frontier" areas, cashing in quickly whenever possible was hard to resist, and the rationale of preserving a resource for the future was especially weak.

Economic integration in the late twentieth century also promoted a rapid "financialization" of the world economy. In the 1970s, the oil producers' cartel (OPEC) brought them vast windfall profits, which they deposited in the world's banks. When states abandoned efforts to regulate capital flows, which many did under the influence of the ideas and pressures of the Reagan-Thatcher era, they made it much easier to make money in finance than in, say, trade or manufacturing. International financial flows dwarfed trade flows after 1980, filling the world's banking systems with cash. This too had its ecological consequences, because banks must lend.

A fair chunk of this cash passed through development banks. The World Bank (founded in 1944), the Inter-American Development Bank (1959), the Asian Development Bank (1965), and a few others were charged with prodding economic development in poor countries. They specialized in lending for specific development projects. In some respects they were successors to the European colonial regimes which ostensibly intended to "develop" economies in Africa and Asia before decolonization. But the banks had far more money. They could borrow huge sums in the flush financial markets of New York, London, and Tokyo and lend it out to poor countries. In keeping with reigning ideas about economic development, these banks tended to invest in infrastructure and energy projects. After 1960 the World Bank was the single largest financier of road building, power plants, oil drilling, coal mining, and dam construction. Until 1987 the development banks paid virtually no attention to the ecological consequences of their lending programs, even those with far-reaching effects such as road building and settlement in Amazonia. The governments borrowing their billions, notably Brazil, India, China, and Indonesia, did not want the banks to worry about environmental effects. They, together with most bank staffs, resisted when the World Bank, after 1987 bowing to American pressures (originating with environmental groups and filtered through Congress), started to require environmental assessment of its projects. Other development banks, not subjected to the same pressures, continued to lend on strictly economic and political criteria into the 1990s. The huge sums—tens of billions of dollars a year—disbursed by the development banks allowed eager states to transform their environments with irrigation schemes, power plants, roads through rainforests, and much more. A discouraging proportion of these projects were ecological fiascos because their promoters gave no thought to ecological contexts.[37]

Conclusion

The changes in energy regime, technology, and economy in the twentieth century were closely linked. Together as clusters of innovation, these changes propelled environmental history, both in pace and direction, in the industrialized world. Their impact elsewhere, while great enough, was limited by the fact that technological change and energy-intensive economies made a fainter imprint. Indeed, the coketown and motown clusters affected many lands only indirectly, through economic linkages with Europe, Japan, and North America. Most people in Mongolia, Borneo, Chad, and Bolivia experienced little change in the way of automobile use, industrialization, and Fordism, but oil and new trans-

[37]The most famous of these projects were the Polonoroeste road and settlement project and the Carajás iron-ore scheme in the Brazilian Amazon, the Singrauli coal mines and power plants in India, the Narmada dams (see Chapter 6) in India, and Indonesian transmigration. On the World Bank, see Wade 1997 and the indictments in Rich 1994. Bayalama 1992 claims that the structural adjustment programs of the International Monetary Fund (IMF) and the World Bank caused environmental damage in Africa, chiefly erosion and desertification.

port technology helped connect them to industrial heartlands in the USSR, Japan, Europe, and the United States, bringing environmental alterations through new crops or intensified resource extraction. The patterns of technological change and energy use shaped the international division of labor, and thus the international distribution of environmental effects. The rich countries, with their energy- and technology-intensive economies, suffered more from air and water pollution, whereas the poor countries, with their low-energy and low-technology economies, got more deforestation, soil erosion, or desertification. Some big and diverse countries, like Russia and China, suffered from all of these effects because their areas were so vast as to allow them to turn their backs on the world economy and thereby reproduce its division of labor within their own borders.

The strongly linked trajectories of energy, technology, and economy together exercised paramount influence over twentieth century environmental history. They were tied less tightly to trends in population and urbanization. And they interlocked, often strongly, with ideological and political currents, which they helped cause and which helped cause them.

11

Nature, to be commanded, must be obeyed.
                    —Francis Bacon, *Novum Organum* (1620)

"Nature, Mr. Allnutt, is what we are put in this world to rise
above."
                    —Katharine Hepburn to Humphrey Bogart in
                                    The African Queen (1951)

The twentieth century witnessed a kaleidoscopic variety of ideologies
and policies. Turbulent times invited reconsideration of old verities. For
environmental history the powerful, prevailing ideas mattered more than
the explicitly environmental ones. Environmental ideas and politics, although part of the new grand equations that governed societies after the
1960s, never came close to dislodging reigning ideas and policies, which
fit so well with the realities of the times. Even when they did not fit, they
had the staying power of incumbents. One reason the environment in the
twentieth century changed so much is because prevailing ideas and politics—from an ecological perspective—changed so little.

Big Ideas

What people thought affected the environment because to some extent it shaped their behavior. And of course, the changing environment played a part in affecting what people thought. Here there are two related points. First, what people thought specifically about the environment, nature, life, and such mattered only very marginally before 1970. Second, at all times, but more so before 1970, other kinds of ideas governed the human behavior that most affected the environment. So this section is divided into two parts: big ideas and environmental ideas.

Big ideas are the ones that somehow succeed in molding the behavior of millions. They are usually ideas about economics and politics. Ideas, like genetic mutations and technologies, are hatched all the time, but most of them quickly disappear for want of followers. Ruthless selection is always at work, but, again like mutations and technologies, the notion of increasing returns to scale often applies. When an idea becomes successful, it easily becomes even more successful: it gets entrenched in social and political systems, which assists in its further spread. It then prevails even beyond the times and places where it is advantageous to its followers. Historians of technology refer to analogous situations as "technological lock-in." For example, the narrow-gauge railway track adopted in the nineteenth century, once it became the standard, could not be replaced even after it prevented improvements that would allow for faster trains. Too much was already invested in the old ways. Ideological lock-in, the staying power of orthodox ideas, works the same way. Big ideas all became orthodoxies, enmeshed in social and political systems, and difficult to dislodge even if they became costly.

The spread of successful ideas is governed mainly by communications technology and politics. Five great changes—language, writing, printing, mass literacy, and electronic transmission—dominated the evolution of communications technology. This cumulation made successful ideas yet more successful, reducing the variety of influential ideas and adding to the sway of the ever fewer winners.[1] Political factors also influenced the

---

[1] It may also mean the winners take over faster, so that ideological shifts (with their attendant social and political changes) happen more rapidly as time goes on. Marxism spread much faster than Buddhism, Christianity, or even Islam.

success of many ideas. Centuries ago, Christianity benefited from states (such as the Roman Empire) devoted to its propagation. In the twentieth century, Anglo-American economic thought spread with help from the prominence achieved by the United States.

At the outset of the century the ideas with mass followings remained the great religions. Their doctrines include various injunctions about nature. The God of the ancient Hebrews enjoins believers to "Be fruitful and multiply, and fill the earth and subdue it; and have dominion over the fish of the sea and over the birds of the air and over every living thing that moves upon the earth" (Genesis 1:26–29). This and other biblical passages[2] inspired an argument to the effect that Christianity, or the Judeo-Christian tradition, uniquely encouraged environmental despoliation. But the record of environmental ruin around the world, even among followers of Buddhism, Taoism, and Hinduism (seen in this argument as creeds more reverent of nature), suggests this is not so: either other religious traditions similarly encouraged predatory conduct, or religions did not notably constrain behavior with respect to the natural world.[3]

The latter proposition makes more sense. Few believers knew more than a smattering of the sacred scriptures. And most of those who did, being human, easily allowed expediency and interest more than the scriptures of religious texts to govern their behavior. Every durable body of scripture is ambiguous, self-contradictory, and amenable to different interpretations to suit different circumstances. Islamic and Hindu societies maintained some sacred groves; observant Jains tried not to kill any animals. But these and other constraints affected environmental outcomes only modestly. In the unusually secular age of the twentieth century, the ecological impact of religions, rarely great, shrank to the vanishing point.

[2]In Genesis 9:1–3, God tells Noah and his sons, "Be fruitful and multiply, and fill the earth. The fear of you and the dread of you shall be upon every beast of the earth, and upon every bird of the air, and upon everything that creeps on the ground and all the fish of the sea; into your hand they are delivered. Every moving thing shall be food for you; and as I gave you the green plants, I give you everything." Similarly, the Qur'an 31:20 says, "Do you not see God has subjected everything that is in the heavens and earth for your use?" Who could ask for clearer license to appropriate all the net primary productivity of the planet?

[3]White 1967 explores the Judeo-Christian tradition's environmental prescriptions. Tuan 1968 shows decisively that Eastern religions did not prevent environmental damage in Chinese history. See also Asquith and Kalland 1997 (on Japanese religions' environmental impacts); Bruun and Kalland 1995; Hou 1997; Livingstone 1994; Toynbee 1972; and Zaidi 1981 (who argues that Islam is especially benign ecologically).

A variation on the Judeo-Christian theme is the notion that Western humanism, rationalism, or the Scientific Revolution uniquely licensed environmental mayhem by depriving nature of its sacred character.[4] While the lucubrations of Erasmus, Descartes, and Francis Bacon probably did not filter into the calculations of peasants, fishermen, or most landowners in the twentieth century or before, there is something to be said for this proposition. Western science helped recast environments everywhere indirectly, by fomenting technological change.[5] Sir Isaac Newton said that if he had seen further than others it was because he stood on the shoulders of giants. Scientists of the twentieth century, such as Haber and Midgley, whose work proved enormously consequential in ecological terms, stood on the shoulders of giants of scientific method who held the notion that the job of science was to unlock the secrets of nature and to deploy scientific knowledge in the service of human health and wealth. This persuasive and pervasive idea legitimated all manner of environmental manipulation wherever modern science took hold. Applied science brought, for example, the chemical industry, which came of age in the mid- to late nineteenth century. By 1990 it had generated some 80,000 new compounds that found routine use, and inevitably also found their way into ecosystems unadapted to them. A small proportion of these, even at tiny concentrations, proved disruptive, poisoning birds and fish, damaging genes, and causing other usually unwelcome effects. Some entered ecosystems at high concentrations, because while the world's chemical industry in 1930 produced only a million tons of organic chemicals, by 1999 the total had grown a thousandfold.[6] Slowly but surely the chemical industry came to play a part in ecology, introducing new selective criteria in biological evolution, namely compatibility with existing chemicals present in the environment. Such a development and others like it were an accidental result of rigorous scientific work over a century or more. In science more than religion, ideas from earlier eras exerted an impact on environmental history in the twentieth century.

Modern political ideas did too. Nationalism, born of the French Revolution, proved an enormously successful idea in the twentieth century.

[4]On this view see Ehrenfeld 1978, Merchant 1980, and Opie 1987.
[5]Science helped shape technology in the West only after 1850 or so. Prior to that, technological change came from tinkerers ignorant of science.
[6]These estimates come from the U.N. Environment Programme (UNEP), cited in Prager 1993:61–2.

It traveled well, across cultures and continents, better than any other European idea, appearing in several guises. It powerfully affected environmental change, although in no single consistent direction.

In some contexts, nationalism served as a spur to landscape preservation. As Europe industrialized quickly after 1880, nostalgic notions of German, Swiss, or English countryside acquired special patriotic overtones. In 1926 an Englishman could write that "the greatest historical monument that we possess, the most essential thing that is England, is the countryside, the market town, the village, the hedgerow, the lanes, the copses, the streams and farmsteads."[7] The Swiss waxed sentimental and patriotic about their mountains and farms, resisting railroads near the Matterhorn and other threatening manifestations of what they often called "Americanism."[8] Germans honed such forms of nationalism to a fine edge, alloyed them with idyllic romanticism, and organized countless countryside preservation societies. Such ideas added a current to Nazism. Himmler's SS (the Nazi special forces) dreamed of converting Poland into a landscape redolent of German tribal origins, with plenty of primeval forest to reflect the peculiarly German love of nature.[9]

Similar equations of national identity with rural righteousness, the sanctity of (our) land, and nature preservation cropped up wherever cities and industrialization spread. Russian populism before 1917; Russian (not Soviet) nationalism after 1917; western Canada's Social Credit movement; D. H. Lawrence's nature worship; the best-selling and Nobel prize-winning Norwegian novelist Knut Hamsun; the intellectual hodge-podge underlying Mediterranean fascism and Japanese militarism;[10] Mao's peasant populism; and all manner of back-to-the-land, antimodern currents—all these reflected political and cultural revulsion at urban and industrial transformations. In the Mediterranean, they provoked some small-scale reforestation schemes, including ones that won Mussolini's favor because he thought they would make Italy colder and thereby make Italians more warlike.[11]

[7]Patrick Abercrombie, quoted in Buller 1992:70.
[8]Walter 1989.
[9]Gröning and Wolschke-Bulmahn 1987a, 1987b, 1991; I thank Daniel Inkelas for pointing me toward this literature. See also Nolte 1966:419–20; Rollins 1997.
[10]Kizaki 1938.
[11]Mussolini also affected Italian ecology with his conviction that the goat was an unsuitable animal for a fascist country. During his regime (1922–1943) goat numbers declined by a third (McNeill 1992b:337–8). General Ioannis Metaxas, Greece's dictator (1936–1941) rhapsodized about forests in a 1939 speech (Metaxas 1969, 2:214–9). On Portuguese reforestation and the Salazar dictatorship, see Brouwer 1995. Rumania's Iron Guard, a fascist group, pledged to "[d]efend the life of the trees and the mountains from further devastation" (Bramwell 1989:162).

In Russia, the connection between nationalism and nature crystallized around the struggle to save Lake Baikal from pollution, a brave fight under Soviet conditions. Lake Baikal, the pearl of Siberia, is the world's deepest lake and one of the oldest. Its biota is unique, containing many species found nowhere else on earth. In its limpid waters some Soviet engineers saw an ideal cleaning fluid well-suited to the needs of the country's military-industrial complex. In 1957, authorities secretly planned a factory for viscose fibers used in jet aircraft tires to be located on Baikal's shores. Members of the Soviet scientific and cultural elite took advantage of the Khrushchev thaw—a time of unusual freedom of expression in Soviet life—to dissent publicly. In the end, vociferous objections failed to prevent two cellulose plants from opening in 1966–1967, by which time new plastics had made the fibers obsolete for jet tires. The dissent did bring an unusual degree of attention, by Soviet standards, to pollution control. The campaign may have helped deter another 1950s plan, one in which nuclear explosions were to open up the southern end of Baikal, raising the water flow through power stations on the Angara River.[12]

In India, Mohandas K. Gandhi (1869–1948), who shared little with Mussolini, the Nazi SS, or the USSR, crystallized a nostalgic nationalist vision of an artisan and peasant India, free from the corrosive influences of modern industry as exemplified by Britain: "God forbid that India should ever take to industrialism after the manner of the West. . . . If an entire nation of 300 million [this was in 1928] took to similar economic exploitation, it would strip the world bare like locusts."[13] Gandhi was exceptional: most Indian nationalists, like Jawaharlal Nehru, wanted an industrial India, locustlike if need be.

The SS did not carry out its plans for Poland, nor did India pursue Gandhi's vision instead of Nehru's. In general, the preservationist, arcadian component in nationalism lost out to a rival one that emphasized power and wealth, and therefore favored industrialization and frontier settlement, regardless of ecological implications. The nationalism unleashed in the Mexican Revolution, for instance, quickly abandoned peasant causes in favor of accelerated industrialization. Argentina and Brazil pursued the same vision, without the revolutions, after 1930. In Japan, nationalism and industrialization were yoked together from the Meiji

[12]See Weiner 1988 and 1999:355–73.

[13]Quoted in Shiva 1991a:17 and in Guha and Martinez-Alier 1997:156. Chapter 8 of the latter book considers Gandhi's environmental thought.

restoration to World War II (1868–1945), and in a more subdued and less militaristic way, after 1945 as well. The vast changes in land use and pollution patterns brought on by industrialization, then, were in part a consequence of nationalisms.

So were the changes provoked by efforts to populate "empty" frontiers. States earned popular support for steps to settle (and establish firm sovereignty over) the Canadian Arctic, Soviet Siberia, the Australian Outback, Brazilian (not to mention Peruvian and Ecuadorian) Amazonia, and the outer islands of Indonesia. Settling and defending such areas involved considerable environmental change, deforestation in some cases, oil infrastructure in others, and road building in nearly all.

Nationalism lurked behind other population policies too, notably pronatalism. Many twentieth-century states sought security in numbers, especially in Europe where birth rates were sagging. Hypernationalist regimes in particular tried to boost birth rates, in France after the humiliation at the hands of Prussia in 1871, in fascist Italy, and in Nazi Germany. The most successful was Rumania, under Nicolae Ceauşescu (1918–1989). In 1965 he set a growth target of 30 million Rumanians by the year 2000, banned all forms of birth control including abortion, and subjected women of childbearing age to police surveillance to make sure they were not shirking their reproductive duties. At the time, abortions outnumbered live births 4 to 1 in Rumania. After 1966, Rumanian maternity wards were deluged, sometimes obliged to wedge two delivering mothers into a single bed. Ceauşescu temporarily reversed the demographic transition and doubled the birth rate, all for the greater glory of Rumania.[14] Other embattled societies, such as Stalin's USSR, Iran after the 1979 revolution, the Syria of Hafiz al-Assad (ruled 1971–2000) also sought to maximize population to safeguard the nation. Nationalism, in its myriad forms and through the multiple policies it inspired, was a crucially important idea for its effects on the environment, especially when its adherents gave this connection no thought.[15]

Communism, another European idea that traveled well, was in some respects the highest form of nationalism. Its political success in Russia and China, in Cuba and Vietnam, depended as much on its promise of independence from foreign domination as on its promise of social justice.

[14]Chesnais 1995:171–73, 177–8. Rumania's orphanages remained overloaded until after Ceauşescu's fall.

[15]On nationalism and pronatalism, see Ipsen 1996 and Quine 1996.

The same ambitions—economic development and state power—that inspired state-sponsored industrialization elsewhere drove the heroic sprints of communist five-year plans.

But communism had other components too. Deep in Marxism is the belief that nature exists to be harnessed by labor. Friedrich Engels believed, like today's most cheerful optimists, that "the productivity of the land can be infinitely increased by the application of capital, labour and science." Karl Marx endorsed the French socialists who urged that the "exploitation of man by man" give way to the "exploitation of nature by man." Explicitly linking communist progress with environmental transformation, the wordsmith V. Zazurbin in 1926 addressed the Soviet Writers Congress:

> Let the fragile green breast of Siberia be dressed in the cement armor of cities, armed with the stone muzzle of factory chimneys, and girded with iron belts of railroads. Let the taiga be burned and felled, let the steppes be trampled. Let this be, and so it will be inevitably. Only in cement and iron can the fraternal union of all peoples, the iron brotherhood of all mankind, be forged.

A Soviet historian, M. N. Pokrovsky, in 1931 foresaw the day "when science and technique have attained to a perfection which we are as yet unable to visualise, [so that] nature will become soft wax in [man's] hands, which he will be able to cast in whatever form he chooses." Any ecological cost could be borne in the quest for such lofty goals.[16]

Communists, especially in the USSR and eastern Europe, liked things big. Ostensibly this was to realize economies of scale, but it became an ideology, a propaganda tactic, and eventually an end in itself. Gigantism most famously affected architecture and statuary but also industry, forestry, and agriculture. The Soviets typically built huge industrial complexes, like Norilsk and Magnitogorsk, concentrating pollution. When the USSR faced a timber shortage in the First Five-Year Plan (1929–1933), millions of prisoners and collective farmworkers were sent to the forests to cut trees as quickly as possible. The resulting deforestation and erosion put sandbars in the Volga, inhibiting traffic on the coun-

---

[16]The quotations in this paragraph are from Ponting 1991:157–8 (Engels and Pokrovsky), Manuel 1995:163 (Marx), and Hillel 1991:294–5 (Zazurbin). The exemplary communist of the Cuban revolution, Che Guevara, told his children: "Grow up to be good revolutionaries. Study hard to be able to dominate the techniques that permit the domination of nature" (quoted in *Washington Post Book World*, 19 October 1997:10).

The coke ovens in Magnitogorsk in the Urals region of the Soviet Union were part of a huge industrialization effort in the 1930s. Stalin thought he needed to industrialize his country as quickly as possible, given the threat he saw arising in Germany and the ideological necessity of an industrial proletariat. He spared no thought for the environment when trying to build communism in the USSR. Indeed, communists everywhere expected that labor could refashion nature to make it serve humanity more completely.

try's chief waterway.[17] In collectivizing agriculture, they created not merely huge farms but huge fields, stretching as far as the eye could see, far larger than necessary to realize efficiencies from mechanization. This maximized wind and water erosion.[18] Gigantism, together with the Marxist enthusiasm for conquering nature, led to the slow death of the Aral Sea, the creation of the world's biggest artificial lake and the world's biggest dam, and countless efforts "to correct nature's mistakes" on the heroic scale.[19]

Communism, at least after its initial consolidation in power, also resisted technological innovation. With fixed production quotas pegged to

[17]Andreev-Khomiakov 1997:29–38.

[18]Oschlies 1985 addresses this theme in Bulgaria. See also Ordos 1991; Jeleček 1988, 1991 (on Czechoslovakia); Stebelsky 1989 (on Ukraine).

[19]The quotation is from Adabashev 1966:110–4, who urged redirecting the Ob southward to create a reservoir larger than the Caspian Sea, melting the Arctic ice cap, and diverting the Japan Current to warm the climate of the far eastern USSR.

Five Year Plans, Soviet and eastern European factory bosses could ill afford to experiment with new technologies. Subsidized energy prices helped to ossify industry in the USSR and eastern Europe, so that, for example, most steel mills in 1990 still used the open-hearth process, an obsolete nineteenth-century invention long since replaced in Japan, Korea, and the West. The political system stymied decarbonization and dematerialization, leaving the communist world with an energy-guzzling, pollution-intensive coketown economy to the end—a fact that helped bring on that end in the Soviet sphere.

Communism aspired to become the universal creed of the twentieth century, but a more flexible and seductive religion succeeded where communism failed: the quest for economic growth. Capitalists, nationalists—indeed almost everyone, communists included—worshiped at this same altar because economic growth disguised a multitude of sins. Indonesians and Japanese tolerated endless corruption as long as economic

Spring sowing in 1955 at the Lenin Collective farm in Krasnodar district, southern Russia. Krasnodar, once home to the Kuban Cossacks, had rich steppe soils and served as a granary in the twentieth century. In the Soviet Union, after the collectivization of agriculture in the 1930s, mechanization combined with ideological commitments to encourage farms with huge open fields, making them acutely susceptible to wind erosion. The tank treads on the tractor were useful in muddy fields.

growth lasted. Russians and eastern Europeans put up with clumsy surveillance states. Americans and Brazilians accepted vast social inequalities. Social, moral, and ecological ills were sustained in the interest of
economic growth; indeed, adherents to the faith proposed that only more
growth could resolve such ills. Economic growth became the indispensable ideology of the state nearly everywhere. How?

This state religion had deep roots in earlier centuries, at least in imperial China and mercantilist Europe. But it succeeded fully only after the
Great Depression of the 1930s. Like an exotic intruder invading disturbed ecosystems, the growth fetish colonized ideological fields around
the world after the dislocations of the Depression: it was the intellectual
equivalent of the European rabbit. After the Depression, economic rationality trumped all other concerns except security. Those who promised
to deliver the holy grail became high priests.

These were economists, mostly Anglo-American economists. They
helped win World War II by reflating and managing the American and
British economies. The international dominance of the United States
after 1945 assured wide acceptance of American ideas, especially in economics, where American success was most conspicuous. Meanwhile the
USSR proselytized within its geopolitical sphere, offering a version of
the growth fetish administered by engineers more than by economists.

American economists cheerfully accepted credit for ending the Depression and managing the war economies. Between 1935 and 1970 they
acquired enormous prestige and power because, or so it seemed, they
could manipulate demand through minor adjustments in fiscal and monetary policy so as to minimize unemployment, avoid slumps, and assure
perpetual economic growth. They infiltrated the corridors of power and
the groves of academe, provided expert advice at home and abroad,
trained legions of acolytes from around the world, wrote columns for
popular magazines—they seized every chance to spread the gospel. Their
priesthood tolerated many sects, but agreed on fundamentals. Their ideas
fitted so well with social and political conditions of the time that in many
societies they locked in as orthodoxy. All this mattered because economists thought, wrote, and prescribed as if nature did not.

This was peculiar. Earlier economists, most notably the Reverend
Thomas Malthus (1766–1834) and W. S. Jevons (1835–1882), tried hard
to take nature into account. But with industrialization, urbanization, and
the rise of the service sector, economic theory by 1935 to 1960 crystallized
as a bloodless abstraction in which nature figured, if at all, as a storehouse
of resources waiting to be used. Nature did not evolve, nor did it twitch

and adjust when tweaked. Economics, once the dismal science, became the jolly science. One American economist in 1984 cheerfully forecast 7 billion years of economic growth—only the extinction of the sun could cloud the horizon. Nobel Prize winners could claim, without risk to their reputations, that "the world can, in effect, get along without natural resources."[20] These were extreme statements but essentially canonical views. If Judeo-Christian monotheism took nature out of religion, Anglo-American economists (after about 1880) took nature out of economics.

The growth fetish, while on balance quite useful in a world with empty land, shoals of undisturbed fish, vast forests, and a robust ozone shield, helped create a more crowded and stressed one. Despite the disappearance of ecological buffers and mounting real costs, ideological lock-in reigned in both capitalist and communist circles. No reputable sect among economists could account for depreciating natural assets. The true heretics, economists who challenged the fundamental goal of growth and sought to recognize value in ecosystem services, remained outside the pale to the end of the century.[21] Economic thought did not adjust to the changed conditions it helped to create; thereby it continued to legitimate, and indeed indirectly to cause, massive and rapid ecological change. The overarching priority of economic growth was easily the most important idea of the twentieth century.

From about 1880 to 1970 the intellectual world was aligned so as to deny the massive environmental changes afoot. While economists ignored nature, ecologists pretended humankind did not exist. Rather than sully their science with the uncertainties of human affairs, they sought out pristine patches in which to monitor energy flows and population dynamics. Consequently they had no political, economic—or ecological—impact.

Environmental Ideas

In contrast to the big ideas of the twentieth century, explicitly environmental thought mattered little before 1970. Acute observers, such as

[20]Robert Solow, 1974 Nobel laureate in economics, quoted in Rees 1992:123. The very long term forecast is by Julian Simon, quoted in Dryzek 1997:48.

[21]These very marginal ideas, ecological economics, are on display in Costanza 1997 and Krishnan et al. 1995. Pioneers in this field are considered in Martinez-Alier 1987.

Aldo Leopold (1887–1948) in the United States, remarked upon changes to forests, wildlife, soils, and biogeochemical flows.[22] Fears of global resource exhaustion, although almost always mistaken, provoked laments and warnings. But the audience was small and the practical results few. Environmental thinking appealed only to a narrow slice of society. Small nature conservation societies arose almost everywhere in the Western world by 1910. Nature preserves and national parks, more or less isolated from economic use, emerged after 1870, first in Australia and North America, where after the near elimination of Aboriginal and Amerindian peoples, there was plenty of open space. These efforts inspired widespread imitation, but in most countries preserves and parks had to be small and had to accommodate existing economic activity. So these developments scarcely slowed the momentum of environmental change. The ideas, however sound and elegantly put, did not mesh with the times.[23] That began to change in the 1960s.

The 1960s were turbulent times. From Mexico to Indonesia and from China to the United States, received wisdom and constituted authority came under fierce attack. Of the many ideas and movements nurtured in these heated conditions, two lasted better than the rest: women's equality and environmentalism. The success of environmentalism (loosely defined as the view that humankind ought to seek peaceful coexistence with, rather than mastery of, nature) depended on many things. In the industrial world, pollution loads and dangerous chemicals had built up quickly in the preceding decades. Wealth had accumulated (and diffused through Fordism) to the point where most citizens could afford to worry about matters beyond money. In a sense, the economic growth of the industrial countries in the era 1945 to 1973 provoked its own antithesis in environmentalism.[24]

Successful ideas require great communicators to bring about wide conversion. The single most effective catalyst for environmentalism was an American aquatic zoologist with a sharp pen, Rachel Carson (1907–1964).

[22]Leopold, yet another son of the Iowa soil (like Henry Wallace and Norman Borlaug), worked for the U.S. Forest Service and wrote *A Sand County Almanac* (1949).

[23]For the history of (Western) environmental thought generally, see Bramwell 1989; Corvol 1987 (on France); Deléage 1992; Glacken 1967; Graaff 1982 (on Holland); Grove 1994; Hermand 1991 and Ditt 1996 (on Germany); Pepper 1996; Teich et al. 1997; Votruba 1993 (on Czechoslovakia); and Worster 1977.

[24]Environmentalism, of course, provoked its own antithesis—claims to the effect that ecological change was minimal, or natural, or for some other reason not a cause for concern.

In 1962 Rachel Carson, a former government biologist, published *Silent Spring*, an attack on the careless use of pesticides in the United States. Through this book, other writings, and the media furor that surrounded her in 1962 and 1963, she helped mobilize American opinion on the issue of pesticides and environmental conservation generally. If modern environmentalism in the United States had a progenitor, it was Rachel Carson. She died of cancer a few months after this photograph was taken in 1963.

While working for the U.S. Bureau of Fisheries (later the Fish and Wildlife Service) she began to publish articles and books, mostly on marine life, that reached wide audiences. In 1962 her salvo against indiscriminate use of pesticides, *Silent Spring*, appeared. She compared the agrochemical companies to the Renaissance Borgias with their penchant for poisoning. This earned her denunciations as a hysterical and unscientific woman from chemical manufacturers and the U.S. Department of Agriculture. The resulting hue and cry got her, and her detractors, onto national television in 1963. But her scientific information, mainly concerning the deleterious effects of DDT and other insecticides upon bird life, was mostly sound and her message successful. After serialization in an influential magazine (*The New Yorker*), her book became a bestseller in several languages. President John F. Kennedy, against the wishes of the USDA, convened a government panel to look into pesticide problems, and its findings harmonized with Carson's. Eventually she had elementary schools named for her and her face graced postage stamps.[25]

In another age Rachel Carson's ideas might have been ignored.[26] Instead she, and hundreds like her, inspired followers and imitators. Millions now found the pollution they had known most of their lives to be unnecessary and intolerable. Earth Day in 1970 mobilized some 20 million Americans in demonstrations against assaults on nature. By the 1980s, anxieties about tropical deforestation, climate change, and the thinning ozone shield added a fillip (and a new focus) to environmentalism. Earth Day in 1990 attracted 200 million participants in 140 countries. American popular music—a global influence—added the environment to its repertoire of subjects.[27] Mainstream religious leaders, from the Dalai Lama to the Greek Orthodox Patriarch (of Istanbul), embraced aspects of environmentalism, as did some fundamentalist religious groups.[28] Big science and its government funders converted too. The United Nations launched

[25]Lear 1997 is a recent biography. Carson's book might be the most important ever written by an American. Admiral Mahan's tome on sea power is a rival.

[26]In a sense they were: warnings of DDT's mischief were published (obscurely) as early as 1946.

[27]Some examples are the songs of Marvin Gaye, Joni Mitchell (a Canadian), and the group Alabama. These artists represented urban and soul music, pop balladeering, and country music, covering most of the spectrum of U.S. popular music.

[28]See Oliver 1992 on the greening of Protestantism. Some fundamentalist Christians regarded it as sinful to visit extinction upon God's creatures, and hence supported biological conservation. The *Economist*, 21 December 1996:108–9, reviews the connections between world religions and environmentalism.

its Man and the Biosphere research program in 1971, and by 1990 most rich countries had global-change science programs. Taken together, by 1998 these amounted to the largest research program in world history.[29]

Between 1960 and 1990 a remarkable and potentially earth-shattering (earth-healing?) shift took place. For millions of people swamps long suited only for draining had become wetlands worth conserving. Wolves graduated from varmints to noble savages. Nuclear energy, once expected to fuel a cornucopian future, became politically unacceptable. Pollution no longer signified industrial wealth but became a crime against nature and society. People held such views with varying emphases and degrees of commitment. Movements based on them proved schismatic in the extreme, but all shared in a common perceptual shift. The package of ideas proved highly successful, to the point where by the late 1980s, oil companies and chemical firms caved in and instructed their public relations staff to construct new "green" identities. While the sincerity of their conversions remained open to doubt, their fig leaves showed that in the realm of ideology, environmentalism had arrived.

This extraordinary intellectual and cultural shift started in the rich countries but emerged almost everywhere. Environmentalism had many faces, each with its own issues and agendas. Where it was once systematically repressed—in some countries of the Soviet bloc ecological data were state secrets—it soon helped topple regimes. In countries as poor as India, vigorous environmental groups emerged by 1973 and coalesced by the 1980s. In poor countries environmentalism normally was entwined in one or another social struggle over water, fish, or wood and had little to do with nature for nature's sake. A 1997 poll found the people most willing to part with their own money to check pollution were Indians, Peruvians, and Chinese.[30] The full meaning of this new current will take decades, conceivably centuries, to reveal itself.[31]

[29]Or so thought officials at the American National Science Foundation: in 1997 there were some 2,000 scientific organizations carrying on global change research or monitoring.

[30]This poll was conducted by Environics International (Toronto) and reported in the Washington Post, 22 November 1997:A15. The populations who most strongly favored giving environmental protection priority over economic growth were those of New Zealand, Canada, Switzerland, Australia, and the Netherlands. Those least inclined to do so were in Ukraine, Nigeria, Poland, and Hungary. See also Guha and Martinez-Alier 1997.

[31]In a philosophical treatise Luc Ferry (1995) argues that deep ecology, a fringe movement within environmentalism, is the first significant challenge to Cartesian thought in 300 years. Deep ecology rejects anthropocentrism and finds intrinsic merit in all forms of life, natural systems, and natural phenomena.

International Politics and War

As with ideas, so with politics. By far the most important political forces for environmental change were inadvertent and unwitting ones. Explicit, conscious environmental politics, while growing in impact after 1970, still operated in the shadow cast by conventional politics. This was true on both the international and national scales.

SECURITY ANXIETY AND ENVIRONMENTAL NONCHALANCE. The dominant characteristic of the twentieth-century international system was its highly agitated state. By the standards of prior centuries, the big economies and populous countries conducted their business with war very much on their minds, especially from about 1910 to 1991. War efforts in the two world wars were all-consuming. Security anxiety between the wars, especially during the long Cold War (1945–1991), was high given the perceived costs of unpreparedness. In this situation, states and societies had strong incentives to maximize their military strength, to industrialize (and militarize) their economies, and after 1945 to develop nuclear weapons. The international system, in Darwinian language, selected rigorously against ecological prudence in favor of policies dictated by short-term security considerations.

Security anxiety had countless environmental ramifications. In France after the defeat of 1870, the army won the power to preserve public and private forests in northeastern France, using them in a reorganized frontier defense system designed to channel German invaders along narrow, well-fortified corridors. (The next German invasion, in 1914, came via Belgium). Many tense borders became de facto nature preserves because ordinary human activities were prohibited (for example, Bulgaria–Greece, the demilitarized zone between North and South Korea, and Iran–USSR). But other border regions became targets for intensive settlement intended, among other goals, to secure sovereignty, and consequently witnessed wide deforestation, as in Brazilian and Ecuadorian Amazonia. Many states built road and rail systems with geopolitical priorities in mind, such as Imperial Russia's Trans-Siberian Railroad, Hitler's autobahns, the U.S. interstate system, and the Karakoram Highway between Pakistan and China. Such major transport systems inevitably affected land-use patterns. Land use occasionally was deliberately changed to promote military transport: In India, the British before 1921 used irrigation to create "horse runs," lush pastures for horse breeding that were intended to keep the India Army supplied with mounts.[32]

By far the largest environmental effect of security anxiety came via the construction of military-industrial complexes. After World War I it became clear that, aside from plenty of young men, the main ingredient of military power was heavy industry. Horses and heroism were obsolete. All the great powers of the twentieth century adopted policies to encourage the production of munitions, ships, trucks, aircraft—and nuclear weapons.

No component of military-industrial complexes enjoyed greater subsidy, protection from public scrutiny, and latitude in its environmental impact than the nuclear weapons business. At least nine countries built nuclear weapons, although only seven admitted doing so (USA, U.K., France, USSR, China, India, and Pakistan). Israel and South Africa developed nuclear weapons while pretending they hadn't.

The American weapons complex involved some 3,000 sites in all. The United States built tens of thousands of nuclear warheads and tested more than a thousand of them. The jewel in this crown was the Hanford Engineering Works, a sprawling bomb factory on the Columbia River in the bone-dry expanse of south-central Washington state. It opened during World War II and built the bomb that destroyed Nagasaki. Over the next 50 years, Hanford released billions of gallons of radioactive wastes into the Columbia and accidentally leaked some more into groundwater. In 1949, shortly after the Soviets had exploded their first atomic bomb, the Americans conducted a secret experiment at Hanford. The fallout detected from the Soviet test had prompted questions about how quickly the Soviets were able to process plutonium. In response, American officials decided to use "green" uranium, less than 20 days out of the reactor, to test their hypotheses about Soviet activities. The Green Run, as it was known to those in on the secret, released nearly 8,000 curies of iodine-131, dousing the downwind region with radiation at levels varying between 80 and 1,000 times the limit then thought tolerable. The local populace learned of these events in 1986, when Hanford became the first of the U.S. nuclear weapons complexes to release documents concerning the environmental effects of weapons production. The Green

[32]On French forests, see Amat 1993. On Amazonia, see Wood and Schmink 1993 and Pichón 1992. On the motives behind the U.S. Interstate Highway System, see Lewis 1997 and Gifford 1998. On the Karakoram Highway, see Ispahani 1989:145–213. On horse runs, see Ali 1988.

Run shows the environmental liberties the Americans took under the influence of Cold War security anxiety.[33]

But that was just the tip of the iceberg. More environmentally serious were the wastes, which in the heat of the Cold War were left for the future to worry about. A half century of weapons production around the United States left a big mess, including tens of millions of cubic meters of long-lived nuclear waste. Partial cleanup is projected to take 75 years and cost $100 billion to $1 trillion, the largest environmental remediation project in history. Full cleanup is impossible. More than half a ton of plutonium is buried around Hanford alone.[34]

The Soviets were more cavalier. Their nuclear program began with Stalin, who wanted atomic weapons as fast as possible, whatever the human and environmental cost. As it happened, the Soviet command economy was rather good at such things: a large nuclear weapons complex arose from nothing in only a few years. The Soviets built about 45,000 warheads and exploded about 715 between 1949 and 1991, mostly at Semipalatinsk (in what is now Kazakhstan) and on the Arctic island of Novaya Zemlya. They used nuclear explosions to create reservoirs and canals and to open mine shafts. In 1972 and 1984 they detonated three nuclear bombs to try to loosen ores from which phosphate (for fertilizer) was derived. They dumped much of their nuclear wastes at sea, mostly in the Arctic Ocean, some of it in shallow water. They scuttled defunct nuclear submarines at sea. Most of the world's known reactor accidents befell the USSR's Northern Fleet, based at Archangel.

The Soviets had only one center for reprocessing used nuclear fuel, at the Mayak complex in the upper Ob River basin in western Siberia, now the most radioactive place on earth. It accumulated 26 metric tons of plutonium, 50 times Hanford's total. From 1948 to 1956 the Mayak complex dumped radioactive waste into the Techa River, an Ob tributary and the sole source of drinking water for 10,000 to 20,000 people. After 1952, storage tanks held some of Mayak's most dangerous wastes, but in 1957 one exploded, raining 20 million curies down onto the neighborhood—about 40 percent of the level of radiation released at Chernobyl. After 1958, liquid wastes were stored in Lake Karachay. In 1967 a drought exposed the lakebed's radioactive sediments to the steppe winds, sprinkling dangerous dust with 3,000 times the radioactivity released at Hiroshima over an area the size of Belgium and onto a half million unsuspecting peo-

---

[33]Gerber 1992.
[34]Fioravanti and Makhijani 1997; USDOE 1995.

ple. By the 1980s, anyone standing at the lakeshore for an hour received a lethal dose of radiation (600 roentgens/hr). A former chairman of the USSR's Supreme Soviet's Subcommittee on Nuclear Safety, Alexander Penyagin, likened the situation at Mayak to 100 Chernobyls. No one knew the extent of contamination in the former USSR because the nuclear complex was so large and so secret. Much of the complex was shut down in the last years of the USSR, but the mess remained and Russia cannot afford much in the way of cleanup.[35]

The lethal residues of the British, French, Chinese, Indian, Pakistani, Israeli, South African (and perhaps a few other) nuclear weapons programs were, mercifully, not on the superpower scale.[36] Taken as a whole, these programs not only burdened posterity with a long-term waste-management obligation, but they also gobbled up nearly a tenth of the commercial energy deployed worldwide after 1940.[37] Future historians will need to be at their best to convey to future generations the Cold War anxiety that led responsible officials to sanction the slipshod production of nuclear weapons and disposal of wastes.

WAR AND THE ENVIRONMENT. Much less was done in war than in the name of war. The twentieth century did not lack for prolonged combat, but most of the environmental changes wrought in combat proved fleeting. Bombers flattened most of Berlin and Tokyo in 1944–1945, but both cities sprang up again within a decade or two. American bombers put some 20 million craters in Vietnam (1965–1973), but vegetation covered most of these wounds, while some eventually served as fishponds.[38] In the war between Japan and China (1937–1945), Chinese Nationalists, hoping to forestall a Japanese advance, destroyed the dikes holding the Huanghe (Yellow River) in 1938. Probably the single most environmentally damaging act of war, it drowned several hundred thousand Chinese (and many thousand Japanese), destroyed millions of hectares of cropland in three provinces, and flooded out 11 cities and 4,000 villages. But the

[35]Cochrane et al. 1993; Nilsen and Bohmer 1994; Nilsen and Hauge 1992; Yablokov 1995. A useful general study of the Soviet nuclear weapons program to 1956 is given by Holloway 1994.

[36]See Danielsson 1990 (on the French in Polynesia); Makhijani et al. 1995 (for a global survey).

[37]Smil 1994:185.

[38]The pioneer vegetation in recolonizing destroyed forests in Vietnam (and elsewhere in Southeast Asia) was mainly *imperata* grass and bamboo. Ecological succession might bring back tropical forest in 100 years (Westing 1980:97–8).

labor of surviving Chinese made good the devastation in a few years. The intense combat on the Western front and at Gallipoli during World War I and the scorched-earth policies of the German-Soviet struggle during World War II brought correspondingly intense environmental devastation. But patient labor and the processes of nature have hidden these scars and assimilated into the surrounding countryside the sites of even the most ferocious battles—except where there has been conscious effort to preserve the battlefields as memorials. In the Gulf War of 1991, Iraqi forces ignited huge oil fires that darkened the skies, and spilled further oil into the shallow and biologically rich Persian Gulf. The atmospheric pall dissipated in a few months when the burning wells were capped, but marine life took (and will take) years to recover. The Gulf War may prove an exception to the rule about the fleeting nature of environmental damage from combat.[39]

While environments governed by irrigation works such as China's were the most vulnerable to war's destruction, deforestation took (and will take) longer to heal. Dryland agriculture recovered quickly from war, on average in about three years. Pasture and grassland often took a little longer, perhaps 10 years. But forests would take a century or three. For centuries warfare had featured forest destruction as policy. Caesar burned Gallic woods. In the twentieth century the prominence of guerrilla tactics meant that war played an unusually large role in deforestation. Many wars of colonial resistance in Africa and Southeast Asia involved guerrilla campaigns. During the Cold War, many of the proxy wars fought in Africa, Asia, and Central America did too. Guerrillas had to hide, and forests provided ideal cover; hence antiguerrilla forces destroyed forest. At times guerrillas did too, often as acts of arson aimed at occupying powers or forces of constituted order.

Twentieth-century technology made forest destruction much easier than in Caesar's (or William Tecumseh Sherman's) day. The French pioneered incendiary bombing of forests in the Rif War (1921–1926), an uprising of Moroccan Berbers against Spanish and French colonial power. Napalm debuted in World War II in flamethrowers and proved its effectiveness

---

[39]Clout 1996; Sobolev 1947; Westing 1990. Remarks on Gallipoli are based on my own observations of 1994. On the Gulf War, see Burger 1997:69–73, Hawley 1992, Hobbs and Radke 1992. Aarsten 1946 reports that 17% of Dutch farmland was destroyed during World War II by saltwater intrusion, but I believe this too was transitory. El-Shobosky and Al-Saedi 1993 report that tank campaigns in deserts (e.g., Egypt, 1941–1943; Kuwait, 1991) broke fragile desert crusts and allowed unusually intense sandstorms.

against forest cover in the Greek Civil War (1944–1949) before becoming a major weapon in the American arsenal in Vietnam. The British inaugurated the use of chemical defoliants in the Malaya insurgency in the 1950s. The Americans used them on a large scale (for example, Agent Orange) in Vietnam. The Soviet-Afghan War begun in 1979 witnessed the use of a variety of high-tech defoliants. These and a hundred cases like them constitute some of the more durable ecological effects of combat.[40]

Outside of combat, war efforts had other ecological impacts. In the karstic limestone of the Veneto Alps, World War I munitions dumps leached copper into groundwater. Eighty years later some springs seemed "to be small mines" of copper. European wheat demand in World War I led to the plowing up of about 6 million hectares (an area the size of West Virginia or Sri Lanka) of grasslands on the American High Plains, and some more in Canada's prairie provinces. This helped prepare the way for the Dust Bowl of the 1930s. The British war effort in World War II consumed about half of Britain's forests. To build Liberty ships in 11 days, as Americans did in Portland, Oregon, during that war, took a lot of electricity, justifying additional hydroelectric dams on the Columbia River, where two big ones had been built in the late 1930s. Frantic drives to raise production of food, fuel, minerals, and other resources surely led to sharp ecological disruptions in every combatant nation, as did crash road- and railroad-building efforts. More recently, belligerents in the civil wars that raged in Cambodia and eastern Myanmar (Burma) financed their campaigns by contracting with Thai logging companies to strip forest areas under their control.[41]

By suppressing normal economic activity, war temporarily reduced some ordinary environmental pressures. Despite depth charges and oil spills in the U-boat campaigns, World War II brought back halcyon days for North Atlantic fish populations, because fishing fleets sat out the war. Industrial emissions slackened because of coal shortages and factory destruction, at least in Europe and Japan. Iraqi land mines in the Kuwaiti desert kept people out and allowed a resurgence of animal and plant life

[40]Demorlaine 1919; McNeill 1992b:260–70; Prochaska 1986; Westing 1990.

[41]Celi 1991 (on copper); Opie 1993:96 (on Great Plains); Kuusela 1994:125 (on British forests). On Southeast Asian wars and timber, see the *Economist*, 17 June 1995:35. War also produced refugees who, when outside networks of efficient transport, might overload local ecosystems in their quest for food and firewood. The 3 million Afghans in northern Pakistan in the 1980s had such effects (Azhar 1985; Allan 1987).

in the 1990s.[42] Combat had its impacts on the environment, occasionally acute but usually fleeting. More serious changes arose from the desperate business of preparing and mobilizing for industrial warfare.

IMPERIALISM, DECOLONIZATION, AND DEMOCRATIZATION. International politics was conducted by means other than war. Here I will touch briefly on only two major themes: first, imperialism and decolonization, and then democratization. As the twentieth century began, Russia, Japan, the United States, and especially the western European powers had embarked on imperial expansions. This often involved the displacement of existing populations, as in South Africa and Algeria. Colonial powers reoriented local economies toward mining and logging, and toward export monocultures of cotton, tea, peanuts, or sisal. Normally these changes were imposed with no thought to environmental consequences: the only goals were to make money for the state and for entrepreneurs, and to assure the mother country ready access to strategic materials. By the 1940s the French and British at least claimed to have local interests at heart when converting as much as possible of Mali to cotton or of Tanganyika to peanut production. But through ecological ignorance they nonetheless brought salinization in the Niger bend region of Mali and turned marginal land into useless hardpan in central Tanganyika.[43]

Decolonization surprisingly changed little of this. New independent regimes often continued the economic policies of their predecessors. Big prestige projects carried on the tradition of colonial environmental manipulation in places such as Ghana, Sudan, and India. Financially weak regimes (such as Indonesia, Papua New Guinea, and Ivory Coast) often sold off timber and minerals as fast as possible, regardless of environmental impacts. Many rulers arrived in office via coup d'état and saw fit to cash in before the next colonel or sergeant followed suit. The decolonization of Soviet Central Asia brought no changes in the water regime that was strangling the Aral Sea. In environmental matters, as in so many respects, independence often proved no more than a change in flags.

Democratization was another matter. A global wind of democratization blew across Greece and Iberia in the 1970s, much of Latin America and

---

[42]Westing 1980:154 (on North Atlantic fish). In Turner et al. 1990 there are chapters on biogeochemical flows that show the impact of World War II. Kuwaiti desert life is reported in *Environment* 35(4):22.

[43]Adams 1992:104; Hogendorn and Scott 1981.

India's first prime minister, Jawaharlal Nehru, envisioned India as an industrial power-house in the world economy. He hoped that with independence from Britain in 1947, India could escape from poverty and weakness through state-sponsored industrialization. That ambition extended to agriculture, which Nehru wanted to modernize with machines and grand-scale irrigation works. Like many leaders of recently decolonized countries, he was in a hurry to make up for lost time under colonial rule. Here he inspects a tractor factory of the International Harvest Company in Chicago in 1949, guided by H. T. Reishus of International Harvester.

East Asia in the 1980s, and parts of eastern Europe and Africa in the 1990s. In some cases, environmental protests helped in modest ways to undermine the legitimacy of authoritarian (for example, Chile) and communist (Poland) regimes. Such regimes had encouraged pollution-intensive economies and ecologically heedless resource extraction in their quests to build state power and maximize economic growth. They normally had kept ecological information under tight control. Democratization broke the hold these regimes enjoyed over information, and brought to light all manner of environmental problems. Those caused by foreigners, the military, or specific factories were often addressed and sometimes resolved. Those caused by the consumption patterns of ordinary citizens often got worse under democracy, as, for instance, when eastern Europe and Russia dropped subsidized public transport in favor of private cars. Moreover the media spotlight shone only on certain kinds of environmental problems, usually those inspiring maximum dread such as industrial accidents and nuclear issues. Slow-moving crises such as soil erosion or biodiversity loss remained hidden in the shadows, uncompelling to the media and the public and entirely irrelevant to politicians attuned to the next election. Thus democracy tended to generate its own characteristic environment.[44]

All these great currents of international politics in the twentieth century—security anxiety, imperialism, decolonization, democratization, and if to a lesser degree war—profoundly shaped the century's environmental history. Almost all of the environmental changes generated by these currents were inadvertent effects of politics and policies designed for other ends. Yet at the same time nations negotiated hundreds of environmental accords, sowing the seeds of what might become a loose regime of environmental governance. That would require a relaxation of the security anxiety that shaped the twentieth century.

Environmental Politics and Policies

In contrast, the politics and policies in which environmental considerations formed a conscious element had modest effects. Environmental politics and policies, as such, began only in the 1960s. Prior to that, local, national, and (on a very limited scale) international laws and treaties regulated some aspects of pollution, land use, fishing, and other issues. Smoke nuisance ordinances go back at least 700 years. Britain established a regulatory agency for a specific source of pollution, the Alkali Inspectorate, in 1865. But all of this was uncoordinated—specific policies and laws for very specific instances.[45] On the international scale, neighboring countries had, from time to time, agreed to restrain fishing or water use. A multilateral agreement in 1911 checked the exploitation, in

---

[44]On democratization and environment, see the contributions to Jänicke and Weidner 1996; Lafferty and Meadowcroft 1996.

[45]An exception might be Peter the Great, who ruled Russia from 1689 to 1725. He introduced laws on wildlife conservation, forest preservation, overfishing, soil conservation, and, in St. Petersburg, water pollution. Much of this was rescinded by Catherine the Great (ruled 1762–1796). See Massey 1992:16–17.

the Bering Sea's Pribilof Islands, of fur seals that Russian, Japanese, Canadian, and American sealers had hunted nearly to extinction between 1865 and 1900. The United States and Canada arrived at a number of wildlife conservation accords before 1916.[46] In the aftermath of World War II, international institutions sprang up left and right, including some concerned with the environment such as the IUCN (International Union for the Conservation of Nature). Others regulated the environment without making it their explicit focus, such as the WHO, FAO (the U.N. Food and Agriculture Organization), and UNESCO (U.N. Educational, Scientific, and Cultural Organization), all of which were born in the years 1945 to 1948. But no coordinated policies or political currents dealt with the environment as such. This changed in the 1960s, a direct result of the tumult in the world of ideas.

Two general phases are discernible in the environmental politics and policies of the late twentieth century. The first began in the mid-1960s and lasted until the late 1970s. In this phase environmental movements and, in some cases, political parties, sprang up in the rich countries. New Zealand's Values Party, born in the late 1960s, was the first explicitly green party but far from the most successful: it splintered after some 15 years on the edges of New Zealand politics. Environmental movements focused mainly on pollution issues but also on fears of resource exhaustion, spurred by OPEC's actions of 1973. Governments responded by creating new agencies charged with protecting the environment as a whole. Sweden (1967) and the United States (1970) led the way. International regimes of cooperation remained very weak, despite the efforts made after the first international conference on the environment in Stockholm (1972). That led to the United Nations Environment Programme (UNEP), headquartered in Nairobi.

In the second phase, beginning around 1980, poorer countries established their own environmental protection agencies, often given the status of ministry. In many cases, such as Nigeria or Russia, environmental laws and policy existed only on paper. In some cases, for example, Angola or Afghanistan, ongoing wars meant there was no environmental politics or policy even on paper. But in India, Brazil, Kenya, and elsewhere, grassroots environmental movements germinated and, whether through civil disobedience or official channels, began to affect national politics. India boasted hundreds of environmental organizations by the 1980s, ranging from scientific research institutes that served as watchdogs—such as New

[46]Dorsey 1998.

Delhi's Centre for Science and Environment—to coalitions composed mainly of peasant women, such as the Chipko Andalan that sought to check logging in the Himalaya. Such movements were often led by women whose lives were most affected by fuelwood shortage (because fuelwood gathering almost everywhere was women's and children's work), soil erosion (where women worked the land, as in much of Africa and the Indian Himalaya), and water pollution (because women fetched water and were responsible for children's health). Kenya's Green Belt movement, dedicated to tree planting, was organized by the National Council of Women of Kenya in 1977. From 1981 to 1987 the movement was led by a woman, a former professor of veterinary anatomy, Wangari Maathai (born 1940). Ordinarily these grassroots environmental movements were embedded in peasant protest or some other social struggle. When strong enough, they won some concessions from governments; when not, they solidified antienvironmental attitudes in the corridors of power, inadvertently inviting elites to equate environmentalism with subversion and treason. The Green Belt movement proved strong enough to make an im-

Wangari Maathai

pact on the land and provoke a backlash: it had planted some 20 million trees in Kenya by 1993, but government spokesmen vilified Maathai and government thugs roughed her up more than once for her efforts.[47]

In the rich countries in this second phase, new concerns added new dimensions to environmental politics: tropical forests, climate change, ozone depletion. In the United States an ideological crusade to roll back environmental regulation (c.1981–1984) boomeranged, as provocative statements by President Reagan's officials served as recruiting devices for environmental pressure groups.[48] Leadership in terms of innovative institutions and planning passed to northern European countries, notably the Netherlands, and to Japan. Green parties entered politics, and in some cases (such as Germany in 1983) to parliaments as well. In 1998 the German Green Party took part in a coalition government, and its members held some important ministries. The Europeans pioneered a consensual politics of environmental moderation, based on corporatist traditions in which government, business, and organized labor hammered out agreements after prolonged bargaining. The Dutch in particular, beginning in 1989, arrived at an integrated national environmental plan, designed to harness the power of the influential ministries and special interests resistant to ecological prudence, such as agribusiness.[49]

The second phase featured unprecedented efforts at international cooperation. Regional and global problems, such as acid rain or ozone depletion, required new institutions, agreements, and regimes of restraint. The Reagan Administration initially scuppered what it could, but found fewer and fewer allies abroad or in Congress. In 1987 (see above), the U.S. Congress helped browbeat the World Bank into environmental awareness. In 1987 the Brundtland Report, the fruit of four years of U.N.-sponsored inquiry into the relationship between environment and economic development, offered intellectual underpinnings for environ-

---

[47]Gadgil and Guha 1995; Guha 1990; Guha and Martinez Alier 1997.

[48]Reagan himself claimed that trees cause most air pollution. His Secretary of the Interior, James Watt, once allowed that environmentalists were not real Americans, and on other occasions suggested that environmentalists should be shot, and (at a congressional hearing) that there was little point to environmental preservation because God was preparing the apocalypse shortly. Donald Hodel, another Reagan Interior Secretary, said that donning hats and sunglasses made more sense than trying to prevent thinning of stratospheric ozone (Rothman 1998:187–9).

[49]On environmental politics c.1970–1995 in rich countries, see Bührs and Bartlett 1994, Broadbent 1998, Cramer 1989, Dalton 1994, de Jongh and Captain 1999, Dede 1993, Diani 1995, Hays 1997, Lee and So 1999, Rothman 1998, Stevis 1993, Villalba 1997, and Votruba 1993.

mental planning, for regimes of restraint, and for the ambition of ecologically sustainable development. The Montreal Protocol (1987) and
subsequent accords showed what good science and diplomacy could do.
Thousands of international environmental agreements were reached
from the mid-1960s onward and many had real effects. Optimistic observers saw in this a nascent "global governance regime" that could address the world's cross-border environmental problems.[50]

The impact of all this, from 1967 on, was considerable in the rich countries. The technically and politically easiest environmental problems were
in fact significantly reduced. Industrial wastewater was cleaned up, with
benefits to the Rhine, the North American Great Lakes, and elsewhere.
Sulfur dioxide emissions waned. Leaded gasoline vanished into history.
Municipal sewage treatment improved. In general, problems that arose
from a single institution or point source were addressed with some success. Initially at least, local solutions such as taller smokestacks merely
shunted ill effects elsewhere. Sometimes more systemic solutions succeeded in their specific task but at the same time deepened other problems. Scrubbers used to control particulate emissions from smokestacks
worsened acid rain. Most truculent of all were those problems that derived from citizen behavior or from diffuse sources. Nitrous oxides from
vehicle exhausts and toxic farm runoff, for instance, continued to mount
in North America and Europe.

Moreover, in most of the rich countries some powerful industries resisted environmental regulation successfully by launching endless lawsuits
or controlling the decisive ministries. This helped prevent serious reform
in transport, energy, and agribusiness, the myriad environmental impacts
of which scarcely abated. The U.S. auto industry fought successfully to
hamstring fuel efficiency standards. The coal industry in Germany retained its giant subsidies. California agribusiness kept getting water at
dirt-cheap prices. More often than not, major decisions affecting the environment remained the province of the powerful ministries—trade, finance, industry, agriculture—rather than of environmental agencies.

Environmental politics ran up against the limits of the possible at the
international level too. Although the United States became more
amenable to international agreements after the late 1980s, it tried hard
to see that in these accords its ox wasn't gored. At the U.N. Conference

---

[50]See, e.g., Tolba and El-Kholy 1992:737–98, and Young 1997. The Brundtland Report's
formal name appears in the Bibliography under WCED 1987. It was directed by Norway's
Prime Minister, Gro Harlem Brundtland.

on Environment and Development in Rio de Janeiro in 1992, the Americans made it clear that U.S. "lifestyles" were not up for negotiation. Other countries matched this stance. Japan proved recalcitrant on whaling prohibitions (as did Norway) and the trade in elephant ivory. Saudi Arabia and other oil producers fought against agreements on carbon emissions. Brazil insisted on its right to develop Amazonia as it wished, regardless of the implications of burning the world's largest rainforest. India and China declined to join the Montreal Protocol and subsequent accords on ozone-destroying CFCs, and in general adamantly refused to compromise their industrial ambitions in the interest of their, or the global, environment. Mexico and many other countries resisted pressures to bring environmental laws into harmony with those of richer nations: countries with more relaxed laws (or enforcement) found multinational firms more eager to invest in new steel mills and chemical plants. While there were many fault lines and alliances in this late-century world of international environmental politics, the main one divided rich from poor. Called, with dubious geography, a north-south confrontation, it crystallized at Rio in 1992 and particularly bedeviled climate-change negotiations, which had achieved only toothless accords up to 1999.

In short, both before and after 1970, for good and for ill, real environmental policy, both on the international and national levels, was made inadvertently, as side effects of conventional politics and policies. Britain managed to reduce its sulfur emissions after 1985 because Margaret Thatcher scuttled the coal industry in her quest to break the political power of trade unionism. Farm subsidies, especially in Japan and Europe, helped create and maintain a chemical-intensive agriculture and dense populations of pigs and cattle, with deleterious consequences.[51] Soviet and Chinese policies reduced the mobility of Central Asian nomads, aggravating overgrazing and desertification.[52] China's collectivization and Cultural Revolution in the 1960s destroyed village-level constraints on marriage and fertility, provoking a gargantuan baby boom that lay behind many aspects of China's manifold environmental crisis in the 1980s and beyond.[53] China's collectivization also helped inspire Tanzania's "villagization" scheme of the 1970s, the largest resettlement plan in African

[51]OECD 1998.
[52]Humphrey and Sneath 1996, vol. 1.
[53]Lee and Feng 1999.

history, and one which led to deep environmental problems.[54] Even in the age of environmental politics and overt environmental policy, real environmental policy almost always derived from other concerns, and traditional politics exercised stronger influence over environmental history.[55]

Conclusion

The grand social and ideological systems that people construct for themselves invariably carried large consequences, for the environment no less than for more strictly human affairs. Among the swirl of ideas, policies, and political structures of the twentieth century, the most ecologically influential probably were the growth imperative and the (not unrelated) security anxiety that together dominated policy around the world. Both were venerable features of the intellectual and political landscape, and both solidified their hold on imaginations and institutions in the twentieth century. Both, but particularly the growth imperative, meshed well with the simultaneous trends and trajectories in population, technology, energy, and economic integration. Indeed, successful (that is, widely adopted) ideas and policies had to mesh with these trends.

Domestic politics in open societies proved mildly more responsive to environmental problems that annoyed citizens than did more authoritarian societies, especially after 1970, but there were clear limits to the ecological prudence that citizens wanted. Regardless of political system, policy makers at all levels from local to international responded more readily to clear and present dangers (and opportunities) than to the more subtle and gradual worries about the environment. The prospect of economic depression or military defeat commanded attention that pollution, deforestation, or climate change could not. More jobs, higher tax revenues, and stronger militaries all appealed, with an immediate lure that cleaner air or diversified ecosystems could not match.

---

[54]Shao 1986 considers the ecological effects of the scheme (1969–1975), which moved 80% of Tanzanians into villages constructed in accordance with the ideas of "African socialism" propounded by Julius Nyerere.

[55]On environmental politics I have consulted Dryzek 1997, Jänicke and Weidner 1996, Karan 1994, McCormick 1991, Mandrillon 1991, Panjari 1997, Price 1994, and Tolba and El-Kholy 1992.

By 1970, however, something new was afoot. The interlocked, mutually supporting (and coevolving) social, ideological, political, economic, and technological systems that we conveniently call industrial society spawned movements that cast doubt on the propriety and prudence of business as usual. Some of these movements demanded the antithesis of industrial society, denouncing technology, wealth, and large-scale organization. Others called for yet more and better technology and organization, and more wealth for those who had least, as solutions to environmental problems. To date these new movements exercise only modest influence over the course of events, but they are still young. When Zhou Enlai, longtime foreign minister of Mao's China and a very worldly man, was asked about the significance of the French Revolution some 180 years after the event, he replied that it was still too early to tell. So it is, after only 35 years, with modern environmentalism.

Environmental change of the scale, intensity and variety witnessed in the twentieth century required multiple, mutually reinforcing causes. The most important immediate cause was the enormous surge of economic activity. Behind that lay the long booms in energy use and population. The reasons economic growth had the environmental implications that it had lay in the technological, ideological, and political histories of the twentieth century. All these histories (and some more that I have omitted) mutually affected one another; they also determined, and in some measure were themselves determined by, environmental history.

Few people paused to contemplate these complexities. In the struggles for survival and power, in the hurly-burly of getting and spending, few citizens and fewer rulers spared a thought for the ecological impacts of their behavior or ideas. Even after 1970, when environmental awareness had hurriedly dawned, easy fables of good and evil dominated public and political discourse. In this context, environmental outcomes continued, as before, to derive primarily from unintended consequences. Many specific outcomes were in a sense accidental. But the general trend of increasing human impact and influence, in the myriad ways described in this book, was no accident. It was, while unintended, strongly determined by the trajectories of human history. So what?

1 2

Epilogue: So What?

It happens then as it does to physicians in the treatment of consumption, which in the commencement is easy to cure and difficult to understand; but when it has neither been discovered in
time nor treated upon a proper principle, it becomes easy to understand and difficult to cure. The same thing happens in state
affairs; by foreseeing them at a distance, which is only done by
men of talents, the evils which might arise from them are soon
cured; but when, from want of foresight, they are suffered to increase to such a height that they are perceptible to everyone,
there is no longer any remedy.

—Machiavelli, *The Prince* (1513)

In 1930 the American physicist and Nobel Prize winner Robert Millikan
(1868–1953) said that there was no risk that humanity could do real harm
to anything so gigantic as the earth.[1] In the same year the American
chemical engineer Thomas Midgley invented chlorofluorocarbons, the
chemicals responsible for thinning the stratospheric ozone layer. Mil-

---

[1]Hobsbawm 1994:534.

likan, although certainly a man "of talents," did not understand what was brewing. What Machiavelli said of affairs of state is doubly true of affairs of global ecology and society. It is nearly impossible to see what is happening until it is inconveniently late to do much about it.

It is impossible to know whether humankind has entered a genuine ecological crisis. It is clear enough that our current ways are ecologically unsustainable, but we cannot know for how long we may yet sustain them, or what might happen if we do. In any case, human history since the dawn of agriculture is replete with unsustainable societies, some of which vanished but many of which changed their ways and survived. They changed not to sustainability but to some new and different kind of unsustainability. Perhaps we can, as it were, pile one unsustainable regime upon another indefinitely, making adjustments large and small but avoiding collapse, as China did during its "3,000 years of unsustainable development."[2] Imperial China, for all its apparent conservatism, was more rat than shark, adopting new food crops, new technologies, shifting its trade relations with its neighbors, constantly adapting—and surviving several crises. However, unsustainable society on the global scale may be another matter entirely, and what China did for millennia the whole world perhaps cannot do for long. If so, then collapse looms, as prophets of the ecological apocalypse so often warn. Perhaps the transition from our current unsustainable regime to another would be horribly wrenching and a fate to be avoided—or at least delayed—at all costs, as beneficiaries of the status quo so regularly claim. We cannot know for sure, and by the time we do know, it will be far too late to do much about it.

The future, even the fairly near future, is not merely unknowable; it is inherently uncertain. Some scenarios are more likely than others, no doubt, but nothing is fixed. Indeed, the future is more volatile than ever before: a greater number of radically different possibilities exist because technology has grown so influential, because ideas spread so rapidly, and because reproductive behavior—usually a variable that changes at a glacial pace—is in rapid flux. Moreover, all these variables are probably more tightly interactive than at most times in the past, so the total system of global society and environment is more uncertain, more chaotic, than ever.

[2]Elvin 1993.

All that said, the probability is that sharp adjustments will be required to avoid straitened circumstances. Many of the ecological buffers—open land, unused water, unpolluted spaces—that helped societies weather difficult times in the past are now gone. The most difficult passages will probably (or better put, least improbably) involve shortage of clean fresh water, the myriad effects of warmer climate, and of reduced biodiversity.

While one cannot say with any confidence what forms an ecological crunch might take, when it might happen, or how severe it might be, it is easier to predict who will have the worst of it. The poor and powerless cannot shield themselves from ecological problems today, nor will they be able to in the future. The wealthy and powerful in the past have normally had the wherewithal to insulate themselves from the effects of pollution, soil erosion, or fishery collapse. Only in a very severe crunch are they likely to face heavy costs. Of course, just who is rich and who is poor is subject to change: consider South Koreans, who in 1960 were on average poorer than Ghanaians, but by the 1990s were among the world's richer populations. This very fact inspires great efforts, individual and collective, to escape poverty and weakness, which efforts often aggravate ecological problems. South Koreans after 1960 paid dearly for their economic miracle, suffering from noxious urban air, toxic industrial effluent in their rivers, and many other disagreeable conditions. But now they are in a much better position than Ghanaians to weather serious ecological dislocations—because they are much richer.

If one accepts the notion that there is a significant chance that more serious ecological problems lie ahead, then, bearing Machiavelli in mind, it is prudent to address the prospect sooner rather than later. My interpretation of modern history suggests that the most sensible things to do are to hasten the arrival of a new, cleaner energy regime and to hasten the demographic transition toward lower mortality and fertility. The former implies concentrated scientific and engineering work, and probably interventions in the marketplace that encourage early retirement of existing energy infrastructures and faster diffusion of new ones. The latter implies furthering the formal education of girls and women in poor countries, because poor countries are where the demographic transition is incomplete, and because female education is the strongest determinant of fertility. There may be other desirable initiatives, such as converting the masses to some new creed of ecological restraint or coaxing rulers into considering time horizons longer than the next election or coup.[3] These are more difficult and less practical, precisely because they are more fundamental.

The reason I expect formidable ecological and societal problems in the future is because of what I see in the past. In this book I have tried to give some measure to the ecological changes experienced in the twentieth century. Only some of them are easily reducible to numbers. They are re-capitulated in Table 12.1, which is intended as a summary measure of en-

TABLE 12.1 THE MEASURE OF THE TWENTIETH CENTURY

Item	Increase Factor, 1890s–1990s
World population	4
Urban proportion of world population	3
Total world urban population	13
World economy	14
Industrial output	40
Energy use	13
Coal production	7
Air pollution	≈5
Carbon dioxide emissions	17
Sulfur dioxide emissions	13
Lead emissions to the atmosphere	≈8
Water use	9
Marine fish catch	35
Cattle population	4
Pig population	9
Horse population	1.1

[3]McGovern 1994 offers an edifying story. Norse Greenland became an extinct society in the 15th century. The Norse settled some enclaves on the coasts of Greenland during the 10th–12th centuries, unusually warm centuries in the Northern Hemisphere. They failed to adapt to a cooling trend, the onset of the Little Ice Age. They remained attached to an economy based on cattle when it would have been prudent to emulate the Inuit and focus on seals

Blue whale population (Southern Ocean only)	0.0025 (99.75% decrease)
Fin whale population	0.03 (97% decrease)
Bird and mammal species	0.99 (1% decrease)
Irrigated area	5
Forest area	0.8 (20% decrease)
Cropland	2

Note: Some of the numbers are more trustworthy than others. Comments on their reliability appear in the text, from which they are all drawn.

vironmental change and some of the factors producing it in the twentieth century. This table ignores expansions that took place entirely after 1900, such as CFC releases or tractor numbers. It ignores expansions for which the baseline of 1900 would produce astronomical coefficients of increase, for example, the world's total of automobiles, the quantities of chemical fertilizer applied, or the tonnage of synthetic chemicals produced. The table is an imperfect measure, but it gives the right impression.

According to the Hebrew Bible, on the fifth day of creation God enjoined humankind to fill and subdue the earth and to have dominion over every living thing. For most of history, our species failed to live up to these (as to so many other) injunctions, not for want of trying so much as for want of power. But in the twentieth century the harnessing of fossil fuels, unprecedented population growth, and a myriad of technological changes made it more nearly possible to fulfill these instructions. The prevailing political and economic systems made it seem imprudent not to try: most societies, and all the big ones, sought to maximize their current formidability and wealth at the risk of sacrificing ecological buffers and tomorrow's resilience. The general policy of the twentieth century was to try to make the most of resources, make Nature perform to the utmost, and hope for the best.

---

and fish. The main reason they did not change their ways, says McGovern, is that Norse Greenland was run by a landed elite that controlled production and trade and saw its position, in particular its privileged access to trade goods from Scandinavia, as dependent upon the perpetuation of an economy based on land and cattle. The tenacious power of the landed elite precluded any effective adjustment to changing ecological circumstances. The lords of Norse Greenland were, in the language of my Preface, committed sharks.

With our new powers we banished some historical constraints on health and population, food production, energy use, and consumption generally. Few who know anything about life with these constraints regret their passing. But in banishing them we invited other constraints in the form of the planet's capacity to absorb the wastes, by-products, and impacts of our actions. These latter constraints had pinched occasionally in the past, but only locally. By the end of the twentieth century they seemed to restrict our options globally. Our negotiations with these constraints will shape the future as our struggles against them shaped our past.

Those responsible for policy tend to take as their frame of reference the world as we know it. This invites them to think of things as they observe and experience them—the regime of perpetual disturbance as I called it in the Preface—as "normal." In fact, in ecological terms, the current situation is an extreme deviation from any of the durable, more "normal," states of the world over the span of human history, indeed over the span of earth history. If we lived 700 or 7,000 years, we would understand this on the basis of experience and memory alone. But for creatures who live a mere 70 years or so, the study of the past, distant and recent, is required to know what the range of possibilities includes, and to know what might endure.

The enormity of ecological change in the twentieth century strongly suggests that history and ecology, at least in modern times, must take one another properly into account. Modern history written as if the life-support systems of the planet were stable, present only in the background of human affairs, is not only incomplete but is misleading. Ecology that neglects the complexity of social forces and dynamics of historical change is equally limited. Both history and ecology are, as fields of knowledge go, supremely integrative. They merely need to integrate with one another.

If and when they do, we will have a better idea of our past, more complete, more compelling, more comprehensible, if perhaps more complicated. We will have a better idea of our present situation, and whether or not it qualifies as a predicament. With these, we will have a better idea of our possible futures. And, with that, we will be better placed to debate and choose among them, and at the very least to avoid the most unpleasant ones. We might then consciously choose a world that would require only irksome adaptations on our part and avoid traumatic ones. We could make our own luck instead of merely trusting to luck. That would distinguish us from rats and sharks—as well as those cyanobacteria of 2 billion years ago.

Bibliography

Aamlid, Dan. 1990. *Forest Decline in Norway* (As: Norwegian Agricultural Advisory Centre).

Aarsten, J. P. von. 1946. "Consequences of the War on Agriculture in the Netherlands," *International Review of Agriculture* 37:55–345, 495–705, and 1085–1235.

Aarts, Wilma, Johan Goudsblom, Kees Schmidt, and Fred Spier. 1995. *Toward a Morality of Moderation* (Amsterdam: Amsterdam School for Social Science Research).

Abdoellah, Oekan S. 1996. "Social and Environmental Impacts of Transmigration: A Case-Study in Barambai, South Kalimantan." In: Christine Padoch and Nancy Lee Peloso, eds., *Borneo in Transition* (Kuala Lumpur: Oxford University Press), 266–79.

Abreu, Mauricio de Almeida. 1988. *Evolução urbana do Rio de Janeiro* (Rio de Janeiro: Instituto de Planejamento Municipal).

Adabashev, I. I. 1966. *Global Engineering* (Moscow: Progress Publishers).

Adam, Paul. 1987. "Les nouvelles pêches maritimes mondiales," *Etudes internationales* 18:7–20.

Adams, Robert M. 1995. *Paths of Fire: An Anthropologist's Inquiry into Western Technology* (Princeton: Princeton University Press).

Adams, W. M. 1992. *Wasting the Rain: Rivers, People and Planning in Africa* (London: Earthscan).

Adler, Robert, Jessica Landman, and Diane Cameron. 1993. *The Clean Water Act 20 Years Later* (Washington, D.C.: Island Press).

Agnihotri, Indu. 1996. "Ecology, Land Use and Colonisation: The Canal Colonies of Punjab," *Indian Economic and Social History Review* 33:37–58.

Ahmed, Sara. 1990. "Cleaning the River Ganga: Rhetoric and Reality," *Ambio* 19:42–5.

Akio, Mishima. 1992. *Bitter Sea: The Human Cost of Minamata Disease* (Tokyo: Kosei).

Alauddin, Mohammad, and Clement Tisdell. 1991. *The Green Revolution and Economic Development: The Process and Its Impact in Bangladesh* (New York: St. Martin's Press).

Albaigues, Joan, M. Aubert, and J. Aubert. 1985. "Footprints of Life and Man." In: Ramón Margalef, ed., *Western Mediterranean* (Oxford: Pergamon Press), 317–52.

Alderton, D. H. M. 1985. "Sediments." In: *Historical Monitoring: A Technical Report* (London: Monitoring and Assessment Research Centre).

Alexander, David. 1993. *Natural Disasters* (New York: Chapman & Hall).

Ali, Imran. 1988. *The Punjab Under Imperialism, 1885–1947* (Princeton: Princeton University Press).

Allan, J. A. 1994. "Overall Perspectives on Countries and Regions." In: Peter Rogers and Peter Lydon, eds. *Water in the Arab World: Perspectives and Prognoses* (Cambridge: Harvard University Press), 65–100.

Allan, N. J. R. 1987. "Impact of Afghan Refugees on the Vegetation Resources of Pakistan's Hindu Kush–Himalaya," *Mountain Research and Development* 7:200–4.

Amanor, Kojo S. 1994. *The New Frontier: Farmer Responses to Land Degradation* (Geneva: U.N. Research Institute for Social Development).

Amat, Jean-Paul. 1993. "Le rôle stratégique de la forêt, 1871–1914: Exemples dans les forêts lorraines," *Revue historique des armées* 1:62–9.

Anagnostakis, S. L., and B. Hillman. 1992. "Evolution of the Chestnut Tree and Its Blight," *Arnoldia* 52:3–10.

Anderson, C. H. 1975. *A History of Soil Erosion by Wind in the Palliser Triangle of Western Canada* (Ottawa: Canada Department of Agriculture, Research Branch).

Anderson, David. 1984. "Depression, Dust Bowl, Demography and Drought: The Colonial State and Soil Conservation in East Africa During the 1930s," *African Affairs* 83:321–43.

Anderson, M. R. 1995. "The Conquest of Smoke: Legislation and Pollution in Colonial Calcutta." In: D. Arnold and R. Guha, eds., *Nature, Culture, Imperialism* (Delhi: Oxford University Press), 293–335.

Anderson, Robert S. 1991. "The Origins of the International Rice Research Institute," *Minerva* 29:61–89.

Andreev-Khomiakov, Gennady. 1997. *Bitter Waters*, Ann E. Healy, trans. (Boulder: Westview Press).

Antoine, Serge. 1993. "18 Pays riverains dans un même bateau," *Peuples méditerranéens* 62–63:255–78.

Argabright, M. Scott, et al. 1996. *Historical Changes in Soil Erosion, 1930–1992: The Northern Mississippi Valley Loess Hills* (Washington: USDA Natural Resources Conservation Service, Historical Notes no. 5).

Arnold, R. W., I. Szabolcs, and V. O. Targulian. 1990. *Global Soil Change* (Laxenburg: International Institute for Applied Systems Analysis).

Asami, T. 1983. "Pollution of Soils by Cadmium." In: J. O. Nriagu, ed., *Changing Metal Cycles and Human Health* (Berlin: Springer-Verlag), 95–111.

———. 1988. "Soil Pollution by Metals from Mining and Smelting Activities." In: W. Salomons and U. Föstner, eds., *Chemistry and Biology of Solid Waste: Dredged Material and Mine Tailings* (Berlin: Springer-Verlag), 143–69.

Ashton, Peter J., and David S. Mitchell. 1989. "Aquatic Plants: Patterns and Modes of Invasion, Attributes of Invading Species and Assessment of Control Programmes." In: J. A. Drake et al., eds., *Biological Invasions: A Global Perspective* (Chichester: Wiley), 111–54.

Ashton, P. J., C. C. Appleton, and P. B. N. Jackson. 1985. "Ecological Impacts and Economic Consequences of Alien Invasive Organisms in Southern African Aquatic Ecosystems." In:

I. A. W. McDonald, F. J. Kruger, and A. A. Ferrar, eds., *The Ecology and Management of Biological Invasions in Southern Africa* (Cape Town: Oxford University Press), 247–57.

Asquith, Pamela, and Arne Kalland. 1997. *Japanese Images of Nature* (Richmond: Curzon Press).

Augelli, John. 1989. "Modernization of Costa Rica's Beef Cattle Economy: 1950–85," *Journal of Cultural Geography* 9:77–90.

Ausubel, Jesse. 1989. "Regularities in Technological Development: An Environmental View." In: Jesse Ausubel and Hedy Sladovich, eds., *Technology and Environment* (Washington: National Academy Press), 70–91.

———. 1996. "The Liberation of the Environment," *Daedalus* 125(3):1–17.

Ausubel, Jesse, and R. Herman, eds. 1988. *Cities and Their Vital Systems Infrastructure: Past, Present, and Future* (Washington: National Academy Press).

Ausubel, Jesse, and Hedy Sladovich, eds. 1989. *Technology and Environment* (Washington: National Academy Press).

Ayeb, Habib. 1996. "Le haut barrage 30 ans après," *Peuples méditerranéens* 74–75:131–46.

Ayers, G. P., and K. K. Yeung. 1996. "Acid Deposition in Hong Kong," *Atmospheric Environment* 30:1581–88.

Ayres, R. U. 1989. "Industrial Metabolism." In: Jesse Ausubel and Hedy Sladovich, eds., *Technology and Environment* (Washington: National Academy Press), 23–49.

Azhar, Said. 1985. "Three Million Uprooted Afghans in Pakistan," *Pakistan Horizon* 38:60–84.

Bahre, Conrad. 1979. *Destruction of the Vegetation of North-Central Chile* (Berkeley: University of California Press).

Baines, Dudley. 1995. *Emigration from Europe, 1815–1930* (Cambridge: Cambridge University Press).

Bairoch, Paul. 1989. "Les trois révolutions agricoles du monde développé: rendements et productivité de 1800 à 1985," *Annales: Economies, sociétés, civilisations* 44(2):317–53.

Barica, J. 1979. "Massive Fish Mortalities Caused by Algal Blooms in Eutrophic Ecosystems," *Symposia Biologica Hungarica,* 19:121–4.

Barker, David, and Duncan MacGregor. 1988. "Land Degradation in the Yallahs Basin, Jamaica: Historical Notes and Contemporary Observations," *Geography* 73:116–24.

Barker, Graeme. 1995. *A Mediterranean Valley: Landscape Archaeology and Annales History in the Biferno Valley* (London and New York: Leicester University Press).

Barrett, Thomas. 1997. "The Terek Cossacks and the North Caucasus Frontier, 1700–1860" (Ph.D. thesis, Georgetown University).

Barrie, L. A. 1986. "Arctic Air Pollution: An Overview of Current Knowledge," *Atmospheric Environment* 20:643–63.

Barrie, L. A., P. Fisher, and R. M. Koerner. 1985. "Twentieth-Century Trends in Arctic Air Pollution Revealed by the Conductivity and Acidity Observations in Snow and Ice in the Canadian High Arctic," *Atmospheric Environment* 19:2055–63.

Bart, François. 1993. *Montagnes d'Afrique, terres paysannes: Le cas du Rwanda* (Bordeaux: Presses Universitaires de Bordeaux).

Bartz, Fritz. 1964. *Die grossen Fischereiräume der Welt* (Weisbaden: Franz Steiner Verlag).

Basu, A. K. 1992. *Ecological and Resource Study of the Ganga Delta* (Calcutta: Bagchi).

Baviskar, Amita. 1995. *In the Belly of the River: Tribal Conflicts over Development in the Narmada Valley* (Delhi: Oxford University Press).

Bayalama, Sylvain. 1992. "The Environment and Structural Adjustment Programs in Africa," *TransAfrica Forum* 9(3):89–99.

Beach, Timothy. 1994. "The Fate of Eroded Soil: Sediment Sinks and Sediment Budgets of Agrarian Landscapes in Southern Minnesota, 1851–1988," *Association of American Geographers, Annals* 84:5–28.

Behre, K.-E., J. Dorjes, and G. Irion. 1985. "A Dated Holocene Sediment Core from the Bottom of the Southern North Sea," *Eiszeitsalter und Gegenwart* 35:9013.

Beinart, William. 1984. "Soil Erosion, Conservation and Ideas about Development: A Southern African Exploration, 1900–1960," *Journal of Southern African Studies* 11:52–83.

Ben Tuvia, Adam. 1983. "The Mediterranean Sea: Biological Aspects." In: Bostwick Ketchum, ed., *Estuaries and Enclosed Seas* (Amsterdam: Elsevier), 239–51.

Berkes, Fikret. 1992. "Success and Failure in Marine Coastal Fisheries of Turkey." In: Daniel W. Bromley, ed., *Making the Commons Work* (San Francisco: Institute for Contemporary Studies), 161–82.

Berkhout, Frans. 1994. "Nuclear Power: An Industrial Ecology That Failed?" In: R. Socolow, C. Andrews, F. Berkhout, and V. Thomas, eds., *Industrial Ecology and Global Change* (Cambridge: Cambridge University Press), 319–27.

Berry, Brian. 1990. "Urbanization." In: B. L. Turner et al., eds., *The Earth as Transformed by Human Action* (New York: Cambridge University Press), 103–19.

Berz, Gerhard. 1990. "Global Warming and the Insurance Industry," *Nature and Resources (English Edition)* 27(1):19–30.

Bethoux, J. P., P. Gentili, J. Raunet, and D. Taillez. 1990. "Warming Trend in the Western Mediterranean Deep Water," *Nature* 347:660–2.

Beveridge, W. I. B. 1993. "Unravelling the Ecology of the Influenza A Virus," *History and Philosophy of the Life Sciences* 15:23–32.

Bevilacqua, Piero. 1989. "Le rivoluzioni dell'acqua: Irrigazione e trasformazioni dell'agricoltura tra Sette e Novecento." In: P. Bevilacqua, ed., *Storia dell'agricoltura italiana in età contemporanea* (Venice: Marsilio), 255–318.

Bezuneh, Mesfin, and Carl C. Mabbs-Zeno. 1984. "The Contribution of the Green Revolution to Social Change in Ethiopia," *Northeast African Studies* 6(3):9–17.

Bianchi, Bruna. 1989. "La nuova pianura: Il paesaggio delle terre bonificate in area padana." In: P. Bevilacqua, ed., *Storia dell'agricoltura italiana in età contemporanea* (Venice: Marsilio), 451–94.

Biraben, J. N. 1979. "Essai sur l'évolution du nombre des hommes," *Population* 34:13–24.

Biswas, Margaret R. 1994. "Agriculture and Environment: A Review, 1972–1992," *Ambio* 23:192–7.

Blaikie, Piers. 1985. *Political Economy of Soil Erosion in Developing Countries* (London: Longman).

Blaxter, Kenneth, and Noel Robertson. 1995. *From Dearth to Plenty: The Modern Revolution in Food Production* (Cambridge: Cambridge University Press).

Bloom, B. R., and C. J. L. Murray. 1992. "Tuberculosis: Commentary on a Reemergent Killer," *Science* 257:1055–64.

Bobak, Martin, and Richard G. A. Feachem. 1995. "Air Pollution and Mortality in Central and Eastern Europe," *European Journal of Public Health* 5:82–6.

Bohm, Georgy, et al. 1989. "Biological Effects of Air Pollution in Sao Paulo and Cubatao," *Environmental Research* 49(2):208–16.

Bonnin, J. R. 1987. "Aménagements hydrauliques avant notre ère." In: Walter O. Wunderlich and J. Egbert Prins, eds., *Water for the Future* (Rotterdam: Balkema), 101–12.

Bonomi, G., A. Calderoni, and R. Mosello. 1979. "Some Remarks on the Recent Evolution of the Deep Italian Subalpine Lakes," *Symposia Biologica Hungarica* 19:87–112.

Boulding, Kenneth E. 1964. *The Meaning of the Twentieth Century* (New York: Harper Colophon).

Bourgeois, Jean, Anton Ervynck, Paul Rondelez, and Michel Gilté. 1996. "De vuilnisblet vertelt Archeologisch onderzoek van modern Gents huishoudelijk afval," *Tijdschridt voor ecologische geschiedenis* 1:46–51.

Bourgeois-Pichat, J. 1988. "Du XXe au XXIe siècle: Europe et sa population après l'an 2000," *Population* 43:9–42.

Boyden, Stephen, Sheelagh Miller, Ken Newcombe, and Beverly O'Neill. 1981. *The Ecology of a City and Its People: The Case of Hong Kong* (Canberra: Australian National University Press).

Bramwell, Anna. 1989. *Ecology in the 20th Century: A History* (New Haven: Yale University Press).

Breburda, Josef. 1990. "Land-Use Zones and Soil Degradation in the Soviet Union." In: Karl-Eugen Wädekin, ed., *Communist Agriculture: Farming in the Soviet Union and Eastern Europe* (London: Routledge), 23–39.

Bresnan, John. 1993. *Managing Indonesia: The Modern Political Economy* (New York: Columbia University Press).

Bridges, Olga, and Jim Bridges. 1996. *Losing Hope: The Environment and Health in Russia* (Aldershot, U.K.: Avebury).

Brimblecombe, Peter. 1995. "History of Air Pollution." In: H. B. Singh, ed., *Composition, Chemistry and Climate of the Atmosphere* (New York: Van Nostrand Reinhold), 1–18.

———. 1987. *The Big Smoke: A History of Air Pollution in London Since Medieval Times* (London: Methuen).

Brimblecombe, Peter, and C. Bowler. 1992. "History of Air Pollution in York, England," *Journal of the Air and Waste Management Association* 42:1562–6.

Brimblecombe, Peter, Trevor Davies, and Martyn Tranter. 1986. "Nineteenth Century Black Scottish Showers," *Atmospheric Environment* 20:1053–57.

Brinkhoff, Thomas. 1999. "Principal Agglomerations and Cities of the World." Web site address: *http://www.citiesin.thecountry.com*.

Brinkmann, Uwe. 1994. "Economic Development and Tropical Disease." In: Mary E. Wilson, Richard Levins, and Andrew Spielman, eds., *Disease in Evolution* (New York: Annals of the New York Academy of Sciences), 740:303–11.

Broadbent, Jeffrey. 1998. *Environmental Politics in Japan: Networks of Power and Protest* (Cambridge: Cambridge University Press).

Brookfield, Harold, Francis Jana Lian, Low Kwai-Sim, and Lesley Potter. 1990. "Borneo and the Malay Peninsula." In: B. L. Turner et al. eds., *The Earth as Transformed by Human Action* (New York: Cambridge University Press), 495–512.

Brouwer, Roland. 1995. "Planting Power: The Afforestation of the Commons and State Formation in Portugal" (Ph.D. thesis, University of Wageningen, Netherlands).

Brown, H. S., R. E. Kasperson, and S. S. Raymond. 1990. "Trace Pollutants." In: B. L. Turner et al. eds., *The Earth as Transformed by Human Action* (New York: Cambridge University Press), 437–54.

Brown, Lester R., et al. 1996. *The State of the World 1996* (New York: Norton).

Brüggemeir, Franz-Josef. 1990. "The Ruhr Basin 1850–1980: A Case of Large-Scale Environmental Pollution." In: Peter Brimblecombe and Christian Pfister, eds., *The Silent Countdown: Essays in European Environmental History* (Berlin: Springer-Verlag), 210–27.

———. 1994. "A Nature Fit for Industry: The Environmental History of the Ruhr Basin, 1840–1990," *Environmental History Review* 18(1):35–54.

Brüggemeir, Franz-Josef, and Thomas Rommelspacher. 1992. *Blauer Himmel über der Ruhr: Geschichte der Umwelt in Ruhrgebiet, 1840–1990* (Essen: Klartext).

Bruun, Ole, and Arne Kalland, eds. 1995. *Asian Perceptions of Nature: A Critical Approach* (Richmond: Curzon Press).

Bryder, Linda. 1988. *Below the Magic Mountain: A Social History of Tuberculosis in Twentieth-Century Britain* (Oxford: Clarendon Press).

Bührs, Tom, and Robert V. Bartlett. 1994. *Environmental Policy in New Zealand* (Oxford: Oxford University Press).

Buller, Henry J. 1992. "Agricultural Change and the Environment in Western Europe." In: Keith Hoggart, ed., *Agricultural Change, Environment and Economy* (London: Mansell), 68–88.

Burger, Joanna. 1997. *Oil Spills* (New Brunswick: Rutgers University Press).

Burmeister, Larry. 1990. "State, Industrialization and Agricultural Policy in Korea," *Development and Change* 21:197–220.

Burney, David. 1996. "Historical Perspectives on Human-Assisted Biological Invasions," *Evolutionary Anthropology* 4(6):216–21.

Burns, G., et al. 1990. "Salinity Threat to Upper Egypt," *Nature* 344:25.

Burrows, Geoff, and Ralph Shlomowitz. 1992. "The Lag in the Mechanization of the Sugarcane Harvest: Some Comparative Perspectives," *Agricultural History* 66(3):61–75.

Busch, Briton C. 1985. *The War against the Seals: A History of the North American Seal Fishery* (Kingston and Montreal: McGill–Queen's University Press).

Busch, Lawrence, et al. 1995. *Making Nature, Shaping Culture: Plant Biodiversity in Global Context* (Lincoln: University of Nebraska Press).

Büschenfeld, Jürgen. 1997. *Flüsse und Kloaken: Umweltfragen im Zeitalter der Industrialisierung (1870–1914)* (Stuttgart: Klett-Cotta).

Butzer, Karl. 1975. "Accelerated Soil Erosion: A Problem in Man-Land Relationships." In: Ian R. Manners and Marvin Mikesell, eds., *Perspectives on Environment* (Washington: Association of American Geographers), 57–78.

Buxton, Ian. 1993. "The Development of the Merchant Ship, 1880–1990," *Mariner's Mirror* 79:71–82.

Caddy, J. F. 1993. "Toward a Comparative Evaluation of Human Impacts on Fishery Ecosystems of Enclosed and Semi-Enclosed Seas," *Reviews in Fisheries Science* 1:57–95.

Caddy, J. F. and J. A. Gulland. 1983. "Historical Patterns of Fish Stocks." *Marine Policy* 7:267–78.

Campbell, John. 1991. "Land or Peasants? The Dilemma Confronting Ethiopian Resource Conservation," *African Affairs* 90:5–21.

Carlton, J. T. 1985. "Transoceanic and Interoceanic Dispersal of Coastal Marine Organisms: The Biology of Ballast Water," *Oceanography and Marine Biology: An Annual Review* 23:313–71.

———. 1989. "Man's Role in Changing the Face of the Ocean: Biological Invasions and Implications for Conservation of Near-Shore Environments," *Conservation Biology* 3:265–73.

———. 1993. "Neoextinctions of Marine Invertebrates," *American Zoologist* 33:499–509.

———. 1996. "Marine Bioinvasions: The Alteration of Marine Ecosystems by Nonindigenous Species," *Oceanography* 9:36–43.

Carlton, J. T., and J. B. Geller. 1993. "Ecological Roulette: The Global Transport of Non-indigenous Marine Organisms," *Science* 261:78–82.

Carré, François. 1982. "Les pêches maritimes dans l'Atlantique du nord-est," *Annales de géographie* 91:173–204.

Cartalis, C., and C. Varotsos. 1994. "Surface Ozone in Athens, Greece, at the Beginning and End of the Twentieth Century," *Atmospheric Environment* 28:3–8.

Carter, F. W. 1993a. "Czechoslovakia." In: F. W. Carter and D. Turnock, eds., *Environmental Problems in Eastern Europe* (London: Routledge), 63–88.

————. 1993b. "Poland." In: F. W. Carter and D. Turnock, eds., *Environmental Problems in Eastern Europe* (London: Routledge), 107–34.

Carter, F. W., and D. Turnock, eds. 1993. *Environmental Problems in Eastern Europe* (London: Routledge).

Caviedes, César N., and Timothy J. Fik. 1992. "The Peru-Chilean Eastern Pacific Fisheries and Climatic Oscillation." In: M. H. Glantz, ed., *Climate Variability, Climate Change, and Fisheries* (Cambridge: Cambridge University Press), 355–76.

Cecchini, Marcella. 1987. "Due missioni tecniche italiane in URSS, 1930–36," *Storia contemporanea* 18:731–65.

Çeçen, Kâzım. 1992. *Sinan's Water Supply System in Istanbul* (Istanbul: T. C. Istanbul Büyük Şehir Belediyesi, Istanbul Su ve Kanalızasyon Idaresi Genel Müdürlüğü).

Celi, M. 1991. "Biospeleologia," *Barbastrijo* 3:1–18.

Centre for Science and Environment. 1982. *The State of India's Environment 1982: A Citizens' Report* (New Delhi: CSE).

Changnon, Stanley, ed. 1994. "The Lake Michigan Diversion at Chicago and Urban Drought" (Final Report to Great Lakes Environmental Research Laboratory, Ann Arbor, Mich.; National Oceanic and Atmospheric Administration, contract no. 50WCNR306047).

Changnon, Stanley, and Joyce Changnon. 1996. "History of the Chicago Diversion and Future Implications," *Journal of Great Lakes Research* 22:100–18.

Chaussade, Jean, and Jean-Pierre Corlay. 1990. *Atlas des pêches et des cultures marines: France, Europe, monde* (Rennes: Editions Ouest France).

Chen, Lincoln C. 1994. "New Diseases—The Human Factor: Commentary." In: Mary E. Wilson, Richard Levins, and Andrew Spielman, eds., *Disease in Evolution* (New York: Annals of the New York Academy of Sciences, vol. 740), 319–324.

Chesnais, Jean-Claude. 1995. *Le crépuscule de l'Occident* (Paris: Laffont).

Cho, Yong Hyu, and Byung Dae Choi. 1995. "The Threat of Polluted Air and the Policies to Control the Problem in the Seoul Metropolitan Area," *International Journal of Public Administration* 18:1725–39.

Chomsky, Aviva. 1996. *West Indian Workers and the United Fruit Company in Costa Rica, 1870–1940* (Baton Rouge: Louisiana State University Press).

Cioc, Mark. 1998. "The Impact of the Coal Age on the German Environment," *Environment and History* 4:105–24.

Cipolla, Carlo. 1978. *The Economic History of World Population* (Harmondsworth: Penguin Books).

Clapp, B. W. 1994. *An Environmental History of Britain* (London: Longman).

Clapp, J. 1994. "The Toxic Waste Trade with Less Industrialised Countries: Economic Linkages and Political Alliances," *Third World Quarterly* 15:505–18.

Clark, David. 1998. "Interdependent Urbanization in an Urban World: An Historical Overview," *Geographical Journal* (U.K.) 164:85–95.

Clark, Martin. 1984. *Modern Italy, 1871–1982* (New York: Longman).

Clarke, John I. 1996. "The Impact of Population on Environment: An Overview." In: Bernardo

Colombo, Paul Demeny, and Max Perutz, eds., *Resources and Population: Natural, Institutional, and Demographic Dimensions of Development* (Oxford: Clarendon Press), 254–68.

Cliff, Andrew, Peter Haggett, and Matthew Smallman-Raynor. 1998. *Deciphering Global Epidemics: Analytical Approaches to the Disease Records of World Cities, 1888–1912* (Cambridge: Cambridge University Press).

Clout, Hugh. 1996. *After the Ruins: Restoring the Countryside of Northern France after the Great War* (Exeter: Exeter University Press).

Cochrane, Thomas B., R. S. Norris, and K. L. Suokko. 1993. "Radioactive Contamination at Chelyabinsk-65, Russia," *Annual Review of Energy and the Environment* 18:507–28.

Cochrane, Willard. 1993. *The Development of American Agriculture: An Historical Perspective* (Minneapolis: University of Minnesota Press).

Cohen, Joel. 1995. *How Many People Can the Earth Support?* (New York: Norton).

Cohen, Mitchell L. 1992. "Epidemiology of Drug Resistance: Implications for a Post-Antimicrobial Era," *Science* 257:1050–55.

Collins, David N. 1991. "Kabinet, Forest, and Revolution in the Siberian Altai to May 1918," *Revolutionary Russia* 4:1–27.

Collins, Robert O. 1990. *The Waters of the Nile* (Oxford: Clarendon Press).

Colten, Craig. 1986. "Industrial Wastes in Southeast Chicago: Production and Disposal, 1870–1970," *Environmental Review* 10:93–105.

———. 1994. "Creating a Toxic Landscape: Chemical Waste Disposal Policy and Practice, 1900–1960," *Environmental History Review* 18:85–116.

Conacher, Brian. 1990. "Salt of the Earth: Secondary Soil Salinization in the Australian Wheat Belt," *Environment* 32(6):4–9, 40–42.

Conniff, Richard. 1990. "You Never Know What the Fire Ant Is Going to Do Next," *Smithsonian* 21(4):48–57.

Conzen, Kathleen Neils. 1990. "Immigrants in Nineteenth-Century Agricultural History." In: Lou Fergeler, ed., *Agriculture and National Development: Views on the Nineteenth Century* (Ames: Iowa State University Press), 303–42.

Cooper, R. C. 1992. "Transboundary Pollution: Sulfur Dioxide Emissions in the Republics of the USSR," *Comparative Economic Studies* 34(2):38–49.

Cooter, Roger. 1993. "War and Modern Medicine." In: W. F. Bynum and Roy Porter, eds., *Companion Encyclopedia of the History of Medicine* (London: Routledge), 1536–73.

Cordell, D. V., J. M. Gregory, and V. Piché. 1996. *Hoe and Wage* (Boulder: Westview Press).

Corner, Paul. 1975. *Fascism in Ferrara, 1915–1925* (Oxford: Oxford University Press).

Corona, Gabriella, and Gino Massullo. 1989. "La terre e le techniche." In: Piero Bevilacqua, ed., *Storia dell'agricoltura italiana in età contemporanea* (Venice: Marsilio), 353–449.

Corvol, A. 1987. "Le discours pré-écologiste." *Revue d'Auvergne* 101:147–57.

Costanza, Robert. 1997. *Frontiers in Ecological Economics* (Cheltenham: Edward Elgar).

Coward, Don Huon. 1988. *Out of Sight: Sydney's Environmental History, 1851–1981* (Canberra: Australian National University Press).

Cox, R. M. 1982. "Smelter Emissions as an Environmental Hazard in Sudbury, Ontario." In: I. Burton, C. D. Fowle and R. S. McCullough, eds., *Living with Risks: Environmental Risk Management in Canada* (Ottawa: Environment Canada Environmental Monograph No. 3), 161–72.

Cox, Thomas R. 1988. "The North American-Japanese Timber Trade." In: John F. Richards and Richard P. Tucker, eds., *World Deforestation in the Twentieth Century* (Durham: Duke University Press), 164–88.

Cramer, Jacqueline. 1989. *De groene golf: Geschiedenis en toekomst van de Nederlandse milieubeweging* (Utrecht: Van Arkel).

Craswell, E. T. 1993. "The Management of World Resources for Sustainable Agricultural Development. In: David Pimentel, ed., *World Soil Erosion and Conservation* (Cambridge: Cambridge University Press), 257–76.

Cronon, William. 1991. *Nature's Metropolis: Chicago and the Great West* (New York: Norton).

Cruse, R. M., and S. C. Gupta. 1991. "Soil Compaction Mechanisms and Their Control." In: R. Lal and F. J. Pierce, eds., *Soil Management for Sustainability* (Ankeny, Iowa: Soil and Water Conservation Society), 19–24.

Curtin, Philip D. 1989. *Death by Migration* (New York: Cambridge University Press).

———. 1993. "Disease Exchange across the Tropical Atlantic," *History and Philosophy of the Life Sciences*, 15:329–56.

Dahlsten, D. L. 1989. "Control of Invaders." In: H. A. Mooney and J. A. Drake, eds., *Ecology of Biological Invasions of North America and Hawaii* (New York: Springer-Verlag), 275–302.

Dalrymple, Dana G. 1974. *Development and Spread of High-Yielding Varieties of Wheat and Rice in the Less Developed Nations* (Washington: USDA).

Dalton, Russell J. 1994. *The Green Rainbow* (New Haven: Yale University Press).

Daniel, Pete. 1977. *Deep'n as It Come: The 1927 Mississippi River Flood* (New York: Oxford University Press).

Daniels, R. B. 1987a. "Saline Seeps in the Northern Great Plains of the USA and Southern Prairies of Canada." In: M. G. Wolman and F. G. A. Fourier, eds., *Land Transformation in Agriculture* (Chichester: Wiley), 381–496.

———. 1987b. "Soil Erosion and Degradation in the Southern Piedmont of the USA." In: M. G. Wolman and F. G. A. Fourier, eds., *Land Transformation in Agriculture* (Chichester: Wiley), 407–28.

Danielsson, Bengt. 1990. "Poisoned Pacific: The Legacy of French Nuclear Testing," *Bulletin of the Atomic Scientists* 46(2):22–31.

Darmstadter, Joel, and Robert W. Fri. 1992. "Interconnections between Energy and Environment: Global Challenges," *Annual Review of Energy and the Environment* 17:45–76.

Dauvergne, Peter. 1997. *Shadows in the Forest: Japan and the Politics of Timber in Southeast Asia* (Cambridge: MIT Press).

David, Christina, and Keijiro Otsuka, ed., 1994. *Modern Rice Technology and Income Distribution in Asia* (Boulder: Lynne Rienner).

Davidson, Cliff I. 1979. "Air Pollution in Pittsburgh: A Historical Perspective," *Journal of the Air Pollution Control Association* 29:1035–41.

Dean, Warren. 1987. *Brazil and the Struggle for Rubber: A Study in Environmental History* (New York: Cambridge University Press).

———. 1995. *With Broadax and Firebrand: The Destruction of the Brazilian Atlantic Forest* (Berkeley: University of California Press).

Debeir, Jean-Claude, Jean-Paul Deléage and Daniel Hémery. 1986. *Servitudes de la puissance* (Paris: Flammarion).

De Bevoise, Ken. 1995. *Agents of Apocalypse: Epidemic Disease in the Colonial Philippines* (Princeton: Princeton University Press).

Dede, Ionna. 1993. *Okologiebewegung in Griechenland und in der Bundesrepublik Deutschland* (Frankfurt: Peter Lang).

de Gruijl, Frank R. 1995. "Impacts of a Projected Depletion of the Ozone Layer," *Consequences* (U.S.) 1(2):12–21.

de Jong, J. 1987. "Water and Land Management in the Netherlands: History, Present Day's Situation and Future." In: W. O. Wunderlich and J. E. Prins, eds., *Water for the Future* (Rotterdam: Balkema), 79–89.

de Jongh, Peter, and Sean Captain. 1999. *Our Common Journey: A Pioneering Approach to Cooperative Environmental Management* (London: Zed Books).

Deléage, J. P. 1992. *Histoire de l'écologie: une science de l'homme et de la nature* (Paris: La Découverte).

Demorlaine, J. 1919. "Importance stratégique des forêts dans la guerre," *Revue des eaux et forêts* 57:25–30.

Dennis, Jerry. 1996. *The Bird in the Waterfall* (New York: HarperCollins).

Derenne, Benoît. 1988. "De la chicotte aux billons: Aperçu des méthodes de lutte contre l'érosion au Rwanda et Burundi du XIXe siècle à nos jours," *Genève-Afrique* 36:46–70.

de Rosa, Luigi. 1989. "Urbanization and Industrialization in Italy," *Journal of European Economic History* 17:469–90.

Deshpande, V. A., K. M. Phadke, and A. L. Aggarwal. 1993. "Trends of Marble Erosion Rates Due to Air Pollution in Indian Urban Centres," *Asian Environment* 15:22–35.

Desowitz, Robert S. 1997. *Who Gave Pinta to the Santa Maria?* (New York: Norton).

Dessus, Benjamin, and François Pharabod. 1990. "Jérémie et Noé: Deux scénarios énergétiques mondiaux à long terme," *Revue de l'énergie* 421:291–307.

De Vos, Antoon, R. H. Manville, and R. G. van Gelder. 1956. "Introduced Mammals and Their Influence on Native Biota," *Zoologica* 41:163–94.

de Walle, F. B., J. Lomme, and M. Nikolopoulou-Tamvakli. 1993a. "General Overview of the Environmental Quality of the Mediterranean Sea." In: F. B. de Walle, M. Nikolopoulou-Tamvakli, and W. J. Heinen, eds., *Environmental Condition of the Mediterranean Sea* (Dordrecht: Kluwer Academic), 34–179.

de Walle, F. B., M. Nikolopoulou-Tamvakli, and W. J. Heinen, eds. 1993b. *Environmental Condition of the Mediterranean Sea* (Dordrecht: Kluwer Academic).

Diani, Mario. 1995. *Green Networks: A Structural Analysis of the Italian Environmental Movement* (Edinburgh: Edinburgh University Press).

di Castri, Francesco. 1989. "History of Biological Invasions with Special Emphasis on the Old World." In: J. A. Drake et al., eds., *Biological Invasions: A Global Perspective* (Chichester: Wiley), 1–30.

Dienes, Leslie, and Theodore Shabad. 1979. *The Soviet Energy System* (New York: Wiley).

Ditt, Karl. 1996. "Nature Conservation in England and Germany, 1900–1970: Forerunner of Environmental Protection?" *Contemporary European History* 5:1–28.

Dobson, A. P., and E. R. Carper. 1996. "Infectious Disease and Human Population History," *BioScience* 46:115–26.

Dobson, A. P., and R. M. May. 1986. "Patterns of Invasion by Pathogens and Parasites." In: H. A. Mooney and J. A. Drake, eds., *Ecology of Biological Invasions of North America and Hawaii* (New York: Springer-Verlag), 58–76.

Dominick, Raymond. 1998. "Capitalism, Communism, and Environmental Protection," *Environmental History* 3:310–32.

Dorsey, Kurkpatrick. 1998. *The Dawn of Conservation Diplomacy: U.S.-Canadian Wildlife Protection Treaties in the Progressive Era* (Seattle: University of Washington Press).

Douglas, Ian. 1990. "Sediment Transfer and Siltation." In: B. L. Turner et al., eds., *The Earth as Transformed by Human Action* (New York: Cambridge University Press), 215–34.

———. 1994. "Human Settlements." In: W. B. Meyer and B. L. Turner, eds., *Changes in Land Use and Land Cover: A Global Perspective* (New York: Cambridge University Press), 149–69.

Dourojeanni, Marc. 1989. "The Environmental Impact of Coca Cultivation and Cocaine Production in the Peruvian Amazon Basin." In: F. R. León and R. Castro de la Mata, eds., *Pasta básica de cocaina: un estudio multidisciplinario* (Lima: Centro de Información y Educación para la Prevención del Abuso de Drogas).

Drake, J. A., H. A. Mooney, F. di Castri, R. H. Groves, F. J. Kruger, M. Rejmánek, and M. Williamson, eds., 1989. *Biological Invasions: A Global Perspective* (Chichester: Wiley).

Dregne, H. E. 1982. "Historical Perspective of Accelerated Erosion and Effect in World Civilization," *Determinants of Soil Loss Tolerance* (American Society of Agronomy, Special Publication no. 45), 1–14.

Drèze, Jean, Meera Samson, and Satyajit Singh, eds. 1997. *The Dam and the Nation: Displacement and Resettlement in the Narmada Valley* (Oxford: Oxford University Press).

Druett, Joan. 1983. *Exotic Intruders: The Introduction of Plants and Animals into New Zealand* (Auckland: Heinemann).

Dryzek, John S. 1997. *The Politics of the Earth* (Oxford: Oxford University Press).

Dudka, S., R. Ponce-Hernandez, and T. C. Hutchinson. 1995. "Current Level of Total Element Concentrations in the Surface Layer of Sudbury's Soils," *Science of the Total Environment* 162:161–72.

Dunnette, David A., and Robert J. O'Brien. 1992. *The Science of Global Change: The Impact of Human Activities on the Environment* (Washington: American Chemical Society).

Dupon, J. F. 1986. *The Effects of Mining on the Environment of High Islands: A Case Study of Nickel Mining in New Caledonia* (Noumea: South Pacific Commission).

Durand, Frédéric. 1993. "Trois Siècles dans l'île du teck: Les politiques forestières aux Indes Neerlandaises (1602–1942)," *Revue française d'histoire d'Outre-Mer* 80:251–305.

Duus, Peter. 1995. *The Abacus and the Sword: The Japanese Penetration of Korea, 1895–1910* (Berkeley: University of California Press).

Earle, Sylvia. 1995. *Sea Change: A Message of the Oceans* (New York: Putnam's).

Edmondson, W. T. 1991. *The Uses of Ecology: Lake Washington and Beyond* (Seattle: University of Washington Press).

Edwards, K. 1993. "Soil Erosion and Conservation in Australia." In: David Pimentel, ed., *World Soil Erosion and Conservation* (Cambridge: Cambridge University Press), 147–70.

Eggleston, Simon, et al. 1992. "Trends in Urban Air Pollution in the United Kingdom during Recent Decades," *Atmospheric Environment* 26B:227–39.

Ehrenfeld, David. 1978. *The Arrogance of Humanism* (New York: Oxford University Press).

Ehrlich, Paul R. 1995. "The Scale of Human Enterprise and Biodiversity Loss." In: John H. Lawton and Robert M. May eds., *Extinction Rates* (Oxford: Oxford University Press), 214–26.

Eisler, Ronald. 1995. "Ecological and Toxicological Aspects of the Partial Meltdown of the Chernobyl Power Plant Reactor." In: David J. Hoffman et al., eds., *Handbook of Ecotoxicology* (Boca Raton, Fla: Lewis).

Elder, Floyd C. 1992. "Acid Deposition." In: David Dunnette and Robert J. O'Brien eds., *The Science of Global Change* (Washington: American Chemical Society), 36–63.

Ellis, Richard. 1991. *Men and Whales* (New York: Knopf).

Elmgren, R. 1989. "Man's Impact on the Ecosystem of the Baltic Sea: Energy Flows Today and at the Turn of the Century," *Ambio* 18:326–32.

Elton, Charles. 1958. *The Ecology of Invasions by Animals and Plants* (London: Chapman & Hall).

El-Shobosky, M. S., and Y. G. Al-Saedi. 1993. "The Impact of the Gulf War on the Arabian Environment," *Atmospheric Environment* 27A:95–108.

Elvin, Mark. 1993. "Three Thousand Years of Unsustainable Development: China's Environment from Archaic Times to the Present," *East Asian History* 6:7–46.

Eng, Pierre van der. 1995. "Stagnation and Dynamic Change in Indonesian Agriculture," *Jahrbuch für Wirtschaftsgeschichte* 1995 1:75–91.

Epstein, Paul R., Timothy Ford, Charles Puccia, and Christina de A. Possas. 1994. Marine Ecosystem Health: Implications for Public Health." In: Mary E. Wilson, Richard Levins, and Andrew Spielman, eds., *Disease in Evolution* (New York: Annals of the New York Academy of Sciences), 740:13–23.

Erickson, Jon. 1995. *The Human Volcano* (New York: Facts on File).

Etemad, Bouda, and Jean Luciani. 1991. *World Energy Production, 1800–1985* (Geneva: Droz).

Ettling, John. 1981. *Germ of Laziness: Rockefeller Philanthropy and Public Health in the New South* (Cambridge: Harvard University Press).

Ewald, Paul W. 1994. *Evolution of Infectious Disease* (New York: Oxford University Press).

Ezcurra, Exequiel. 1990a. "The Basin of Mexico." In: B. L. Turner et al., eds., *The Earth as Transformed by Human Action* (New York: Cambridge University Press), 577–88.

———. 1990b. *De las chinampas a la megalópolis: el medio ambiente en la Cuenca de México* (Mexico City: Fondo de Cultura Económica).

Ezcurra, Exequiel, and Marisa Mazari-Hiriart. 1996. "Are Megacities Viable: A Cautionary Tale from Mexico City," *Environment* 38(1):6–15 and 26–32.

Falkenmark, Malin. 1996. "Approaching the Ultimate Constraint: Water Shortage in the Third World." In: Bernardo Colombo, Paul Demeny, and Max Perutz, eds., *Resources and Population* (Oxford: Clarendon Press), 70–81.

Fang, J., and Z. Xie. 1994. "Deforestation in Preindustrial China: The Loess Plateau Region as an Example," *Chemosphere* 29:983–99.

Fang, Shu-hwei, and Hsiung-wen Chen. 1996. "Air Quality and Pollution Control in Taiwan," *Atmospheric Environment* 30:735–41.

Fanos, Alfy Morcos. 1995. "The Impact of Human Activities on the Erosion and Accretion of the Nile Delta," *Journal of Coastal Research* 11:821–33.

Farley, John. 1991. *Bilharzia: A History of Tropical Medicine* (Cambridge: Cambridge University Press).

Feeny, David. 1988. "Agricultural Expansion and Forest Depletion in Thailand, 1900–1975." In: John F. Richards and Richard P. Tucker, eds., *World Deforestation in the Twentieth Century* (Durham: Duke University Press), 112–43.

Fenner, Frank. 1993. "Smallpox: Emergence, Global Spread, and Eradication," *History and Philosophy of the Life Sciences* 15:397–420.

Fernando, C. H., and Juraj Holcik. 1991. "Some Impacts of Fish Introductions into Tropical Freshwaters." In: P. S. Ramakrishnan, ed., *Ecology of Biological Invasion in the Tropics* (New Delhi: International Scientific Publishers), 103–29.

Ferry, Luc. 1995. *The New Ecological Order* (Chicago: University of Chicago Press).

Feshbach, Murray. 1995. *Ecological Disaster: Cleaning Up the Hidden Legacy of the Soviet Regime* (New York: Twentieth Century Fund Press).

Feshbach, Murray, and Albert Friendly. 1992. *Ecocide in the USSR* (New York: Basic Books).

Findley, Roger. 1988. "Pollution Control in Brazil," *Ecology Law Quarterly* 15:1–68.

Fioravanti, Marc, and Arjun Makhijani. 1997. *Containing the Cold War Mess: Restructuring the Environmental Management of the U.S. Nuclear Weapons Complex* (Takoma Park, Md.: Institute for Energy and Environmental Research).

Fitter, R. S. R. 1946. *London's Natural History* (London: Collins).

Fitzgerald, Deborah. 1986. "Exporting American Agriculture: The Rockefeller Foundation in Mexico," *Social Studies of Science* 16:457–83.

———. 1990. *The Business of Breeding: Hybrid Corn in Illinois, 1890–1940* (Ithaca: Cornell University Press).

———. 1996. "Blinded by Technology: American Agriculture in the Soviet Union, 1928–1932," *Agricultural History* 70:459–86.

Fontana, Vincenzo. 1981. *Il nuovo paesaggio dell'Italiana giolittiana* (Bari: Laterza).

Ford, John. 1971. *The Role of Trypanosomiases in African Ecology: A Study of the Tsetse Fly Problem* (Oxford: Clarendon Press).

Foweraker, Joe. 1981. *The Struggle for Land* (Cambridge: Cambridge University Press).

Francks, Penelope. 1984. *Technology and Agricultural Development in Pre-War Japan* (New Haven: Yale University Press).

Frank, Philipp. 1947. *Einstein, His Life and Times* (New York: Knopf).

Freeman, O. 1993. "Perspectives and Prospects." In: Douglas Helms and Douglas Bowers, eds., *The History of Agriculture and the Environment* (Washington: Agricultural History Society), 3–11.

Freemantle, Michael. 1995. "An Acid Test for Europe," *Chemical and Engineering News* 73(18):10–17.

French, Hilary. 1997. "Learning from the Ozone Experience." In: Lester Brown, Christopher Flavin, Hilary French, eds., *State of the World* (New York: Norton), 151–71.

Freund, Peter, and George Martin. 1993. *The Ecology of the Automobile* (Montreal: Black Rose Books).

Friedrich, D., and D. Müller. 1984. "Rhine." In: B. A. Whitton, ed., *Ecology of European Rivers* (Oxford: Blackwell Scientific), 265–316.

Friedrich, Monika. 1993. "Die Aktivitten des Deutschen Stickstoff-Syndikats in Gypten, 1924–1939," *Zeitschrift für Unternehmensgeschichte* 38:26–48.

Frost, Ruth S. 1934. "The Reclamation of the Pontine Marshes," *Geographical Review* (U.S.) 24:584–95.

Fuhrer, Erwin. 1990. Forest Decline in Central Europe: Additional Aspects of Its Cause," *Forest Ecology and Management* 37:249–57.

Gadgil, Madhav, and Ramachandra Guha. 1995. "Ecological Conflicts and the Environmental Movement in India," *Development and Change* 25:101–35.

———. 1995. *Ecology and Equity: The Use and Abuse of Nature in Contemporary India* (London: Routledge).

Galat, David L., and Ann G. Frazier, eds. 1995. *Overview of River-Floodplain Ecology in the Upper Mississippi River Basin.* Vol. 3 of *Science for Floodplain Management into the 21st Century,* John Kelmelis, ed. (Washington: U.S. Government Printing Office).

Gallagher, Nancy E. 1990. *Egypt's Other Wars: Epidemics and the Politics of Public Health* (Syracuse: Syracuse University Press).

Gambell, Ray. 1993. "International Management of Whales and Whaling: An Historical Review of the Regulation of Commercial and Aboriginal Subsistence Whaling," *Arctic* 46(2):97–107.

Garbrecht, Günther. 1987. "Irrigation Throughout History—Problems and Solutions." In: W. O. Wunderlich and J. E. Prins, eds., *Water for the Future* (Rotterdam: Balkema), 3–17.

Gardner, Gary. 1997. "Preserving Global Cropland," in Lester Brown et al., eds., *State of the World 1997* (New York: Norton), 42–59.

Garfias, J., and R. González. 1992. "Air Quality in Mexico City." In: David Dunnette and Robert O'Brien, eds., *The Science of Global Change* (Washington: American Chemical Society), 149–61.

Garnett, G. P., and E. C. Holmes. 1996. "The Ecology of Emergent Infectious Disease," *Bio-Science* 46(2):127–35.

Garrett, Laurie. 1994. *The Coming Plague* (New York: Farrar, Straus & Giroux).

Garza Villareal, Gustavo. 1985. *El proceso de industrialización en la ciudad de México, 1821–1970* (Mexico City: El Colegio de México).

Geertz, Clifford. 1963. *Agricultural Involution* (Berkeley: University of California Press).

GEMS [Global Environmental Monitoring System]. 1989. *Global Freshwater Quality: A First Assessment,* Michel Meybeck, Deborah Chapman, and Richard Helmer, eds. (Oxford: Blackwell Scientific Publications).

Gerasimov, I. P., and A. M. Gindin. 1977. "The Problem of Transforming Runoff from Northern and Siberian Rivers to the Arid Regions of the European USSR, Soviet Central Asia, and Kazakhstan." In: Gilbert F. White, ed., *Environmental Effects of Complex River Development* (Boulder: Westview Press), 59–70.

Gerber, Michele S. 1992. *On the Home Front: The Cold War Legacy of the Hanford Nuclear Site* (Lincoln: University of Nebraska Press).

German Advisory Council on Global Change. 1995. *World in Transition: The Threat to Soils* (Berlin: Economica Verlag).

Ghose, N. C., and C. B. Sharma. 1989. *Pollution of Ganga River: Ecology of Mid-Ganga Basin* (New Delhi: Ashish Publishing House).

Giblin, James L. 1990. "Trypanosomiasis Control in African History: An Evaded Issue?" *Journal of African History* 31:59–80.

———. 1992. *The Politics of Environmental Control in Northeastern Tanzania, 1840–1940* (Philadelphia: University of Pennsylvania Press).

Gifford, J. 1998. *Planning the Interstate Highway System* (Boulder: Westview Press).

Gilbert, O. L. 1991. *The Ecology of Urban Habitats* (London: Chapman & Hall).

Gilhaus, Ulrike. 1995. *'Schmerzenskinder der Industrie': Umveltverschmutzung, Umweltpolitik und sozialer Protest im Industriezeitalter in Westfalen, 1845–1914* (Paderborn: Schöningh).

Gilmartin, David. 1994. "Scientific Empire and Imperial Science: Colonialism and Irrigation Technology in the Indus Basin," *Journal of Asian Studies* 53:1127–49.

Glacken, Clarence. 1967. *Traces on the Rhodian Shore* (Berkeley: University of California Press).

Glantz, Michael H., ed. 1992. *Climate Variability, Climate Change and Fisheries* (Cambridge: Cambridge University Press).

Gleick, Peter, ed. 1993. *Water in Crisis* (New York: Oxford University Press).

———, ed. 1999. *The World's Water 1998–99* (Washington: Island Press).

Goldman, Marshall. 1972. *The Spoils of Progress: Environmental Pollution in the Soviet Union* (Cambridge: MIT Press).

Goran, Morris. 1967. *The Story of Fritz Haber* (Norman: University of Oklahoma Press).

Gorman, Martha. 1993. *Environmental Hazards: Marine Pollution* (Santa Barbara: ABC-Clio).

Gorres, M., B. Frenzel, and H. Kempter. 1995. "Das Hochmoor als Archiv: der Elementgehalt des Torfes spiegelt die Luftverschutzung im Mittelalter und in der Romerzeit," *Telma* 35:129–141.

Goubert, Jean-Pierre. 1989. *The Conquest of Water: The Advent of Health in the Industrial Age* (Princeton: Princeton University Press).

Goudie, Andrew. 1985. "Man, Maker of Landscapes." *Geographical Magazine* 57:12–16.

Graaff, M. G. H. A. de. 1982. "Milieuvervuiling: een oud proobleem," *Spiegel Historiael* 17:86–96.

Graedel, Thomas. 1990. "Regional Environmental Forces: A Methodology for Prediction." In: John L. Helms, ed., *Energy: Production, Consumption, and Consequences* (Washington: National Academy Press), 85–110.

Graedel, Thomas, and Paul Crutzen. 1989. "The Changing Atmosphere," *Scientific American* 261(3):58–68.

———. 1990. "Atmospheric Trace Constituents." In: B. L. Turner et al., eds., *The Earth as Transformed by Human Action* (New York: Cambridge University Press), 295–312.

Graedel, Thomas, et al. 1995. "Global Emissions Inventories of Acid-related Compounds," *Water, Air and Soil Pollution* 85:25–36.

Graetz, Dean. 1994. "Grasslands." In: W. B. Meyer and B. L. Turner, eds., *Changes in Land Use and Land Cover: A Global Perspective* (New York: Cambridge University Press), 125–48.

Graf, William L. 1985. *The Colorado River: Instability and Basin Management* (Washington: Association of American Geographers).

Graham, Michael. 1956. "Harvests of the Seas." In: William L. Thomas, ed., *Man's Role in Changing the Face of the Earth* (Chicago: University of Chicago Press, 2 vols.), 2:487–503.

Grenon, Michel, and Michel Batisse, eds. 1989. *Futures for the Mediterranean Basin* (Oxford: Oxford University Press).

Grepperud, Sverre. 1996. "Population Pressure and Land Degradation: The Case of Ethiopia," *Journal of Environmental Economics and Management* 30:18–33.

Grigg, David. 1992. *The Transformation of Agriculture in the West* (Oxford: Basil Blackwell).

Grimmett, M. Ross, and Kim Currie. 1991. "The Chemistry of Air Pollution: An Overview," *New Zealand Journal of Geography* 91:5–12.

Gröning, Gert, and Joachim Wolschke-Bulmahn. 1987a. *Die Liebe zur Landschaft. Teil III. Der Drang nach Osten* (Munich: Minerva).

———. 1987b. "Politics, Planning and the Protection of Nature," *Planning Perspectives* 2:127–48.

———. 1991. "1 September 1939. Der Überfall auf Polens als Ausgangspunkt 'totaler' Landespflege," *Raumplanung* 46/47:149–53.

Grove, A. T. 1985. *The Niger and Its Neighbors* (Rotterdam: Balkema).

Grove, Richard. 1990. "Colonial Conservation, Ecological Hegemony and Popular Resistance: Towards a Global Synthesis." In: J. M. Mackenzie, ed., *Imperialism and the Natural World* (Manchester: Manchester University Press), 15–50.

———. 1994. *Green Imperialism: Colonial Scientists, Ecological Crises, and the History of Environmental Concern, 1600–1860* (Cambridge: Cambridge University Press).

Groves, R. H., and J. J. Burdon, eds. 1986. *Ecology of Biological Invasions* (Cambridge: Cambridge University Press).

Grübler, Arnulf. 1994. "Industrialization as a Historical Phenomenon." In: R. Socolow, C. Andrews, F. Berkhout, and V. Thomas, eds., *Industrial Ecology and Global Change* (Cambridge: Cambridge University Press), 43–68.

Grübler, Arnulf, and Nebojša Nakičenović. 1991. *Evolution of Transport Systems: Past and Future* (Laxenburg: International Institute for Applied Systems Analysis).

Guayacochea de Onofrí, Rosa. 1987. "Urbanismo y salubridad en la ciudad de Mendoza (1880–1916)," *Revista de historia de América y Argentina* 14:171–202.

Gugler, Joseph, ed. 1996. *The Urban Transformation of the Developing World* (Oxford: Oxford University Press).

Guha, Ramachandra. 1990. *The Unquiet Woods: Ecological Change and Peasant Resistance in the Himalaya* (Berkeley: University of California Press).

Guha, Ramachandra, and J. Martinez-Alier. 1997. *Varieties of Environmentalism: Essays North and South* (London: Earthscan).

Gujja, B., and A. Finger-Stich. 1996. "What Price Prawn? Shrimp Aquaculture's Impact on Asia," *Environment* 38(7):12–15, 33–9.

Gunst, P. 1990. "Die Mechanisierung der Ungarischen Landwirtschaft bis 1945," *Etudes historiques hongroises* 1990(3):237–50.

Gutman, P. S. 1994. "Involuntary Resettlement in Hydropower Projects," *Annual Review of Energy and the Environment* 19:189–210.

Guzman, Francisco, Maria Ruiz, and Elizabeth Vega. 1996. "Air Quality in Mexico City," *Science* 271:1040–42.

Gytarsky, M. L., R. T. Karaban, I. M. Nazarov, T. I. Sysygina and M. V. Chemeris. 1995. "Monitoring of Forest Ecosystems in the Russian Subarctic: Effects of Industrial Pollution," *Science of the Total Environment* 164:57–68.

Haas, Peter. 1990. *Saving the Mediterranean: The Politics of International Environmental Cooperation* (New York: Columbia University Press).

Haberer, Klaus. 1991. "Die Belastung des Rheins mit Schadstoffen," *Geographische Rundschau* 43:334–41.

Hager, Carol J. 1995. *Technological Democracy: Bureaucracy and Citizenry in the German Energy Debate* (Ann Arbor: University of Michigan Press).

Hahn, Peter L. 1991. *The United States, Great Britain, and Egypt, 1945–1956* (Chapel Hill: University of North Carolina Press).

Hall, Jane V. 1995. "Air Quality Policy in Developing Countries," *Contemporary Economic Policy* 13:77–85.

Hannah, Lee, et al. 1994. "A Preliminary Inventory of Human Disturbance of World Ecosystems," *Ambio* 23:246–50.

Hannesson, Rognvaldur. 1994. "Trends in Fishery Management." In: Eduardo Loayza, ed., *Managing Fishery Resources* (Washington: World Bank), 91–6.

Hanley, Susan B. 1987. "Urban Sanitation in Preindustrial Japan," *Journal of Interdisciplinary History* 18:1–26.

Hardjono, J. M. 1977. *Transmigration in Indonesia* (Kuala Lumpur: Oxford University Press).
———. 1988. "The Indonesian Transmigration Program in Historical Perspective," *International Migration* 26:427–38.

Hardoy, Jorge, Diana Mitlin, and David Satterthwaite. 1993. *Environmental Problems in Third World Cities* (London: Earthscan).

Harrison, Paul. 1992. *The Third Revolution: Population, Environment and a Sustainable World* (Harmondsworth: Penguin Books).

Hartwell, Robert. 1967. "A Cycle of Economic Change in Imperial China: Coal and Iron in Northeast China, 750–1350," *Journal of the Economic and Social History of the Orient/Journal d'Histoire economique et sociale de l'Orient* 10:102–59.

Harvard Working Group on New and Resurgent Diseases. 1996. "Globalization, Development, and the Spread of Disease." In: *The Case against the Global Economy,* Jerry Mander and Edward Goldsmith, eds. (San Francisco: Sierra Club Books), 160–70.

Hashimoto, Michio. 1989. "History of Air Pollution Control in Japan." In: H. Nishimura, ed., *How to Conquer Air Pollution: A Japanese Experience* (Amsterdam: Elsevier), 1–94.

Hashimoto, Y., et al. 1994. "Atmospheric Fingerprints of East Asia, 1986–1991," *Atmospheric Environment* 28:1437–45.

Haub, Carl. 1995. "How Many People Ever Have Lived on Earth?" *Population Today* 23(2):4–5.

Hawley, T. M. 1992. *Against the Fires of Hell: The Environmental Disaster of the Gulf War* (New York: Harcourt Brace Jovanovich).

Hayami, Yujiro. 1975. *A Century of Agricultural Growth in Japan* (Tokyo: University of Tokyo Press).

Hayami, Yujiro, and Saburo Yamada. 1991. *The Agricultural Development of Japan* (Tokyo: University of Tokyo Press).

Hays, J. N. 1998. *The Burdens of Disease: Epidemics and Human Response in Western History* (New Brunswick: Rutgers University Press).

Hays, Samuel P. 1997. *Explorations in Environmental History* (Pittsburgh: University of Pittsburgh Press).

Hazell, Peter B. R., and C. Ramasamy. 1991. *The Green Revolution Reconsidered: The Impact of High-yielding Rice Varieties in South India* (Baltimore: Johns Hopkins University Press).

Headrick, Daniel. 1988. *The Tentacles of Progress: Technology Transfer in the Age of Imperialism, 1850–1940* (New York: Oxford University Press).

Headrick, Daniel. 1990. "Technological Change." In: B. L. Turner et al., eds., *The Earth as Transformed by Human Action* (New York: Cambridge University Press), 55–68.

Headrick, Rita. 1994. *Colonialism, Health and Illness in French Equatorial Africa, 1885–1935* (Atlanta: African Studies Association Press).

Hein, Laura. 1990. *Fueling Growth: The Energy Revolution and Economic Policy in Postwar Japan* (Cambridge: Harvard University Press).

Heine, Klaus. 1983. "Outline of Man's Impact on the Natural Environment in Central Mexico," *Jahrbuch für Geschichte von Staat, Wirtschaft und Gesellschaft Lateinamerikas* 20:121–31.

Helms, Douglas. 1992. *Readings in the History of the Soil Conservation Service* (Washington: USDA Soil Conservation Service, Historical Notes no. 1).

Hempel, Lamont C. 1997. "Population in Context: A Typology of Environmental Driving Forces," *Population and Environment* 18:439–61.

Herman, R., S. A. Ardakhani, and Jesse Ausubel. 1989. "Dematerialization." In: Jesse Ausubel and Hedy Sladovich, eds., *Technology and Environment* (Washington: National Academy Press), 50–69.

Hermand, Jost. 1991. *Grüne Utopien in Deutschland: Zur Geschichte des ökologischen Bewusstseins* (Frankfurt: Fischer).

Herz, O. 1989. "La politique française de prévention de la pollution atmosphérique." In: L. J. Brasser and W. L. Mulder, eds., *Man and His Ecosystem* (Amsterdam Elsevier), 1–7.

Hewa, Soma. 1992. *Colonialism, Tropical Disease and Imperial Medicine: Rockefeller Philanthropy in Sri Lanka* (Lanham, Md.: University Press of America).

Heywood, V. H., and R. T. Watson, eds. 1995. *Global Biodiversity Assessment* (Cambridge: Cambridge University Press).

Hilborn, Ray. 1990. "Marine Biota." In: B. L. Turner et al., eds., *The Earth as Transformed by Human Action* (New York: Cambridge University Press), 371–85.

Hillel, Daniel. 1991. *Out of the Earth: Civilization and the Life of the Soil* (Berkeley: University of California Press).

———. 1994. *Rivers of Eden: The Struggle for Water and the Quest for Peace in the Middle East* (New York: Oxford University Press).

Ho, Kong Chong. 1997. "From Port City to City-State: Forces Shaping Singapore's Built Environment." In: Won Bae Kim, Mike Douglass, Sang-Chuel Choe, and Kong Chong Ho, eds., *Culture and the City in East Asia* (Oxford: Clarendon Press), 212–33.

Hobbs, Peter V., and Lawrence F. Radke. 1992. "Airborne Studies of the Smoke from the Kuwait Oil Fires," *Science* 256:987–91.

Hobsbawm, Eric. 1994. *The Age of Extremes* (New York: Pantheon).

Hodgson, Dominic, and Nadine Johnston. 1997. "Inferring Seal Populations from Lake Sediments," *Nature* 387:30–1.

Hogendorn, J. S., and K. M. Scott. 1981. "The East African Groundnut Scheme: Lessons of a Large-Scale Agricultural Failure," *African Economic History* 10:81–115.

Hohenberg, Paul, and Lynn Hollen Lees. 1985. *The Making of Urban Europe, 1000–1950* (Cambridge: Harvard University Press).

Holdren, John P. 1991. "Energy in Transition." In: *Energy for Planet Earth: Readings from Scientific American Magazine* (New York: Freeman), 119–30.

Holland, Heinrich D., and Ulrich Petersen. 1995. *Living Dangerously* (Princeton: Princeton University Press).

Holloway, David. 1994. *Stalin and the Bomb: The Soviet Union and Atomic Energy, 1939–1956* (New Haven: Yale University Press).

Hong, Sungmin, Jean-Pierre Candelone, Clair C. Patterson, and Claude F. Boutron. 1996. "History of Ancient Copper Smelting Pollution during Roman and Medieval Times Recorded in Greenland Ice," *Science* 272:246–8.

Hooke, Roger L. 1994. "On the Efficacy of Humans as Geomorphic Agents," *GSA Today (Geological Society of America)* 4:217–25.

Hoppe, Kirk A. 1997. "Lords of the Fly: Environmental Images, Colonial Science and Social Engineering in British East African Sleeping Sickness Control, 1903–1963" (Ph.D. thesis, Boston University).

Hoshino, Yoshiro. 1992. "Japan's Post-Second World War Environmental Problems." In: Jun Ui, ed., *Industrial Pollution in Japan* (Tokyo: U.N. University Press), 64–76.

Hou, Wenhui. 1997. "Reflections on Chinese Traditional Ideas of Nature," *Environmental History* 2:482–93.

Houghton, R. A. 1994. "The Worldwide Extent of Land-Use Change," *BioScience* 44:305–13.

Houghton, R. A., D. S. Lefkowitz, and D. L. Skole. 1991. "Changes in the Landscape of Latin America between 1850 and 1985. I. Progressive Loss of Forests," *Forest Ecology and Management* 38:143–72.

Howell, P. P., and J. A. Allan, eds. 1994. *The Nile: Sharing a Scarce Resource* (Cambridge: Cambridge University Press).

Humphrey, Caroline, and David Sneath. 1996. *Culture and Environment in Inner Asia* (Cambridge: White Horse Press, 2 vols.).

Hunter, J. M., et al. 1993. *Parasitic Diseases in Water Resource Development* (Geneva: World Health Organization).

Hurley, Andrew, ed. 1997. *Common Fields: An Environmental History of St. Louis* (St. Louis: Missouri Historical Society Press).

Husar, Rudolf B., and Janja Djukic Husar. 1990. "Sulfur." In: B. L. Turner et al., eds., *The Earth as Transformed by Human Action* (New York: Cambridge University Press), 409–21.

Hutchinson, John F. 1985. "Tsarist Russia and the Bacteriological Revolution," *Journal of the History of Medicine and Allied Sciences* 40:420–39.

Hvidt, Martin. 1995. "Water Resource Planning in Egypt." In: Eric Watkins, ed., *The Middle Eastern Environment* (Cambridge: St. Malo Press), 90–100.

Hyndman, David. 1994. *Ancestral Rain Forests and the Mountain of Gold: Indigenous Peoples and Mining in New Guinea* (Boulder: Westview Press).

Iliffe, John. 1995. *Africans* (Cambridge: Cambridge University Press).

Illinois Department of Energy and Natural Resources. 1994. *The Changing Illinois Environment: Critical Trends* (Springfield, Ill.: IDENR, 7 vols.).

Ioffe, Grigory, and Tatyana Nefedova. 1997. *Continuity and Change in Rural Russia* (Boulder: Westview).

IPCC [Intergovernmental Panel on Climate Change]. 1996. *Climate Change 1995* (Cambridge: Cambridge University Press, 3 vols.).

Ipsen, Carl. 1996. *Dictating Demography: The Problem of Population in Fascist Italy* (Cambridge: Cambridge University Press).

Ispahani, Mahnaz. 1989. *Roads and Rivals: The Political Uses of Access in the Borderlands of Asia* (Ithaca: Cornell University Press).

Ives, Jack, and Bruno Messerli. 1989. *The Himalayan Dilemma* (London: Routledge).

Jacks, G. V. 1958. "The Influence of Man on Soil Fertility," *Annual Report of the Smithsonian Institution 1957* (Washington: U.S. Government Printing Office).

Jacks, G. V., and R. O. Whyte. 1939. *The Rape of the Earth: A World Survey of Soil Erosion* (London: Faber & Faber).

Jackson, Kenneth T. 1985. *Crabgrass Frontier: The Suburbanization of the United States* (New York: Oxford University Press).

Jahnke, Hans, Dieter Kirschke, and Johannes Lagemann. 1987. *The Impact of Agricultural Research on Tropical Africa* (Washington: World Bank).

Jaksic, F. M., and E. R. Fuentes. 1991. "Ecology of a Successful Invader: The European Rabbit in Central Chile." In: R. H. Groves and F. di Castri, eds., *Biogeography of Mediterranean Invasions* (Cambridge: Cambridge University Press), 273–83.

Janetos, Anthony C. 1997. "Do We Still Need Nature?" *Consequences* (U.S.) 3(1):17–25.

Jänicke, M., and H. Weidner, eds. 1996. *National Environmental Politics* (Berlin: Springer).

Jarosz, Lucy. 1993. "Defining and Explaining Tropical Deforestation: Shifting Cultivation and Population Growth in Colonial Madagascar (1895–1940)," *Economic Geography* 69:366–79.

Jayal, N. D. 1985. "Destruction of Water Resources—The Most Critical Ecological Crisis of East Asia," *Ambio* 14:95–98.

Jedrej, M. C. 1983. "The Growth and Decline of a Mechanical Agriculture Scheme in West Africa," *African Affairs* 82:541–58.

Jeleček, Leoš. 1988. "Some Thoughts on Historical Geography of Environmental Changes: Development of Agricultural Landscape of Czech Lands in Historical Perspective." In: V. V. Annenkov and L. Jeleček, eds., *Historical Geography of Environmental Changes* (Prague: Institute of Czechoslovak and World History), 351–80.

———. 1991. "Některé ekologické souvisloti vývoje zemědelské krajiny v zemedělstvi v českých zemich," *Ceský Casopis Historický* 89:375–94.

Jennings, Bruce. 1988. *Foundations of International Agricultural Research* (Boulder: Westview).

Jiang, Gaoming. 1996. "Tree Ring Analysis for Determination of Pollution History of Chengde City, North China," *Journal of Environmental Sciences* (China) 8:77–85.

Johnson, Richard. 1988. "Malaria and malaria control in the USSR, 1917–41" (Ph.D. dissertation, History Department, Georgetown University).

Jones, David C. 1987. *Empire of Dust: Settling and Abandoning the Prairie Dry Belt* (Edmonton: University of Alberta Press).

Judson, Sheldon. 1968. "Erosion of the Land, or What's Happening to Our Continents," *American Scientist* 56(4):356–74.

Kalland, Arne, and Brian Moeran. 1992. *Japanese Whaling: End of an Era?* (London: Curzon Press).

Kandler, Otto, and J. L. Innes. 1995. "Air Pollution and Forest Decline in Central Europe," *Environmental Pollution* 90:171–80.

Kaplan, Temma. 1981. "Class Consciousness and Community in Nineteenth-Century Andalusia," *Political Power and Social Theory* 2:21–57.

Karan, P. P. 1994. "Environmental Movements in India," *Geographical Journal* (U.K.) 84:32–41.

Karnosky, D. F. 1979. "Dutch Elm Disease: A Review of the History, Environmental Implications, Control, and Research Needs," *Environmental Conservation* 6:311–22.

Kasting, James. 1998. "The Carbon Cycle, Climate, and the Long-Term Effects of Fossil Fuel Burning," *Consequences* (U.S.) 4(1):15–27.

Katsoulis, Basil. 1996. "The Relationship between Synoptic, Mesoscale and Microscale Meteorological Parameters during Poor Air Quality Events in Athens, Greece," *Science of the Total Environment* 181:13–24.

Katsoulis, Basil, and J. M. Tsangaris. 1994. "The State of the Greek Environment in Recent Years," *Ambio* 23:274–9.

Katsouyanni, K., et al. 1990. "Air Pollution and Cause Specific Mortality in Athens," *Journal of Epidemiology and Community Health* 44:321–4.

Kaufman, Les, and Kenneth Mallory, eds. 1993. *The Last Extinction* (Cambridge: MIT Press).

Kay, Jane Holtz. 1997. *Asphalt Nation* (New York: Crown).

Keskinler, B., B. İpekoğlu, U. Danış, F. Acar and O. Özbay. 1994. "Hava kırlığının Erzurum'da tarihi yapıtlara etkisi," *Turkish Journal of Engineering and Environmental Sciences* 18:169–74.

Kettani 1993. Citation mislaid.

Keyfitz, N. 1966. "How Many People Have Lived on the Earth?" *Demography* 3:581–2.

Khalil, M. A. K., and R. A. Rasmussen. 1995. "The Changing Composition of the Earth's Atmosphere." In: H. B. Singh, ed., *Composition, Chemistry and Climate of the Atmosphere* (New York: Van Nostrand Reinhold), 50–87.

Khan, Farieda. 1997. "Soil Wars," *Environmental History* 2:439–59.

Khoshoo, T. N., and K. G. Tejwani. 1993. "Soil Erosion and Conservation in India." In: David Pimentel, ed., *World Soil Erosion and Conservation* (Cambridge: Cambridge University Press), 109–146.

King, Carolyn. 1984. *Immigrant Killers: Introduced Predators and the Conservation of Birds in New Zealand* (Auckland: Oxford University Press).

Kishk, Mohammed A. 1986. "Land Degradation in the Nile Valley," *Ambio* 15:226–30.

Kitagishi, Kazuko, and Ichiro Yamane, eds. 1981. *Heavy Metal Pollution in Soils of Japan* (Tokyo: Japan Scientific Societies Press).

Kizaki, Harumichi. 1938. *Keikan yori Mitaru Nihon Seishin* [Japanese Spirit from the Perspective of Nature] (Tokyo: Kokudosha).

Kjekshus, Helge. 1977. *Ecology Control and Economic Development in East African History* (Berkeley: University of California Press).

Klarer, J., and B. Moldan, eds. 1997. *The Environmental Challenges for Central European Economies in Transition* (Chichester: Wiley).

Klein, Herbert S. 1995. "European and Asian Migration to Brazil." In: Robin Cohen, ed., *The Cambridge Survey of World Migration* (New York: Cambridge University Press), 208–14.

Klein Robbenhaar, J. F. I. 1995. "Agro-Industry and the Environment: The Case of Mexico in the 1990s," *Agricultural History* 69:395–412.

Klidonas, Y. 1993. "The Quality of the Atmosphere in Athens," *Science of the Total Environment* 129:83–94.

Klige, R. K., Liu Hong, and A. O. Selivanov. 1996. "Regime of the Aral Sea during Historical Time," *Water Resources* (Russia; English translation) 23:375–80.

Klimm, Lester E. 1956. "Man's Ports and Channels." In: William L. Thomas, ed., *Man's Role in Changing the Face of the Earth* (Chicago: University of Chicago Press, 2 vols.), 2:522–41.

Klumpp, A., M. Domingos, and G. Klumpp. 1996. "Assessment of the Vegetation Risk by Fluoride Emissions from Fertiliser Industries at Cubatão, Brazil," *Science of the Total Environment* 192:219–28.

Knapen, Han. 1998. "Lethal Diseases in the History of Borneo: Mortality and the Interplay between Disease Environment and Human Geography." In: Victor T. King, ed., *Environmental Challenges in South-East Asia* (Richmond: Curzon Press for Nordic Institute of Asian Studies), 69–94.

Knox, George. 1994. *The Biology of the Southern Ocean* (Cambridge: Cambridge University Press).

Kobori, Iwao, and Michael H. Glantz, eds. 1998. *Central Eurasian Water Crisis* (Tokyo: U.N. University Press).

Kock, K. -H. 1995. "Walfang und Walmanagement in den Polarmeeren," *Historisch-Meereskundliches Jahrbuch* 3:7–34.

Konvitz, Josef. 1985. *The Urban Millennium* (Carbondale: Southern Illinois University Press).

Kosambi, Meera. 1986. *Bombay in Transition: The Growth and Social Ecology of a Colonial City, 1880–1980* (Stockholm: Almquist and Wiksell International).

Kostin, V. 1986. "Protection of the Sea of Azov," *Ambio* 15(5):350.

Kotamarthi, V. R., and G. R. Carmichael. 1990. "The Long Range Transport of Pollutants in the Pacific Rim Region," *Atmospheric Environment* 24A:1521–34.

Kotkin, Stephen. 1995. *Magnetic Mountain: Stalinism as a Civilization* (Berkeley: University of California Press).

Kotov, Vladimir, and Elena Nikitina. 1996. "Norilsk Nickel," *Environment* 38(9):6–11, 32–37.

Krause, Richard M. 1992. "Origin of Plagues: Old and New," *Science* 257:1073–8.

Krishnan, A. Radha, and Malcom Tull. 1994. "Resource Use and Environmental Management in Japan, 1890–1990," *Australian Economic History Review* 34:3–23.

Krishnan, Rajaram, Jonathan M. Harris, and Neva Goodwin, eds. 1995. *A Survey of Ecological Economics* (Washington: Island Press).

Kucera, V., and S. Fitz. 1995. "Direct and Indirect Air Pollution Effects on Materials Including Cultural Monuments," *Water, Air and Soil Pollution* 85:153–65.

Kudo, Akira, and Shojiro Miyahara. 1991. "A Case History; Minamata Mercury Pollution in Japan—From Loss of Human Lives to Decontamination," *Water Science and Technology* 23:283–90.

Kummer, David. 1991. *Deforestation in the Postwar Philippines* (Chicago: University of Chicago Press).

———. 1994. "Environmental Degradation in the Uplands of Cebu," *Geographical Review* (U.S.) 84:266–76.

Kunitz, Stephen J. 1994. *Disease and Social Diversity: The European Impact on Non-European Health* (New York: Oxford University Press).

Kuusela, Kullervo. 1994. *Forest Resources in Europe, 1950–1990* (Cambridge: Cambridge University Press).

Lacy, Rodolfo, ed. 1993. *La calidad del aire en el valle de México* (Mexico City: Colegio de México).

Lafferty, William M., and James Meadowcroft, eds. 1996. *Democracy and the Environment: Problems and Prospects* (Cheltenham: Edward Elgar).

Laj, Paolo, Julia Palais, and Haralder Sigurdsson. 1992. "Changing Impurities to the Greenland Ice Sheet over the Last 250 Years," *Atmospheric Environment* 26A:2627–40.

Lal, R. 1990. "Soil Erosion and Land Degradation: The Global Risks," *Advances in Soil Science* 11:129–72.

Lal, R., and F. J. Pierce. 1991a. "The Vanishing Resource." In: R. Lal and F. J. Pierce, eds., *Soil Management for Sustainability* (Ankeny, Iowa: Soil and Water Conservation Society), 1–6.

Lal, R., and F. J. Pierce, eds. 1991b. *Soil Management for Sustainability* (Ankeny, Iowa: Soil and Water Conservation Society).

Lambert, Audrey M. 1971. *The Making of the Dutch Landscape: An Historical Geography of the Netherlands* (London: Seminar Press).

Landau, Matthew. 1992. *Introduction to Aquaculture* (Chichester: Wiley).

Larsson, Ulf, Ragnor Elmgren, and Fredrik Wulff. 1985. "Eutrophication and the Baltic Sea: Causes and Consequences," *Ambio* 14(1):9–14.

Lawton, John H., and Robert M. May, eds. 1995. *Extinction Rates* (Oxford: Oxford University Press).

Lay, M. G. 1992. *Ways of the World: A History of the World's Roads and the Vehicles That Used Them* (New Brunswick: Rutgers University Press).

Leakey, Richard, and Roger Lewin. 1995. *The Sixth Extinction* (New York: Doubleday).

Lear, Linda. 1997. *Rachel Carson* (New York: Henry Holt).

Lee, James Z., and Wang Feng. 1999. *Malthusian Mythology and Chinese Reality: The Population History of One Quarter of Humanity, 1700–2000* (Cambridge: Harvard University Press).

Lee, Yok-shiu, and Alvin Y. So. 1999. *Asia's Environmental Movements* (Armonk, N.Y.: M. E. Sharpe).

Lelek, Antonin. 1989. "The Rhine River and Some of Its Tributaries Under Human Impact in the Last Two Centuries." In: D. P. Dodge, ed., *Proceedings of the International Large River Symposium* (Ottawa: Department of Fisheries and Oceans, Special Publication of Fisheries and Aquatic Sciences no. 106), 469–87.

Le Lourd, Philippe. 1977. "Oil Pollution in the Mediterranean Sea," *Ambio* 6:317–20.

Lents, James, and William Kelly. 1993. "Cleaning the Air in LA," *Scientific American* 269(4):32–9.

Leontidou, Lila. 1990. *The Mediterranean City in Transition* (Cambridge: Cambridge University Press).

Levang, P., and O. Sevin. 1989. "80 ans de Transmigration en Indonésie," *Annales de géographie* 98:538–66.

Levinson, Arik. 1992. *Efficient Environmental Regulation: Case Studies of Urban Air Pollution* (Washington: World Bank).

Levy, Stuart. 1992. *The Antibiotic Paradox* (New York: Plenum Press).

Lewis, Martin W. 1992. *Wagering the Land: Ritual, Capital, and Environmental Degradation in the Cordillera of Northern Luzon, 1900–1986* (Berkeley: University of California Press).

Lewis, Tom. 1997. *Divided Highways: Building the Interstate Highways, Transforming American Life* (New York: Viking).

Liao, I-chu, and Chung-zen Shyu. 1992. "Evaluation of Aquaculture in Taiwan: Status and Constraints." In: J. B. Marsh, ed., *Resources and Environment in Asia's Marine Sector* (Washington: Taylor & Francis), 185–98.

Lincoln, W. Bruce. 1994. *The Conquest of a Continent: Siberia and the Russians* (New York: Random House).

Linden, Olof. 1990. "Human Impact on Tropical Coastal Zones," *Nature and Resources* 26(4):3–11.

Little, Charles. 1995. *The Dying of the Trees* (New York: Viking).

Livi-Bacci, Massimo. 1992. *A Concise History of World Population* (Oxford: Basil Blackwell).

Livingstone, David N. 1994. "The Historical Roots of Our Ecological Crisis: A Reassessment," *Fides et Historia* 26:38–55.

Locke, G., and K. K. Bertine. 1986. "Magnetite in Sediments as an Indication of Coal Combustion," *Applied Geochemistry* 1:345–56.

Logan, T. J. 1990. "Chemical Degradation of Soil," *Advances in Soil Sciences* 11:187–221.

Long, John L. 1981. *Introduced Birds of the World* (Sydney: Reed).

Lonkiewicz, B., W. Strykowski, and Z. Pryzborski. 1987. "Poland." In: E. G. Richards, ed., *Forestry and Forest Industries: Past and Future* (Dordrecht: Martinus Nijhoff), 363–71.

Louis, William Roger, and Roger Owen, eds. 1989. *Suez 1956* (Oxford: Clarendon Press).

Low, Allan. 1985. *Agricultural Development in Southern Africa* (London: James Currey).

Low, Kwai Sim, and Yat Hoong Yip. 1984. "An Overview of Past Researches and Contemporary Issues on Urbanization and Ecodevelopment in Malaysia with Special Reference to Kuala Lumpur." In: Yip and Low, eds., *Urbanization and Ecodevelopment* (Kuala Lumpur: University of Malaya Institute of Advanced Studies), 11–37.

Lower, A. R. M. 1973. *Great Britain's Woodyard: British America and the Timber Trade* (Montreal and Kingston: McGill–Queen's University Press).

Lowi, Miriam. 1993. *Water and Power: The Politics of a Scarce Resource in the Jordan River Basin* (Cambridge: Cambridge University Press).

Lowman, Gwen, and Rita Gardner. 1996. "Conference Report: Petroleum and Nigeria's Environment," *Geographical Journal* (U.K.) 162:358–9.

Lubin, Nancy. 1995. "Uzbekistan." In: Philip R. Pryde, ed., *Environmental Resources and Constraints in the Former Soviet Republics* (Boulder: Westview), 289–306.

Lucas, AnElissa. 1982. *China's Medical Modernization* (New York: Praeger)

Ludwig, Donald, Ray Hilborn, and Carl Walters. 1993. "Uncertainty, Resource Exploitation, and Conservation: Lessons from History," *Science* 260:17–36.

Ludyansky, M. L., Derek McDonald, and David MacNeill. 1993. "Impact of the Zebra Mussel, a Bivalve Invader," *BioScience* 43:533–44.

Lumsden, Malvern. 1975. "Conventional Warfare and Human Ecology," *Ambio* 4:223–8.

Lupton, F. G. H. 1987. "History of Wheat Breeding." In: F. G. H. Lupton, ed., *Wheat Breeding* (London: Chapman & Hall), 51–70.

L'vovich, Mark, and Gilbert F. White. 1990. "Use and Transformation of Terrestrial Water Systems." In: B. L. Turner et al., eds., *The Earth as Transformed by Human Action* (New York: Cambridge University Press), 235–252.

Lyons, Maryinez. 1992. *The Colonial Disease: A Social History of Sleeping Sickness in Northern Zaire, 1900–1940* (Cambridge: Cambridge University Press).

MacCleery, Douglas W. 1994. *American Forests: A History of Resilience and Recovery* (Durham: Forest History Society).

MacDonald, Calum, Eric Hampton, and Owen Harrop. 1993. "The Changing Face of Glasgow's Air Quality," *Clean Air* 22:233–7.

Macfarlane, Alan. 1997. *The Savage Wars of Peace: England, Japan, and the Malthusian Trap* (Oxford: Blackwell).

Mack, R. N. 1986. "Alien Plant Invasions into the Intermountain West: A Case History." In: H. A. Mooney and J. A. Drake, eds., *Ecology of Biological Invasions of North America and Hawaii* (New York: Springer-Verlag), 191–213.

———. 1989. "Temperate Grasslands Vulnerable to Plant Invasions: Characteristics and Consequences." In: J. A. Drake et al., eds., *Biological Invasions: A Global Perspective* (Chichester: Wiley), 155–79.

MacKellar, F. Landis, Wolfgang Lutz, A. J. McMichael, and Astri Suhrke. 1998. "Population and Climate Change." In: Steve Rayner and Elizabeth Malone eds., *Human Choice and Climate Change* (Columbus: Battelle Press, 4 vols.), 1:89–193.

Mackenzie, J. M., ed. 1990. *Imperialism and the Natural World* (Manchester: Manchester University Press).

Maddison, Angus. 1995. *Monitoring the World Economy, 1820–1992* (Paris: OECD Development Centre).

Maddox, Gregory, James Giblin, and Isaria Kimambo, eds. 1996. *Custodians of the Land: Ecology and Culture in the History of Tanzania* (London: James Currey).

Mageed, Yahia Abdel. 1994. "The Central Region: Problems and Perspectives," In: Peter Rogers and Peter Lydon, eds., *Water in the Arab World* (Cambridge: Harvard University Press, 1994), 101–20.

Majkowski, Slawomir. 1994. "Oberschlesien, ein ökologisches Katastrophengebiet," *Osteuropa-Wirtschaft* 39:310–34.

Majumdar, S. K., E. W. Miller, and J. J. Cahir, eds. 1991. *Air Pollution: Environmental Issues and Health Effects* (Easton: Pennsylvania Academy of Sciences).

Makhijani, Arjun, Howard Hu, and Katherine Yih. 1995. *Nuclear Wastelands: A Global Guide to Nuclear Weapons Production and Its Health and Environmental Effects* (Cambridge: MIT Press).

Malle, Karl-Geert. 1996. "Cleaning Up the River Rhine," *Scientific American* 274(1):70–75.

Mamane, Y. 1987. "Air Pollution Control in Israel during the First and Second Century," *Atmospheric Environment* 21:1861–63.

Mander, Jerry, and Edward Goldsmith, eds. 1996. *The Case Against the Global Economy* (San Francisco: Sierra Club Books).

Mandrillon, Marie-Hélène. 1991. "Les voies du politique en URSS: L'exemple de l'écologie," *Annales: Economies, sociétés, civilisations* 46(6):1375–98.

Mangelsdorf, Paul C. 1974. *Corn: Its Origin, Evolution, and Improvement* (Cambridge: Belknap Press).

Mann, Michael E., Raymond S. Bradley and Malcolm K. Hughes. 1998. "Global-Scale Temperature Patterns and Climate Forcing over the Past Six Centuries," *Nature* 392:779–87.

Manning, Richard. 1995. *Grassland* (New York: Viking).

Mannion, A. M. 1995. *Agriculture and Environmental Change* (Chichester: Wiley).

Mantis, Homer T., and Christos C. Repapis. 1992. "Assessment of the Potential for Photochemical Air Pollution in Athens: A Comparison of Emissions and Air-Pollutant Levels in Athens with Those in Los Angeles," *Journal of Applied Meteorology* 31:1467–76.

Manuel, Frank E. 1995. *A Requiem for Karl Marx* (Cambridge: Harvard University Press).

Manzanova, G. V., and A. K. Tulokhonov. 1994. "Traditsii i novatsii v razvitii sel'skogo khoziaistva zabaikal'ia," *Vostok* 1:100–13.

Marchak, M. Patricia. 1995. *Logging the Globe* (Montreal and Kingston: McGill–Queen's University Press).

Marchetti, Roberto, and Attilio Rinaldi. 1989. "Le condizioni del Mare Adriatico." In: Giovanni Melandri, ed., *Ambiente italia* (Turin: ISEDI), 33–7.

Martinez-Alier, J. 1987. *Ecological Economics* (Oxford: Blackwell).

Mason, B. J. 1992. *Acid Rain: Its Causes and Its Effects on Inland Waters* (Oxford: Oxford University Press).

Massey, John Stewart. 1992. *The Nature of Russia* (New York: Cross River Press).

Mather, Alexander S. 1990. *Global Forest Resources* (London: Belhaven Press).

Matson, Stacey, and E. Lynn Miller. 1991. "Air Pollution and Its Effects on Buildings and Monuments." In: Majumdar, S. K., E. W. Miller, and J. J. Cahir, eds., *Air Pollution: Environmental Issues and Health Effects* (Easton: Pennsylvania Academy of Sciences), 242–54.

Maurer, G. 1968. "Les paysans de Haut Rif Central," *Revue de géographie du Maroc* 14:3–70.

May, Robert M., John H. Lawton, and Nigel Stork. 1995. "Assessing Extinction Rates." In: John Lawton and Robert May, eds., *Extinction Rates* (Oxford: Oxford University Press), 1–24.

McCann, James C. 1997. "The Plow and the Forest: Narratives of Deforestation in Ethiopia, 1840–1992," *Environmental History* 2:138–59.

McCormick, John. 1991. *Reclaiming Paradise: The Global Environmental Movement* (Bloomington: Indiana University Press).

McCreery, David. 1989. "Tierra, trabajo y conflicto en San Juan Ixcoy, Huehuetenango, 1890–1940," *Anales de la Academia de Geografía e Historia de Guatemala* 63:101–12.

McEvedy, Colin, and Richard Jones. 1978. *Atlas of World Population History* (Harmondsworth: Penguin).

McEvoy, Arthur P. 1986. *The Fisherman's Problem: Ecology and Law in the California Fisheries, 1860–1980* (New York: Cambridge University Press).

McGovern, Thomas. 1994. "Management for Extinction in Norse Greenland." In: Carol Brumley, ed., *Historical Ecology* (Santa Fe: School of American Research Press).

McKean, Margaret. 1981. *Environmental Protest and Citizen Politics in Japan* (Berkeley: University of California Press).

———. 1985. "The Evolution of Japanese Images of the Environment," *Internationales Asienforum* 16:25–48.

McNeill, J. R. 1988. "Deforestation in the Araucaria Zone of Southern Brazil, 1900–1983." In: J. F. Richards and R. P. Tucker, eds., *World Deforestation in the Twentieth Century* (Durham: Duke University Press), 1–32.

———. 1992a. "Kif in the Rif: An Historical and Ecological Perspective on Marijuana, Markets, and Manure in Northern Morocco," *Mountain Research and Development* 12:389–92.

———. 1992b. *The Mountains of the Mediterranean: An Environmental History* (New York: Cambridge University Press).

McNeill, William H. 1976. *Plagues and Peoples* (New York: Doubleday).

Meade, Robert H., T. R. Yuzyk, and T. J. Day. 1990. "Movement and Storage of Sediment in Rivers of the United States and Canada." In: M. G. Wolman and H. C. Riggs, eds., *Surface Water Hydrology* (Boulder: Geological Society of America), 255–80.

Melosi, Martin V. 1981. *Garbage in the Cities: Refuse, Reform and the Environment, 1880–1980* (College Station: Texas A&M University Press).

———. 1985. *Coping with Abundance: Energy and Environment in Industrial America* (Philadelphia: Temple University Press).

———. 1990. "Cities, Technical Systems, and the Environment," *Environmental History Review* 14:45–64.

———. 1993. "The Place of the City in Environmental History," *Environmental History Review* 17:1–23.

Menzies, Nicholas. 1996. "Forestry." In: Joseph Needham, ed., *Science and Civilization in China. VI. Biology and Biological Technology*, pt. 3 (Cambridge: Cambridge University Press), 540–689.

Mercer, Alex. 1990. *Disease, Mortality and Population in Transition: Epidemiological-*

Demographic Change in England since the Eighteenth Century as Part of a Global Phenomenon (Leicester: Leicester University Press).

Merchant, Carolyn. 1980. *The Death of Nature: Women, Ecology, and the Scientific Revolution* (San Francisco: Harper & Row).

Metaxas, Ioannis. 1969. *Logoi kai skepseis* [Speeches and Thoughts] (Athens: Ikaros, 2 vols.).

Meybeck, Michel. 1979. "Concentrations des eaux fluviales en éléments majeurs et apports en solution aux océans," *Revue de géologie dynamique et géographie physique* 221:215–46.

Meybeck, Michel, and Richard Helmer. 1989. "The Quality of Rivers: From Pristine Stage to Global Pollution," *Palaeogeography, Palaeoclimatology, Palaeoecology* 75:283–309.

Meyer, William. 1996. *Human Impact on the Earth* (New York: Cambridge University Press).

Michaux, Jacques, Gilles Cheylan, and Henri Croset. 1990. "Of Mice and Men." In: F. di Castri, A. J. Hansen, and M. Debussche, eds., *Biological Invasions in Europe and the Mediterranean Basin* (Dordrecht: Kluwer Academic), 263–84.

Michel, Aloys Arthur. 1967. *The Indus Rivers: A Study of the Effects of Partition* (New Haven: Yale University Press).

Micklin, Philip. 1995. "Turkmenistan." In: Philip R. Pryde, ed., *Environmental Resources and Constraints in the Former Soviet Republics* (Boulder: Westview), 275–88.

Micklin, Philip, and William D. Williams, eds. 1995. *The Aral Sea Basin* (Berlin: Springer-Verlag).

Miège, J.-L. 1989. "The French Conquest of Morocco." In: J. A. de Moor and H. L. Wesselring, eds., *Imperialism and War* (Leiden: E. J. Brill), 201–17.

Mielke, Howard W., Jana C. Anderson, and Kenneth J. Berry. 1983. "Lead Concentrations in Inner-City Soils as a Factor in the Child Lead Problem," *American Journal of Public Health* 73:1366–69.

Mignon, Christian. 1981. *Campagnes et paysans de l'Andalousie méditerranéenne* (Clermont-Ferrand: Faculté des Lettres et Sciences Humaines).

Milliman, J. D., and R. H. Meade. 1983. "World-Wide Delivery of River Sediments to the Oceans," *Journal of Geology* 91:1–21.

Milliman, John D., Yun-shan Qin, Moi-E Ren, and Yoshiki Saito. 1987. "Man's Influence on the Erosion and Transport of Sediments by Asian Rivers: The Yellow River (Huanghe) Example," *Journal of Geology* 95:751–62.

Mills, Edward L., Joseph H. Leach, James T. Carlton, and Carol L. Secor. 1993. "Exotic Species in the Great Lakes: A History of Biotic Crises and Anthropogenic Introductions," *Journal of Great Lakes Research* 19:1–54.

Mitchell, B. R. 1978. *European Historical Statistics, 1750–1970* (New York: Columbia University Press).

———. 1993. *International Historical Statistics: The Americas 1750–1988* (New York: Stockton Press).

———. 1995. *International Historical Statistics: Africa, Asia, & Oceania, 1750–1988* (New York: Stockton Press).

Miura, Toyohiko. 1975. *Taiki osen kara mita kankyo hakai no rekishi* [History of Environmental Destruction from the Perspective of Air Pollution] (Tokyo: Rodo Kagaku Sosho).

Moldan, Bedřich. 1997. "Czech Republic." In: J. Klarer and B. Moldan, eds., *The Environmental Challenge for Central European Economies in Transition* (Chichester: Wiley), 107–29.

Molina Buck, J. S. 1993. "Soil Erosion and Conservation in Argentina." In: David Pimentel, ed., *World Soil Erosion and Conservation* (Cambridge: Cambridge University Press), 171–92.

Molinelli-Cancellieri, Lucia. 1995. *Boues Rouges: La Corse dit non* (Paris: L'Harmattan).

Monastersky, Richard. 1994. "Earthmovers: Humans Take Their Place Alongside Wind, Water and Ice," *Science News* 146:432–33.

Monteiro, Salvador, and Leonel Kaz. 1992. *Floresta atlântica* (Rio de Janeiro: Edições Alumbramento).

Moore, Curtis. 1995. "Poisons in the Air," *International Wildlife* 25:38–45.

Moore, T. R. 1979. "Land Use and Erosion in the Machakos Hills," *Association of American Geographers, Annals* 69:419–31.

Morris-Suzuki, Tessa. 1994. *The Technological Transformation of Japan* (New York: Cambridge University Press).

Morse, Stephen S., ed. 1993. *Emerging Viruses* (New York: Oxford University Press).

Moser, Henri. 1894. *L'irrigation en Asie centrale* (Paris: Editions Scientifique).

Moulin, Anne Marie. 1992. "La métaphore vaccine: De l'inoculation à la vaccinologie," *History and Philosophy of the Life Sciences* 14:271–97.

Moussiopoulos, N., H. Power and C. A. Brebbia, eds. 1995. *Urban Pollution*. Vol. 3 of *Air Pollution III* (Southampton, U.K.: Computational Mechanics Publishers).

Munch, Peter. 1993. *Stadthygiene im 19. und 20. jahrhundert* (Göttingen: Vandenhoeck und Ruprecht).

Murley, Loveday, ed. 1995. *Clean Air around the World* (Brighton: International Union of Air Pollution Prevention and Environmental Protection Associations).

Murray, Christopher J. L., and Alan D. Lopez eds. 1996, *The Global Burden of Disease* (Cambridge: Harvard School of Public Health).

Murphy, Frederick A. 1994. "New, Emerging, and Reemerging Infectious Diseases," *Advances in Virus Research* 43:1–52.

Murphy, Frederick A., and Neal Nathanson. 1994. "The Emergence of New Virus Diseases: An Overview," *Virology* 5:87–102.

Myers, K. 1986. "Introduced Vertebrates in Australia, with an Emphasis on Mammals." In: R. H. Groves and J. J. Burdon, eds., *Ecology of Biological Invasions* (Cambridge: Cambridge University Press), 120–36.

Myers, Norman. 1993. "Population, Environment, and Development." *Environmental Conservation* 20:205–16.

Nagy, Laszló. 1988. "A Duna hasznosítása," *Földrajzi Közlemények* 36(1–2):55–60.

Naiman, R. J., J. M. Melillo, and E. J. Hobbie. 1986. "Ecosystem Alteration of Boreal Forest Streams by Beaver *(Castor canadiensis),*" *Ecology* 67:1254–69.

Nakahara, Hiroyuki. 1992. "Japanese Efforts in Marine-Ranching Development." In: J. B. Marsh, ed., *Resources and Environment in Asia's Marine Sector* (Washington: Taylor & Francis), 199–218.

Nakicenovic, Nebojsa. 1996. "Freeing Energy from Carbon," *Daedalus* (U.S.) 125(3):95–112.

Natural Resources Defense Council. 1996. *Breath-Taking: Premature Mortality Due to Particulate Air Pollution in 239 American Cities* (Washington: NRDC).

Nemecek, Sasha. 1995. "When Smog Gets in Your Eyes," *Scientific American* 273 (July):29–30.

Neu, Harold C. 1992. "Crisis in Antibiotic Resistance," *Science* 257:1064–73.

Neuvy, Guy. 1991. *L'homme et l'eau dans le domaine tropical* (Paris: Masson).

Nielsen, Svend. 1988. "Dansk Landbrug, 1788–1988," *Arv og Eje* 1988:7–62.

Nilsen, Knut Erik, and Frederic Hauge. 1992. "Mayak: The Most Radioactive Polluted Place on Earth," Bellona Foundation (Oslo), Report no. 1:92.

Nilsen, Thomas, and Nils Bohmer. 1994. *Sources to Radioactive Contamination in Murmansk and Arkanghel'sk Counties* (Oslo: Bellona Foundation).

Nimura, Kazuo. 1997. *The Ashio Riot of 1907: A Social History of Mining in Japan* (Durham: Duke University Press).

Nir, Dov. 1983. *Man, A Geomorphological Agent* (Dordrecht: Reidel).

Nishigaki, S., and M. Harada. 1975. "Methylmercury and Selenium in Umbilical Cords of Inhabitants of the Minamata Area," *Nature* 258:324–5.

Nishimura, H., ed. 1989. *How to Conquer Air Pollution: A Japanese Experience* (Amsterdam: Elsevier).

Nolte, Ernst. 1966. *Three Faces of Fascism* (New York: Holt, Rinehart & Winston).

Northrup, David. 1995. *Indentured Labor in the Age of Imperialism, 1834–1922* (New York: Cambridge University Press).

Norwich, John Julius. 1991. "Venice in Peril," *Proceedings of the Royal Institution of Great Britain* 63:243–67.

Notehelfer, F. G. 1975. "Japan's First Pollution Incident," *Journal of Japanese Studies* 1:351–84.

NRC [National Research Council]. 1989. *Irrigation-induced Water Quality Problems: What Can Be Learned from the San Joaquin Valley Experience* (Washington: NRC).

———. 1992. *Restoration of Aquatic Ecosystems* (Washington: National Academy Press).

———. 1993a. *Soil and Water Quality: An Agenda for Agriculture* (Washington: National Academy Press).

———. 1993b. *Sustainable Agriculture and the Environment in the Humid Tropics* (Washington: National Academy Press).

———. 1996. *The Bering Sea Ecosystem* (Washington: National Academy Press).

Nriagu, J. O. 1990a. "Global Metal Pollution," *Environment* 32(7):7–11, 28–33.

———. 1990b. "The Rise and Fall of Leaded Gasoline," *Science of the Total Environment* 92:13–28.

———. 1994. "Industrial Activity and Metal Emissions." In: R. Socolow, C. Andrews, F. Berkhout, and V. Thomas, eds., *Industrial Ecology and Global Change* (Cambridge: Cambridge University Press), 277–85.

———. 1996. "History of Global Metal Pollution," *Science* 272 (12 April 1996):223–4.

Nriagu, Jerome, Champak Jinabhai, Rajen Naidoo, and Anna Coutsoudis. 1996a. "Atmospheric Lead Pollution in KwaZulu/Natal, South Africa," *Science of the Total Environment* 191:69–76.

Nriagu, J. O., Mary Blankson, and Kwamena Ocran. 1996b. "Childhood Lead Poisoning in Africa," *Science of the Total Environment* 181:93–100.

Nuccio, Richard, and Angelina Ornelas. 1987. *Developing Disasters: Mexico's Environment and the United States* (Washington: World Resources Institute).

Obeng, L. E. 1977. "Should Dams Be Built? The Volta Lake Example," *Ambio* 6:46–50.

OECD [Organization for Economic Cooperation and Development]. 1995. *Motor Vehicle Pollution* (Paris: OECD).

———. 1998. *The Environmental Effects of Reforming Agricultural Policies* (Paris: OECD).

OECD Nuclear Energy Agency. 1995. "Chernobyl Ten Years On: Radiological and Health Impact." Unpublished report available at *http://www.nea.fr/html/rp/chernobyl/allchernobyl.html.*

Ogawa, Y. 1991. "Economic Activity and the Greenhouse Effect," *Energy Journal* 12:23–6.

Ogutu-Ohwayo, R. 1990. "The Decline of Native Fishes of Lakes Victoria and Kyoga (East Africa) and the Impact of Introduced Species," *Environmental Biology of Fishes* 27:81–96.

Okada, Yuko. 1993. "Kosaka kozan engai mondai to hantai undo, 1901–17," [The movement against smoke pollution at the Kosaka copper mine, 1901–17] *Shakai-Keizai Shigaku* 56:59–89.

Oldstone, Michael B. A. 1998. *Viruses, Plagues and History.* (New York: Oxford University Press).

Oliver, Harold H. 1992. "The Neglect and Recovery of Nature in the Twentieth-Century Protestant Thought," *American Academy of Religion: Journal* 60:379–404.

Olson R., D. Binkley, and M. Böhm, eds. 1992. *The Response of Western Forests to Air Pollution* (New York: Springer-Verlag).

Olson, R. A. 1987. "The Use of Fertilizers and Soil Amendments." In: M. G. Wolman and F. G. A. Fourier, eds., *Land Transformation in Agriculture* (Chichester: Wiley), 203–26.

Omrane, Mohammed Maceur. 1991. "La croissance de l'agglomération de Tunis, et ses conséquences sur l'utilisation de l'eau." In: *L'eau et la ville* [no editor] (Tours: Centre d'Études et de Recherches URBAMA), 163–72.

Ondiege, Peter. 1996. "Land Tenure and Soil Conservation." In: Calestous Juma and J. B. Ojwang, eds., *In Land We Trust: Environment, Private Property and Constitutional Change* (Nairobi: Initiatives Publishers), 117–42.

Opie, John. 1987. "Renaissance Origins of the Environmental Crisis," *Environmental Review* 11:2–17.

———. 1993. *Ogallala: Water for a Dry Land* (Lincoln: University of Nebraska Press).

Ordos, J. 1991. "Landnutzung und Umweltprobleme in der Slowakischen Republik," *Österreichische Osthefte* 33:697–716.

Orhonlu, Cengiz. 1984. *Osmanlı İmperatorluğunda Şehircilik ve Ulaşim* (Izmir: Ticaret Matbaacılık T.A.S.).

Osaghae, Eghosa E. 1995. "The Ogoni Uprising," *African Affairs* 94:325–44.

Oschlies, Wolf. 1985. *Aus Sorge um "Mutter Erde": Unweltschutz und Okologiediskussion in Bulgarien* (Cologne: Bundesinstitut für Ostwissenschaftliche und Internationale Studien).

Ostrom, Elinor. 1992. "The Rudiments of a Theory of the Origins, Survival and Performance of Common-Property Institutions." In: Daniel W. Bromley, ed., *Making the Commons Work* (San Francisco: Institute for Contemporary Studies), 293–318.

Öziş, Ünal. 1987. "Historical Parallels in the Water Supply Development of Rome and Istanbul," In: W. O. Wunderlich and J. E. Prins, eds., *Water for the Future* (Rotterdam: Balkema), 35–44.

Outwater, Alice. 1996. *Water: A Natural History* (New York: Basic Books).

Palloni, Alberto. 1994. "The Relation between Population and Deforestation: Methods for Drawing Causal Inferences from Macro and Micro Studies." In: Lourdes Arizpe, Priscilla Stone, and David Major, eds., *Population and Environment: Rethinking the Debate* (Boulder: Westview), 125–65.

Panjari, Rance K. L. 1997. *The Earth Summit at Rio* (Boston: Northeastern University Press).

Pantazopoulou, A., et al. 1995. "Short-Term Effects of Air Pollution on Hospital Emergency Outpatient Visits and Admissions in the Greater Athens, Greece Area," *Environmental Research* 69:31–6.

Papaioanniou, J. 1967. "Air Pollution in Athens," *Ekistics* 24:72–80.

Pascon, Paul, and Jacques Berque 1978. *Structures sociales du Haut-Atlas* (Paris: Presses Universitaires de France).

Pattas, K., Z. Samaras, N. Moussiopoulos, and K.-H. Zierock. 1994. *Policy for Reduction of Traffic Related Emissions in the Greater Athens Area* (Luxembourg: European Commission).

Paul, Adam. 1987. "Les nouvelles pêches maritimes mondiales," *Etudes internationales* 18:7–20.

Pauly, D., and V. Christensen. 1995. "Primary Production Required to Sustain Global Fisheries," *Nature* 374:255–7.

Payne, Roger. 1995. *Among Whales* (New York: Delta Books).

Pelekasi, Katerina, and Michalis Skourtos. 1991. "Air Pollution in Greece: An Overview," *Ekistics* 58:135–55.

———. 1992. *He Atmosphairiki rypansi stin Ellada* (Athens: Papazisi).

Peluso, Nancy Lee. 1992. *Rich Forests, Poor People: Resource Control and Resistance in Java* (Berkeley: University of California Press).

Pepper, David. 1996. *Modern Environmentalism* (London: Routledge).

Peterson, D. J. 1993. *Troubled Lands: The Legacy of Soviet Environmental Destruction* (Boulder: Westview).

Petts, Geoff. 1990a. "Forested River Corridors: A Lost Resource." In: Denis Cosgrove and Geoff Petts, eds., *Water, Engineering and Landscape* (London: Belhaven), 12–34.

Petts, Geoff. 1990b. "Water, Engineering and Landscape: Development, Protection and Restoration." In: Denis Cosgrove and Geoff Petts, eds., *Water, Engineering and Landscape* (London: Belhaven), 188–208.

Pfister, Christian, ed. 1995. *Das 1950er Syndrom: Das Weg in die Konsumgesellschaft* (Bern: Verlag Paul Haupt).

Phillips, Steven. 1994. *The Soil Conservation Service Responds to the 1993 Midwest Floods* (Washington: USDA Soil Conservation Service).

Pichón, Francisco J. 1992. "Agricultural Settlement and Ecological Crisis in the Ecuadorian Amazon Frontier," *Policy Studies Journal* 20:662–678.

Pick, James B., and Edgar W. Butler. 1997. *Mexico Megacity* (Boulder: Westview).

Pilkey, Orrin, and Katharine Dixon. 1996. *The Corps and the Shore* (Washington: Island Press).

Pimentel, David, ed. 1993. *World Soil Erosion and Conservation* (Cambridge: Cambridge University Press).

Pimentel, David, and G. H. Heichel. 1991. "Energy Efficiency and Sustainability of Farming Systems." In: R. Lal and F. J. Pierce, eds., *Soil Management for Sustainability* (Ankeny, Iowa: Soil and Water Conservation Society), 113–24.

Pimentel, David, and Hugh Lehman, eds. 1993. *The Pesticide Question: Environment, Economics, and Ethics* (New York: Chapman & Hall).

Pimentel, David, et al. 1993. "Soil Erosion and Agricultural Productivity." In: Pimentel, ed., *World Soil Erosion and Conservation* (Cambridge: Cambridge University Press), 277–92.

Pimentel, David, et al. 1995. "Environmental and Economic Costs of Soil Erosion and Conservation Benefits," *Science* 267:1117–23.

Pineo, Ronn F. 1996. *Social and Economic Reform in Ecuador: Life and Work in Guayaquil* (Gainesville: University Press of Florida).

Pineo, Ronn F., and James A. Baer, eds. 1998. *Cities of Hope: People, Protests, and Progress in Urbanizing Latin America, 1870–1930* (Boulder: Westview).

Pinon, Pierre, and Stephane Yerasimos. 1994. "Istanbul, acquedoti, cisterne, fontane e dighe," *Rassegna* 57:54–9.

Planhol, X. de. 1969. "Le déboisement de l'Iran," *Annales de géographie* 73:625–35.

Platt, Harold A. 1991. *The Electric City: Energy and the Growth of the Chicago Area, 1880–1930* (Chicago: University of Chicago Press).

———. 1995. "Invisible Gases: Smoke, Gender and the Redefinition of Environmental Policy in Chicago, 1900–1920," *Planning Perspectives* 10:67–97.

Pletcher, Jim. 1991. "Ecological Deterioration and Agricultural Stagnation in Eastern Province, Zambia," *Centennial Review* 35:369–388.

Polyakov, Alexei, and Igor Ushkalov. 1995. "Migrations in Socialist and Post-Socialist Russia." In: Robin Cohen, ed., *The Cambridge Survey of World Migration* (Cambridge: Cambridge University Press), 490–95.

Pomeranz, Kenneth. 1993. *The Making of a Hinterland: State, Society and Economy in Inland North China, 1853–1937* (Berkeley: University of California Press).

Ponting, Clive. 1991. *A Green History of the World* (New York: Penguin).

Population Reference Bureau. 1996. *Population Data Sheet* (Washington: PRB).

Por, Francis Dov. 1978. *Lessepsian Migration: The Influx of Red Sea Biota into the Mediterranean by Way of the Suez Canal* (Berlin: Springer-Verlag).

——. 1990. "Lessepsian Migration: An Appraisal and New Data," *Bulletin de l'Institut Océanographique de Monaco*, special no. 7:1–7.

Porter, Roger. 1997. *The Greatest Benefit to Mankind: A Medical History of Humanity from Antiquity to the Present* (London: HarperCollins).

Postel, Sandra. 1992. *Last Oasis* (New York: Norton).

——. 1996. "Forging a Sustainable Water Strategy." In: Lester Brown et al., *State of the World 1996* (Washington: Worldwatch Institute).

——. 1999. *Pillar of Sand: Can the Irrigation Miracle Last?* (New York: Norton).

Postel, Sandra, Gretchen Daily, and Paul Ehrlich. 1996. "Human Appropriation of Renewable Fresh Water," *Science* 271:785–88.

Potter, Lesley. 1996. "Forestry in Contemporary Indonesia." In: J. T. Lindblad, ed., *Historical Foundations of a National Economy in Indonesia, 1890s–1990s* (Amsterdam: Royal Netherlands Academy of Sciences), 369–84.

Powell, R. J., and L. M. Wharton. 1982. "Development of the Canadian Clean Air Act," *Journal of the Air Pollution Control Association* 32(1):62–5.

Prager, Herman. 1993. *Global Marine Environment* (Lanham: University Press of America).

Precoda, Norman. 1991. "Requiem for the Aral Sea," *Ambio* 20(3–4):109–114.

Price, Marie. 1994. "Ecopolitics and Environmental Nongovernmental Organizations in Latin America," *Geographical Journal* (U.K.) 84:42–59.

Prince, Hugh. 1997. *Wetlands of the American Midwest* (Chicago: University of Chicago Press).

Prochaska, David. 1986. "Fire on the Mountain: Resisting Colonialism in Algeria." In: D. Crummey, ed., *Banditry, Rebellion and Social Protest in Africa* (London: James Currey)

Quine, Maria Sophia. 1996. *Population Politics in Twentieth-Century Europe* (London: Routledge).

Quinn, M.-L. 1988. "Tennessee's Copper Basin: A Case for Preserving an Abused Landscape," *Journal of Soil and Water Conservation* 43:140–44.

——. 1989. "Early Smelter Sites: A Neglected Chapter in the History and Geography of Acid Rain in the United States," *Atmospheric Environment* 23:1281–92.

Rackham, Oliver, and Jennifer Moody. 1996. *The Making of the Cretan Landscape* (Manchester: Manchester University Press).

Raghavan, G. S. D. V., P. Alvo, and E. McKyes. 1990. "Soil Compaction in Agriculture: A View toward Managing the Problem," *Advances in Soil Sciences* 11:1–36.

Randrianarijaona, Philemon. 1983. "The Erosion of Madagascar," *Ambio* 12:308–11.

Ranger, Terence. 1992. "Plagues of Beasts and Men: Prophetic Responses to Epidemic in Eastern and Southern Africa." In: Terence Ranger and Paul Slack, eds., *Epidemics and Ideas* (Cambridge: Cambridge University Press), 241–68.

Rao, Radhakrishna. 1989. "Water Scarcity Haunts World's Wettest Place," *Ambio* 18(5):300.

Raskin, P. D. 1995. "Methods for Estimating the Population Contribution to Environmental Change," *Ecological Economics* (Netherlands) 15:225–33.

Rasmussen, Wayne. 1982. "The Mechanization of Agriculture," *Scientific American* 247(2):67–75.

Ravenstijn, Wim. 1997. *De zegenrijke heren der wateren: Irrigatie en staat op Java, 1832–1942* (Delft: Delft University Press).

Raviglione, Mario, Dixie E. Snider, and Arata Kocki. 1995. "Global Epidemiology of Tuberculosis," *Journal of the American Medical Association* 273:220–6.

Reddy, Amulya K. N., and José Goldemberg. 1991. "Energy for the Developing World." In: *Energy for Planet Earth: Readings from Scientific American Magazine* (New York: Freeman), 58–71.

Redford, Kent. 1992. "The Empty Forest," *BioScience* 42:412–22.

Rees, William E. 1992. "Ecological Footprints and Appropriated Carrying Capacity: What Urban Economics Leaves Out," *Environment and Urbanization* 4:121–30.

Regier, Henry A., and John L. Goodier. 1992. "Irruption of Sea Lamprey in the Upper Great Lakes." In: Michael H. Glantz, ed., *Climate Variability, Climate Change and Fisheries* (New York: Cambridge University Press), 192–212.

Reihelt, Günther. 1986. *Lasst den Rhein leben* (Düsseldorf: Girardet).

Reij, Chris, Ian Scoones, and Camilla Toulmin, eds. 1996. *Indigenous Soil and Water Conservation in Africa* (London: Earthscan).

Reinhard, Marcel, André Armengaud, and Jacques Dupâquier. 1968. *Histoire générale de la population mondiale* (Paris: Editions Montchrestien).

Relph, E. C. 1987. *The Modern Urban Landscape: 1880 to the Present* (Baltimore: Johns Hopkins University Press).

Ren, M., and X. Zhu. 1994. "Anthropogenic Influences on Changes in the Sediment Load of the Yellow River, China, during the Holocene," *Holocene* 4(3):314–20.

Ren, Mei-e, and Jesse Walker. 1998. "Environmental Consequences of Human Activity on the Yellow River Delta, China," *Physical Geography* 19(5):421–32.

Repetto, R. 1986. "Soil Loss and Population Pressure on Java," *Ambio* 15:14–18.

Repetto, R., and Thomas Holmes. 1983. "The Role of Population in Resource Depletion in Developing Countries," *Population and Development Review* 9:609–32.

Restrepo, Ivan, ed. 1992. *La contaminación atmosférica en México* (Mexico City: Comisión Nacional de Derechos Humanos).

ReVelle, Penelope, and Charles ReVelle. 1992. *The Global Environment: Securing a Sustainable Future* (Boston: Jones & Bartlett).

Rich, Bruce. 1994. *Mortgaging the Earth: The World Bank, Environmental Impoverishment, and the Crisis of Development* (Boston: Beacon Press).

Richards, J. F. 1990a. "Agricultural Impacts in Tropical Wetlands: Rice Paddies for Mangroves in South and Southeast Asia." In: M. Williams, ed., *Wetlands: A Threatened Landscape* (Oxford: Basil Blackwell), 217–33.

———. 1990b. "Land Transformation." In: B. L. Turner et al., eds., *The Earth as Transformed by Human Action* (New York: Cambridge University Press), 163–78.

Richards, J. F., and Elizabeth P. Flint. 1990. "Long-Term Transformations in the Sundarbans Wetlands Forests of Bengal," *Agriculture and Human Values* 7:17–33.

Ripley, Earle A., Robert E. Redmann and Adele A. Crowder. 1996. *Environmental Effects of Mining* (Delray Beach, Fl.: St. Lucie Press).

RIVM [Rijkinstituut voor Volksgezondheid en Milieu] 1997. *A Hundred Year (1890–1990)*

Database for Integrated Environmental Assessments, C. G. M. Klein Goldewijk and J. J. Battjes, eds. (Bilthoven: RIVM).

RIVM/UNEP. [RIVM and U.N. Environment Programme]. 1997. *The Future of the Global Environment,* A. J. Bakkes and J. W. van Woerden., eds. (Bilthoven: RIVM).

Roberts, Bryan. 1994. "Urbanization and the Environment in Developing Countries." In: Lourdes Arizpe, M. Priscilla Stone, and David Major, eds., *Population and Environment: Rethinking the Debate* (Boulder: Westview), 303–36.

Roberts, Neil. 1989. *The Holocene: An Environmental History* (Oxford: Basil Blackwell).

Robertson, C. J. 1938. "Agricultural Regions of the North Italian Plain," *Geographical Review* (U.S.) 28:573–96.

Rodhe, Henning. 1989. "Acidification in a Global Perspective," *Ambio* 18:155–60.

Rodhe, H., et al. 1995. "Global Scale Transport of Acidifying Compounds," *Water, Air and Soil Pollution* 85:37–50.

Rollins, William H. 1997. *A Greener Vision of Home: Cultural Politics and Environmental Reform in the German Heimatschutz Movement, 1904–1918* (Ann Arbor: University of Michigan Press).

Romero Lankao, Patricia. 1999. *Obra hidráulica de la ciudad de México y su impacto socioambiental (1880–1990)* (Mexico City: Instituto Mora).

Rosen, Christine M. 1995. "Businessmen against Pollution in Nineteenth-Century Chicago," *Business History Review* 69:351–97.

Rosenzweig, Cynthia, and Daniel Hillel. 1995. "Potential Impacts of Climate Change on Agriculture and Food Supply," *Consequences* (U.S.) 1(2):22–32.

Rosner, David, and Gerald Markowitz. 1985. "A 'Gift from God'? The Public Health Controversy over Leaded Gasoline during the 1920s," *Journal of Public Health Policy* 75:344–52.

Rothman, Hal K. 1998. *The Greening of a Nation* (Fort Worth: Harcourt Brace).

Rothschild, Brian J. 1996. "How Bountiful Are Ocean Fisheries," *Consequences* (U.S.) 2(1):14–25.

Rouse, Hunter. 1963. *History of Hydraulics* (New York: Dover Press).

Rozanov, Boris G., Viktor Targulian, and D. S. Orlov. 1990. "Soils." In: B. L. Turner et al., eds., *The Earth as Transformed by Human Action* (New York: Cambridge University Press), 203–14.

Ruddle, K. 1987. "The Impact of Wetland Reclamation." In: M. G. Wolman and F. G. A. Fournier, eds., *Land Transformation in Agriculture* (Chichester: Wiley), 171–201.

Ryabchikov, A. R. 1975. *The Changing Face of the Earth* (Moscow: Progress Publishers).

SADCC [South African Development Coordination Conference]. 1987. *History of Soil Conservation in the SADCC Region* (Maseru, Lesotho: SADCC Soil and Water Conservation and Land Utilization Programme).

Sadkovich, J. J. 1996. "The Indispensable Navy." In: N. A. M. Rodger, ed., *Naval Power in the Twentieth Century* (Annapolis: Naval Institute Press), 66–76.

Said, Rushdi. 1993. *The River Nile: Geology, Hydrology, and Utilization* (Oxford: Pergamon Press).

Salau, Fatai Kayode. 1996. "Nigeria." In: M. Jänicke and H. Weidner, eds., *National Environmental Policies* (Berlin: Springer-Verlag), 257–78.

Salmon, Lynn, et al. 1995. "Source Contributions to Airborne Particle Deposition at the Yungang Grottoes, China," *Science of the Total Environment* 167:33–47.

Salstein, David. 1995. "Mean Properties of the Atmosphere." In: H. B. Singh, ed., *Composition, Chemistry and Climate of the Atmosphere* (New York: Van Nostrand Reinhold), 19–49.

Santiago, Myrna. 1997. "Huasteca Crude: Indians, Ecology and Labor in the Mexican Oil Industry, Northern Veracruz, 1900–1938" (Ph.D dissertation, University of California, Berkeley).

———. 1998. "Rejecting Progress in Paradise: Huastecs, the Environment, and the Oil Industry in Veracruz, Mexico, 1900–1935," *Environmental History* 3:169–88.

Satake, Ken'ichi, Atushi Tanaka, and Katsuhiko Kimura. 1996. "Accumulation of Lead in Tree Trunk Bark Pockets as Pollution Time Capsules," *Science of the Total Environment* 181:25–30.

Savchenko, V. K. 1995. *The Ecology of the Chernobyl Disaster* (Paris: UNESCO).

Scheraga, J. D. 1986. "Pollution in Space: An Economic Perspective," *Ambio* 15(5):358–60.

Schlager, Neil. 1994. *When Technology Fails* (Detroit: Gale Research).

Schröder, R. 1987. "Decline of Swamps in Lake Constance," *Symposia Biologica Hungarica* 19:43–8.

Schulze, R. H. 1993. "The 20-Year History of the Evolution of Air Pollution Control Legislation in the United States," *Atmospheric Environment* 27B(1):15–22.

Schurr, S. H. 1984. "Energy Use, Technical Change and Productive Efficiency: An Economic-Historical Interpretation," *Annual Review of Energy and the Environment* 9:409–25.

Schwarz, Harry E., Jacque Emel, W. J. Dickens, Peter Rogers, and John Thompson. 1990. "Water Quality and Flows." In: B. L. Turner et al., eds., *The Earth as Transformed by Human Action* (New York: Cambridge University Press), 253–70.

Shaler, Nathaniel Southgate. 1905. *Man and the Earth* (New York: Fox, Duffield).

Shao, John. 1986. "The Villagization Program and the Disruption of Ecological Balance in Tanzania," *Canadian Journal of African Studies* 20:219–39.

Sharma, Rita, and Thomas T. Poleman, eds. 1993. *The New Economics of India's Green Revolution: Income and Employment Diffusion in Uttar Pradesh* (Ithaca: Cornell University Press).

Shaw, Glenn. 1995. "The Arctic Haze Phenomenon," *Bulletin of the American Meteorological Society* 76:2403–13.

Shcherbak, Yuri M. 1996. "Ten Years of the Chornobyl Era," *Scientific American* 274(4):44–9.

Sheail, John. 1997. "The Sustainable Management of Industrial Watercourses: An English Historical Perspective," *Environmental History* 2:197–215.

Shen, Tun-Li, P.J. Woolridge, and M. J. Molina. 1995. "Stratospheric Pollution and Ozone Depletion." In: B. H. Singh, ed., *Composition, Chemistry and Climate of the Atmosophere* (New York: Van Nostrand), 394–442.

Shen, Xiao-ming, et al. 1996. "Childhood Lead Poisoning in China," *Science of the Total Environment* 181:101–9.

Sherlock, R. L. 1931. *Man's Influence on the Earth* (London: T. Butterworth).

Shiklomanov, I. A. 1990. "Global Water Resources," *Nature and Resources (English Edition)* 26(3):34–43.

———. 1993. "World Fresh Water Resources." In: Peter Gleick, ed., *Water in Crisis* (New York: Oxford University Press), 13–24.

Shiva, Vandana. 1991a. *Ecology and the Politics of Survival* (New Delhi: U.N. University Press).

———. 1991b. *The Violence of the Green Revolution* (London: Zed Books).

Shoji, Kichiro, and Masuro Sugai. 1992. "The Ashio Copper Mine Pollution Case: The Origins of Environmental Destruction." In: Jun Ui, ed., *Industrial Pollution in Japan* (Tokyo: U.N. University Press), 18–64.

Shope, Robert, and Alfred S. Evans. 1993. "Assessing Geographic and Transport Factors, and Recognition of New Viruses." In: Stephen S. Morse, ed., *Emerging Viruses* (New York: Oxford University Press, 1993), 109–19.

Showers, Kate. 1989. "Soil Erosion in the Kingdom of Lesotho: Origins and Colonial Response, 1830s–1950s," *Journal of Southern African Studies* 15(2):263–86.

Showers, Kate, and Gwendolyn M. Malahlela. 1992. "Oral Evidence in Historical Environmental Impact Assessment: Soil Conservation in Lesotho in the 1930s and 1940s," *Journal of Southern African Studies* 18:276–98.

Sievert, James. 1996. "Construction and Destruction of Nature in Italy, 1860–1914" (Ph.D. dissertation, University of California, Santa Cruz).

Sifakis, Nicholas T. 1991. "Air Pollution in Athens: Similarities of Findings with Remote Sensing Methods in 1967 and 1987," *Ekistics* 58:164–66.

Sikiotis, D., and P. Kirkitsos. 1995. "The Adverse Effects of Nitrates on Stone Monuments," *Science of the Total Environment* 171:173–82.

Silversides, Ross C. 1997. *Broadaxe to Flying Shear: The Mechanization of Forest Harvesting East of the Rockies* (Ottawa: National Museum of Science and Technology).

Simberloff, Daniel. 1996. "Impacts of Introduced Species in the United States," *Consequences* (U.S.) 2(2):13–22.

Simmons, Jack, and Gordon Biddle, eds. 1997. *Oxford Companion to British Railway History* (Oxford: Oxford University Press).

Simon, Joel. 1997. *Endangered Mexico: An Environment on the Edge* (San Francisco: Sierra Club Books).

Simonian, Lane. 1995. *Defending the Jaguar: A History of Conservation in Mexico* (Austin: University of Texas Press).

Singh, B. H., ed. 1995. *Composition, Chemistry, and Climate of the Atmosphere* (New York: Van Nostrand Reinhold).

Sit, Victor F. S. 1995. *Beijing: The Nature and Planning of a Chinese Capital City* (Chichester: Wiley).

Smil, Vaclav. 1984. *The Bad Earth: Environmental Degradation in China* (Armonk: M. E. Sharpe).

———. 1990. "Nitrogen and Phosphorus." In: B. L. Turner et al., eds., *The Earth as Transformed by Human Action* (New York: Cambridge University Press), 423–36.

———. 1993. *China's Environmental Crisis* (Armonk and London: M. E. Sharpe).

———. 1994. *Energy in World History* (Boulder: Westview).

Smith, Bernard, and Clark Baillie. 1985. "Erosion in the Savannas," *Geographical Magazine* 57(3):137–43.

Smith, David. 1995. "Kazakhstan." In: Philip R. Pryde, ed., *Environmental Resources and Constraints in the Former Soviet Republics* (Boulder: Westview), 251–74.

Snyder, Lynne Page. 1994. " 'The Death-dealing Smog over Donora, Pennsylvania': Industrial Air Pollution, Public Health Policy, and the Politics of Expertise, 1948–1949," *Environmental History Review* 18:117–133.

Sobolev, S. S. 1947. "Protecting the Soils of the USSR," *Journal of Soil and Water Conservation* 2:123–32.

Society for General Microbiology. 1995. *Fifty Years of Antimicrobials* (New York: Cambridge University Press).

Solbrig, Otto, and Dorothy Solbrig. 1994. *So Shall You Reap: Farming and Crops in Human Affairs* (Washington: Island Press).

Solomon, Susan Gross, and John F. Hutchinson, eds. 1990. *Health and Society in Revolutionary Russia* (Bloomington: Indiana University Press).

Sonnenfeld, David A. 1992. "Mexico's 'Green Revolution' 1940–1980: Toward an Environmental History," *Environmental History Review* 16(4):28–52.

Sørensen, Bent. 1995. "History of, and Recent Progress in, Wind-Energy Utilization," *Annual Review of Energy and the Environment* 20:387–424.

Sorocos, Eustache P. 1985. *La morphologie social du Pirée à travers son évolution* (Athens: National Social Research Centre).

Sparks, D. L. 1984. Namibia's Coastal and Marine Development Potential," *African Affairs* 83:477–96.

Speer, Lisa, et al. 1997. *Hook, Line, and Sinking: The Crisis in Marine Fisheries* (New York: Natural Resources Defense Council).

Spelsberg, Gerd. 1984. *Rauchplage: Hundert Jahre Saurer Regen* (Aachen: Alano Verlag).

Spielman, Andrew. 1994. "The Emergence of Lyme Disease and Human Babesiosis in a Changing Environment." In: Mary E. Wilson, Richard Levins, and Andrew Spielman, eds., *Disease in Evolution* (New York: Annals of the New York Academy of Sciences, vol. 740), 146–56.

Spinage, C. A. 1962. "Rinderpest and Faunal Distribution Patterns," *African Wildlife* 16:55–60.

Sreenivasan, A. 1991. "Transfers of Freshwater Fishes into India." In: P. S. Ramakrishnan, ed., *Ecology of Biological Invasion in the Tropics* (New Delhi: International Scientific Publishers), 131–8.

Srinanda, K. V. 1984. "Air Pollution Control and Air Quality Management in Malaysia," *Kaijan Malaysia* 2:67–91.

Stanley, D. J. 1996. "Nile Delta: Extreme Case of Sediment Entrapment on a Delta Plain and Consequent Coastal Land Loss," *Marine Geology* 129:189–95.

––––––. 1997. "Degradation of the Nile Delta," *Environmental Review* 4(10):1–7.

Stanley, D. J., and A. G. Warne. 1993. "Nile Delta: Recent Geological Evolution and Human Impact," *Science* 260:628–34.

Stanners, David, and Philippe Bourdeau. 1995. *Europe's Environment: The Dobříš Assessment* (Copenhagen: European Environment Agency).

Stanton, B. F. 1998. "Agriculture." In: Richard W. Bulliet, ed., *The Columbia History of the 20th Century* (New York: Columbia University Press), 345–80.

Stark, Malcolm. 1987. "Soil Erosion Out of Control in Southern Alberta," *Canadian Geographic* 107 (June–July):16–25.

Starr, Chauncey. 1990. "Implications of Continuing Electrification." In: John L. Helms, ed., *Energy: Production, Consumption, and Consequences* (Washington: National Academy Press), 52–71.

––––––. 1996. "Sustaining the Human Environment: The Next 200 Years," *Daedalus* (U.S.) 125(3):235–53.

St. Clair, David J. 1986. *The Motorization of American Cities* (Westport: Praeger).

Stebelsky, Ihor. 1989. "Soil Management in Ukraine: Responding to Environmental Degradation," *Canadian Slavonic Papers/Revue Canadienne des Slavistes* 31(3–4):247–66.

Stern, A. C. 1982. "History of Air Pollution Legislation in the United States," *Journal of the Air Pollution Control Association* 32(1):44–61.

Stern, David I., and Robert K. Kaufman. 1996. "Estimates of Global Anthropogenic Methane Emissions, 1860–1993," *Chemosphere* 33:159–76.

Stevis, Dimitris. 1993. "Political Ecology in the Semi-Periphery: Lessons from Greece," *International Journal of Urban and Regional Research* 17:85–97.

Stewart, John Massey. 1992. *The Nature of Russia* (New York: Cross River Press).

Stocking, Michael. 1985. "Soil Conservation Policy in Colonial Africa." In: Douglas Helms and Susan Flader, eds., *The History of Soil and Water Conservation* (Washington: Agricultural History Society), 46–59.

Stoett, Peter J. 1997. *The International Politics of Whaling* (Vancouver: University of British Columbia Press).

Stolberg, Michael. 1994. *Ein Recht auf saubere Luft? Umweltkonflikte am Beginn des Indus-triezeitalters* (Erlangen: Fischer Verlag).

Stout, Glen, and William Ackermann. 1987. "Past and Future Water Systems for Chicago." In: W. O. Wunderlich and J. E. Prins, eds., *Water for the Future* (Rotterdam: Balkema), 201–10.

Stradling, David. 1996. "Civilized Air: Coal Smoke and Environmentalism in America, 1880–1920" (Ph.D. dissertation, University of Wisconsin).

Stradling, David, and Peter Thorsheim. 1999. "The Smoke of Great Cities: British and American Efforts to Control Air Pollution, 1860–1914," *Environmental History* 4:6–31.

Sundaramoorthy, S., R. Kannapan, E. Vedagiri, and J. Upton. 1991. "Experiences in Sewage Treatment in Madras, India." In: M. D. F. Haigh and C. P. James, eds., *Water and Environmental Management: Design and Construction of Works* (New York: Ellis Norwood), 453–65.

Sykora, K. V. 1990. "History of the Impact of Man on the Distribution of Plant Species." In: F. di Castri, A. J. Hansen, and M. Debussche, eds., *Biological Invasions in Europe and the Mediterranean Basin* (Dordrecht: Kluwer Academic), 37–50.

Tam, On Kit. 1985. *China's Agricultural Modernization: The Socialist Mechanization Scheme* (Beckenham: Croom Helm).

Tarr, Joel A. 1996. *The Search for the Ultimate Sink: Urban Pollution in Historical Perspective* (Akron: University of Akron Press).

Tarr, Joel A., and Carl Zimring. 1997. "The Struggle for Smoke Control in St. Louis." In: Andrew Hurley, ed., *Common Fields: An Environmental History of St. Louis* (St. Louis: Missouri Historical Society Press), 199–220.

Teich, Mikulas, Roy Porter, and Bo Gustaffson, eds., 1997. *Nature and Society in Historical Context* (Cambridge: Cambridge University Press).

Tenner, Edward. 1996. *Why Things Bite Back: Technology and the Revenge of Unintended Consequences* (New York: Knopf).

Tersch, F. 1987. "Austria." In: E. G. Richards, ed., *Forestry and the Forest Industries: Past and Future* (Dordrecht: Martinus Nijhoff), 214–225.

Thandi, Shinder Singh. 1994. "Strengthening Capitalist Agriculture: The Impact of Overseas Remittances in Rural Central Punjab in the 1970s," *International Journal of Punjab Studies* 1:239–70.

Thellung, A. 1915. "Pflanzwanderungen unter dem Einfluss des Menschen," *Beiblat Nr. 116 zu den Botanischen Jahrbüchern* 53:37–66.

Thomas, V. M. 1995. "The Elimination of Lead in Gasoline," *Annual Review of Energy and the Environment* 20:301–24.

Thompson, Harry V., and Carolyn M. King. 1994. *The European Rabbit: The History and Biology of a Successful Colonizer* (Oxford: Oxford University Press).

Thukral, E. G. 1992. *Big Dams, Displaced People: Rivers of Sorrow, Rivers of Change* (New Delhi: Sage Publishers).

Thumerelle, Pierre-Jean. 1996. *Les populations du monde* (Paris: Editions Nathan).

Tiffen, Mary, Michael Mortimore, and Francis Gichuki. 1994. *More People, Less Erosion: Environmental Recovery in Kenya* (New York: Wiley).

Tolba, Mostafa K., and Osama A. El-Kholy, eds. 1992. *The World Environment, 1972–1992* (London: Chapman & Hall).

Tønnessen, J. N., and A. O. Johnsen. 1982. *The History of Modern Whaling* (Berkeley: University of California Press).

Totman, Conrad. 1989. *The Green Archipelago: Forestry in Preindustrial Japan* (Berkeley: University of California Press).

Toynbee, A. J. 1965. *Hannibal's Legacy* (Oxford: Oxford University Press, 2 vols.).

———. 1972. "Religious Background of the Present Environment Crisis: A Viewpoint," *International Journal of Environmental Studies* 3:141–6, 4:157–8.

Trafas, Kazimierz. 1991. *Luftverschmutzung in Südpolen* (Vienna: Österreichisches Ost- und Südeuropa-Institut).

Travis, J. 1993. "Invader Threatens Black, Azov Seas," *Science* 262:1366–7.

Treadgold, Donald W. 1957. *The Great Siberian Migration* (Princeton: Princeton University Press).

Trefil, James. 1994. *A Scientist in the City* (New York: Doubleday).

Trimble, Stanley W. 1985. "Perspectives on the History of Soil Erosion Control in the Eastern United States," *Agricultural History* 59:162–180.

Troadec, Jean-Paul. 1989. "The Mutation of World Fisheries." In: Edward L. Miles, ed., *Management of World Fisheries* (Seattle: University of Washington Press), 1–18.

Trudgill, S. T., et al. 1990. "Rates of Stone Loss at St. Paul's Cathedral, London," *Atmospheric Environment* 24B:361–3.

Tsuru, Shigeto. 1989. "History of Pollution Control Policy." In: S. Tsuru and Helmut Weidmer, eds., *Environmental Policy in Japan* (Berlin: Sigma), 15–42.

———. 1993. *Japan's Capitalism: Creative Defeat and Beyond* (Cambridge: Cambridge University Press).

Tuan, Yi-fu. 1968. "Discrepancies between Environmental Attitude and Behaviour: Examples from Europe and China," *Canadian Geographer* 3:175–91.

Tucker, Richard P. (forthcoming). *An Embarassment of Riches: The United States and the Ecological History of the Tropical World* (Berkeley: University of California Press).

Tuncel, Semra, and Sevgi Üngör. 1996. "Rain Water Chemistry in Ankara, Turkey," *Atmospheric Environment* 30:2721–27.

Tuncer, G. T., S. G. Tuncel, G. Tuncel, and T. I. Balkas. 1993. "Metal Pollution in the Golden Horn, Turkey: Contribution of Natural and Anthropogenic Components since 1913," *Water Science and Technology* 28:59–64.

Turco, Richard P. 1997. *Earth under Siege: From Air Pollution to Global Change* (Oxford: Oxford University Press).

Türkiye Çevre Sorunları Vakfı 1991. *Türkiye'nin Çevre Sorunları* (Ankara: Türkiye Çevre Sorunları Vakfı).

Turner, R. E., and N. N. Rabalais. 1991. "Changes in Mississippi River Quality This Century: Implications for Coastal Food Webs," *BioScience* 41(3):140–7.

Turner, B. L. et al., eds. 1990. *The Earth as Transformed by Human Action* (New York: Cambridge University Press).

Twain, Mark. 1899. *Following the Equator* (New York: Harper & Brothers, 2 vols.).

Ui, Jun., ed. 1992a. *Industrial Pollution in Japan* (Tokyo: U.N. University Press).

———. 1992b. "Minimata Disease." In Jun Ui, ed., *Industrial Pollution in Japan* (Tokyo: U.N. University Press), 103–32.

U.N. Economic Commission for Europe. 1992. *Impacts of Long-Range Transboundary Air Pollution* (New York: United Nations).

UNEP [U.N. Environment Programme]. 1997. *Global Environment Outlook* (New York: Oxford University Press).

UNFAO [U.N. Food and Agriculture Organization]. 1997. *The State of World Fisheries and Agriculture* (Rome: UNFAO).

U.S. Department of the Interior, Bureau of Mines. 1995. *Minerals Yearbook. Vol. III* (Washington: U.S. Government Printing Office).

USDOE [U.S. Department of Energy]. 1995. *Estimating the Cold War Mortgage* (Washington: USDOE).

USEPA [U.S. Environmental Protection Agency]. 1995. *National Air Pollutant Emission Trends, 1900–1994* (Research Triangle Park, N.C.: USEPA).

U.S. Global Change Research Program. 1998. *Our Changing Planet* (Washington: USGCRP).

USOTA [U.S. Office of Technology Assessment]. 1993. *Harmful Non-Indigenous Species in the United States* (Washington: USOTA).

Vance, James E. 1990. *The Continuing City: Urban Morphology in Western Civilization* (Baltimore: Johns Hopkins University Press).

van den Bosch, Robert, P. S. Messenger, and A. P. Gutierrez. 1982. *An Introduction to Biological Control* (New York: Plenum Press).

Van der Weijden, C. H., and J. J. Middleburg. 1989. "Hydrogeochemistry of the River Rhine: Long Term and Seasonal Variability, Elemental Budgets, Base Levels, and Pollution," *Water Research* 23:1247–66.

van Lier, H. N. 1991. "Historical Land Use Changes: The Netherlands." In: F. M. Brouwer, A. J. Thomas, and M. J. Chadwick, eds., *Land Use Changes in Europe* (Dordrecht: Kluwer Academic), 379–402.

Van Urk, G. 1984. "Lower Rhine-Meuse." In: B. A. Whitton, ed., *Ecology of European Rivers* (Oxford: Blackwell Scientific), 437–68.

Varady, Robert G. 1989. "Land Use and Environmental Change in the Gangetic Plain: Nineteenth-Century Human Activity in the Banaras Region." In: Sandra Freitag, ed., *Culture and Power in Banaras: Community, Performance, and Environment, 1880–1980* (Berkeley: University of California Press), 229–45.

Vasey, Daniel E. 1992. *An Ecological History of Agriculture, 10,000 B.C.–A.D. 10,000* (Ames: Iowa State University Press).

Vassilopoulos, M., and M. Nikopoulou-Tamvakli. 1993. "Greek Mediterranean Environment." In: F. B. de Walle, M. Nikopoulou-Tamvakli, and W. J. Heinen, eds., *Environmental Condition of the Mediterranean Sea* (Dordrecht: Kluwer Academic), 425–501.

Vendrov, S. L., and A. B. Avakyan. 1977. "The Volga River." In: Gilbert F. White, ed., *Environ-mental Effects of Complex River Development* (Boulder: Westview Press), 23–38.

Vennetier, Pierre. 1988. "Cadre de vie urbain et problèmes de l'eau en Afrique noire," *Annales de géographie* 92:171–94.

Vesely, J., et al. 1993. "The History and Impact of Air Pollution at Certovo Lake, Southwestern Czech Republic," *Journal of Paleolimnology* 8:211–31.

Vileisis, Ann. 1997. *Discovering the Unknown Landscape: A History of American Wetlands* (Washington: Island Press).

Villalba, Bruno. 1997. "La genèse inachevée des Verts," *Vingtième siècle* 53:85–97.

Viras, L. G., A. G. Paliatsos, A. G. Fotopoulos. 1995. "Nine-Year Trend of Air Pollution by CO in Athens, Greece," *Environmental Monitoring and Assessment* 40:203–14.

Vitale, Luís. 1983. *Hacia una historia del ambiente en América Latina* (Caracas: Nueva Sociedad).

Vitousek, P. M., P. R. Ehrlich, A. R. Ehrlich, P. A. Matson. 1986. "Human Appropriation of the Products of Photosynthesis," *BioScience* 36:368–73.

Vitousek, P. M., C. M. D'Antonio, L. L. Loope, and R. Westbrooks. 1996. "Biological Invasions as Global Environmental Change," *American Scientist* 84:468–78.

Vizcaíno Murray, Francisco. 1975. *La contaminación en México* (Mexico City: Fondo de Cultura Económica).

Vizcarra Andreu, M. A. 1989. "A Case of a Quarter Century's SO₂ Pollution." In: L. J. Brasser and W. C. Mulder, eds., *Man and His Ecosystem. Proceedings of the 8th World Clean Air Congress* (The Hague: Elsevier), 25–30.

Volin, Lazar. 1970. *A Century of Russian Agriculture* (Cambridge: Harvard University Press).

von Broembsen, Sharon L. 1989. "Invasions of Natural Ecosystems by Plant Pathogens." In: J. A. Drake et al., eds., *Biological Invasions: A Global Perspective* (Chichester: Wiley), 77–83.

von Maydell, H.-Z., and H. Ollmann. 1987. "Federal Republic of Germany." In: E. G. Richards, ed., *Forestry and the Forest Industries: Past and Future* (Dordrecht: Martinus Nijhoff), 152–64.

Votruba, Ladislav. 1993. "K Dĕjinám Péče o Přírodu," *Dĕjiny vĕd a Techniky (Prague)* 26:1–6.

Wade, Robert. 1997. "Greening the Bank: The Struggle over the Environment, 1970–1995." In: John Lewis, Richard Webb, and Devesh Kapur, eds., *The World Bank: Its First Half Century* (Washington: Brookings Institution, 2 vols.), 2:611–734.

Walker, Jesse. 1990. "The Coastal Zone." In: B. L. Turner et al., eds., *The Earth as Transformed by Human Action* (New York: Cambridge University Press), 271–94.

Walker, H. J. 1984. "Man's Impact on Shorelines and Nearshore Environments: A Geomorphological Perspective," *Geoforum* 15:395–417.

Waller, Richard D. 1990. "Tsetse Fly in Western Narok, Kenya," *Journal of African History* 31:81–101.

Walsh, James. 1992. "Adoption and Diffusion Processes in the Mechanisation of Irish Agriculture," *Irish Geography* 25:33–53.

Walsh, Michael. 1990. "Global Trends in Motor Vehicle Use and Emissions," *Annual Review of Energy and the Environment* 15:217–43.

Walter, François. 1989. "Attitudes towards the Environment in Switzerland, 1880–1914," *Journal of Historical Geography* 15:287–299.

Waterbury, John. 1979. *Hydropolitics of the Nile* (Syracuse: Syracuse University Press).

Watts, Susan, and Samiha El Katsha. 1997. "Irrigation, Farming and Schistosomiasis: A Case Study in the Nile Delta," *International Journal of Environmental Health Research* 7:101–13.

WCED [World Commission on Environment and Development]. 1987. *Our Common Future* (Oxford: Oxford University Press).

Weeber, K. W. 1990. *Smog über Attika: Umveltverhalten im Altertum* (Zurich: Artemis).

Weiner, Douglas. 1988. "The Changing Face of Soviet Conservation." In: Donald Worster, ed., *The Ends of the Earth* (New York: Cambridge University Press), 252–73.

———. 1999. *A Little Corner of Freedom: Russian Nature Protection from Stalin to Gorbachev* (Berkeley: University of California Press).

Weiss, D., B. Whitten, and D. Leddy. 1972. "Lead Content of Human Hair (1871–1971)," *Science* 178:69–70.

Wen, Dazhong. 1993. "Soil Erosion and Conservation in China." In: David Pimentel, ed., *World Soil Erosion and Conservation,* (Cambridge: Cambridge University Press), 63–86.

Wernick, Iddo K., Robert Herman, Shekhar Govind, and Jesse Ausubel. 1996. "Materialization and Dematerialization: Measures and Trends," *Daedalus* (U.S.) 125(3):171–98.

Westing, Arthur P. 1980. *Warfare in a Fragile World: Military Impact on the Human Environment* (London: Taylor & Francis).

———. 1981. "Note on How Many Humans Have Ever Lived," *BioScience* 31:523–4.

———. 1990. *Environmental Hazards of War: Releasing Dangerous Forces in an Industrialized World* (Newbury Park, Calif.: Sage Publishers).

Westoby, Jack C. 1989. *Introduction to World Forestry* (Oxford: Blackwell).

Whitcombe, Elizabeth. 1995. "The Environmental Costs of Irrigation in British India: Waterlogging, Salinity and Malaria." In: David Arnold and Ramachandra Guha, eds., *Nature, Culture, Imperialism* (Delhi: Oxford University Press), 237–59.

White, Gilbert F. 1988. "The Environmental Effects of the High Dam at Aswan," *Environment* 30(7):4–11, 34–40.

White, Lynn. 1967. "The Historical Roots of Our Ecologic Crisis," *Science* 155:1203–7.

White, Richard. 1995. *The Organic Machine* (New York: Hill & Wang).

Whitehand, J. W. R. 1987. *The Changing Face of Cities* (Oxford: Basil Blackwell).

———. 1992. *The Making of the Urban Landscape* (Oxford: Basil Blackwell).

Whitlow, J. R. 1988. "Soil Erosion—A History," *Zimbabwe Science News* 22:83–5.

WHO [World Health Organization]. 1996. *World Health Report 1996* (Geneva: WHO).

WHO/UNEP [World Health Organization and United Nations Environment Programme]. 1992. *Urban Air Pollution in the Megacities of the World* (Oxford: Basil Blackwell).

Williams, Michael. 1988. "The Death and Rebirth of the American Forest: Clearing and Reversion in the United States, 1900–1980." In: John F. Richards and Richard P. Tucker, eds., *World Deforestation in the Twentieth Century* (Durham: Duke University Press), 211–29.

———. 1990a. "Agricultural Impacts in Temperate Wetlands." In: M. Williams, ed., *Wetlands: A Threatened Landscape* (Oxford: Blackwell), 181–216.

———. 1990b. "Understanding Wetlands." In: M. Williams, ed., *Wetlands: A Threatened Landscape* (Oxford: Blackwell), 1–41.

———. 1990c. "Forests." In: B. L. Turner et al. eds., *The Earth as Transformed by Human Action* (New York: Cambridge University Press), 179–202.

———. 1994. "Forests and Tree Cover." In: W. B. Meyer and B. L. Turner, eds., *Changes in Land Use and Land Cover: A Global Perspective* (New York: Cambridge University Press), 97–124.

Williamson, Mark. 1996. *Biological Invasions* (London: Chapman & Hall).

Wills, Christopher. 1996. *Yellow Fever, Black Goddess: The Co-Evolution of People and Plagues* (Reading, Mass.: Addison-Wesley).

Wilson, Edward O. 1992. *The Diversity of Life* (Cambridge: Harvard University Press).

Wilson, John P., and Christine Ryan. 1988. "Landscape Change in the Lake Simcoe-Couchiching Basin, 1800–1983," *Canadian Geographer/Géographe Canadien* 32:206–22.

Wilson, Mary E., Richard Levins, and Andrew Spielman, eds. 1994. *Disease in Evolution: Global Changes and the Emergence of Infectious Diseases* (New York: N.Y. Academy of Sciences).

Winslow, Donna. 1993. "Mining and the Environment in New Caledonia: The Case of Thio." In: Michael C. Howard, ed., *Asia's Environmental Crisis* (Boulder: Westview), 111–134.

Woischnik, Alwine. 1992. *Die spanische Ökologiebewegung* (Frankfurt: Peter Lang).

Wood, Charles, and Marianne Schmink. 1993. "The Military and the Environment in the Brazilian Amazon," *Journal of Political and Military Sociology* 21:81–105.

Wood, Leslie B. 1982. *The Restoration of the Tidal Thames* (Bristol: Adam Higher).

Worster, Donald. 1977. *Nature's Economy: A History of Ecological Ideas* (San Francisco: Sierra Club Books).

———. 1979. *Dust Bowl* (New York: Oxford University Press).

———. 1985. *Rivers of Empire* (New York: Pantheon).

Worthington, E. Barton. 1983. *The Ecological Century: A Personal Appraisal* (Oxford: Clarendon Press).

WRI [World Resources Institute]. 1996. *World Resources 1996–1997* (New York: Oxford University Press).

———. 1997. *The Last Frontier Forests* (Washington: WRI).

Wunderlich, Walter O., and J. Egbert Prins, eds. 1987. *Water for the Future* (Rotterdam: Balkema).

Xu, Guohua, and L. J. Peel, eds. 1991. *The Agriculture of China* (Oxford: Oxford University Press).

Yablokov, A. V. 1995. "The Protection of Nature: Lessons and Problems from Russia," *Science of the Total Environment* 175:1–8.

Yamamoto, Tadashi, and Hajime Imanishi. 1992. "Use of Shared Stocks in the Northwest Pacific Ocean with Particular Reference to Japan and the USSR." In: J. B. Marsh, ed., *Resources and Environment in Asia's Marine Sector* (Washington: Taylor & Francis), 13–40.

Yeoh, Brenda S. A. 1993. "Urban Sanitation, Health and Water Supply in Late Nineteenth and Eartly Twentieth Century Colonial Singapore," *South East Asia Research* 2:143–172.

Yeung, Yue-man. 1997. "Geography in the Age of Mega-Cities," *International Social Science Journal* 151:91–104.

Yip, Ka-che. 1995. *Health and National Reconstruction in Nationalist China: The Development of Modern Health Services, 1928–1937* (Ann Arbor: Association for Asian Studies, Monograph and Occasional Paper Series no. 50).

Young, Ann R. M. 1996. *Environmental Change in Australia since 1788* (Melbourne: Oxford University Press).

Young, Oran, ed. 1997. *Global Governance: Drawing Insights from the Environmental Experience* (Cambridge: MIT Press).

Zaidi, I. H. 1981. "On the Ethics of Man's Interaction with the Environment: An Islamic View," *Environmental Ethics* 3:35–47.

Zemskov, V. 1991. "Kulakskaya ssylka v 30-e gody," *Sotsiologicheskie issledovaniya* 10:3–21.

Zimmerer, Karl. 1993. "Soil Erosion and Labor Shortages in the Andes with Special Reference to Bolivia 1953–91," *World Development* 21:1659–73.

Zirnstein, Gottfried. 1994. *Ökologie und Umwelt in der Geschichte* (Marburg: Metropolis-Verlag).

Zolberg, Aristide. 1997. "Global Movements, Global Walls: Responses to Migration, 1885–1925." In: Wang Gungwu, ed., *Global History and Migrations* (Boulder: Westview), 279–307.

Zon, Raphael, and William N. Sparhawk. 1923. *Forest Resources of the World* (New York: McGraw-Hill), 2 vols.

Credits

Courtesy of the Library of Congress: pp. 33, 69, 71, 73, 130, 158, 240, 241, 294, 299, 300, 317.

Courtesy of the U.S. National Archives: pp. 88, 182, 185, 237, 308, 309, 333, 334.

Courtesy of the American Chemical Society: p. 112.

CORBIS/Bettmann-UPI: pp. 67, 70, 338, 348.

CORBIS/Adrian Arbib: p. 351.

Index